NATIONAL ACADEMIES *Sciences Engineering Medicine*

NATIONAL ACADEMIES PRESS
Washington, DC

Constructing Valid Geospatial Tools for Environmental Justice

Committee on Utilizing Advanced Environmental Health and Geospatial Data and Technologies to Inform Community Investment

Board on Earth Sciences and Resources

Board on Environmental Sciences and Toxicology

Division on Earth and Life Studies

Board on Mathematical Sciences and Analytics

Division on Engineering and Physical Sciences

Consensus Study Report

NATIONAL ACADEMIES PRESS   500 Fifth Street, NW   Washington, DC 20001

This activity was supported by a grant between the National Academy of Sciences and the Bezos Earth Fund. Any opinions, findings, conclusions, or recommendations expressed in this publication do not necessarily reflect the views of any organization or agency that provided support for the project.

International Standard Book Number-13: 978-0-309-71200-2
International Standard Book Number-10: 0-309-71200-9
Digital Object Identifier: https://doi.org/10.17226/27317
Library of Congress Control Number: 2024949404

This publication is available from the National Academies Press, 500 Fifth Street, NW, Keck 360, Washington, DC 20001; (800) 624-6242 or (202) 334-3313; http://www.nap.edu.

Printed in the United States of America.

Suggested citation: National Academies of Sciences, Engineering, and Medicine. 2024. *Constructing Valid Geospatial Tools for Environmental Justice*. Washington, DC: The National Academies Press. https://doi.org/10.17226/27317.

The **National Academy of Sciences** was established in 1863 by an Act of Congress, signed by President Lincoln, as a private, nongovernmental institution to advise the nation on issues related to science and technology. Members are elected by their peers for outstanding contributions to research. Dr. Marcia McNutt is president.

The **National Academy of Engineering** was established in 1964 under the charter of the National Academy of Sciences to bring the practices of engineering to advising the nation. Members are elected by their peers for extraordinary contributions to engineering. Dr. John L. Anderson is president.

The **National Academy of Medicine** (formerly the Institute of Medicine) was established in 1970 under the charter of the National Academy of Sciences to advise the nation on medical and health issues. Members are elected by their peers for distinguished contributions to medicine and health. Dr. Victor J. Dzau is president.

The three Academies work together as the **National Academies of Sciences, Engineering, and Medicine** to provide independent, objective analysis and advice to the nation and conduct other activities to solve complex problems and inform public policy decisions. The National Academies also encourage education and research, recognize outstanding contributions to knowledge, and increase public understanding in matters of science, engineering, and medicine.

Learn more about the National Academies of Sciences, Engineering, and Medicine at **www.nationalacademies.org**.

Consensus Study Reports published by the National Academies of Sciences, Engineering, and Medicine document the evidence-based consensus on the study's statement of task by an authoring committee of experts. Reports typically include findings, conclusions, and recommendations based on information gathered by the committee and the committee's deliberations. Each report has been subjected to a rigorous and independent peer-review process and it represents the position of the National Academies on the statement of task.

Proceedings published by the National Academies of Sciences, Engineering, and Medicine chronicle the presentations and discussions at a workshop, symposium, or other event convened by the National Academies. The statements and opinions contained in proceedings are those of the participants and are not endorsed by other participants, the planning committee, or the National Academies.

Rapid Expert Consultations published by the National Academies of Sciences, Engineering, and Medicine are authored by subject-matter experts on narrowly focused topics that can be supported by a body of evidence. The discussions contained in rapid expert consultations are considered those of the authors and do not contain policy recommendations. Rapid expert consultations are reviewed by the institution before release.

For information about other products and activities of the National Academies, please visit www.nationalacademies.org/about/whatwedo.

COMMITTEE ON UTILIZING ADVANCED ENVIRONMENTAL HEALTH AND GEOSPATIAL DATA AND TECHNOLOGIES TO INFORM COMMUNITY INVESTMENT

HARVEY J. MILLER (*Co-chair*), The Ohio State University
ERIC TATE (*Co-chair*), Princeton University
SUSAN ANENBERG, George Washington University
LAUREN BENNETT, Esri, Inc.
JAYAJIT CHAKRABORTY, University of California, Santa Barbara
IBRAHEEM KARAYE, Hofstra University
MARCOS LUNA, Salem State University
BHRAMAR MUKHERJEE (NAM), University of Michigan
KATHLEEN SEGERSON (NAS), University of Connecticut
MONICA E. UNSELD, Until Justice Data Partners
WALKER WIELAND, California Environmental Protection Agency

Study Staff

SAMMANTHA MAGSINO, Study Director
CLIFFORD DUKE, Director, Board on Environmental Studies and Toxicology
DEBORAH GLICKSON, Director, Board on Earth Sciences and Resources
MICHELLE SCHWALBE, Director, Board on Mathematical Sciences and Analytics
ANTHONY DePINTO, Program Officer
SARAH HARTMAN, Mirzayan Fellow (2023)
MILES LANSING, Senior Program Assistant (*since July 2023*)
OSHANE ORR, Senior Program Assistant (*until July 2023*)
BRYAN RUFF, Senior Program Assistant (*since May 2024*)

Reviewers

This Consensus Study Report was reviewed in draft form by individuals chosen for their diverse perspectives and technical expertise. The purpose of this independent review is to provide candid and critical comments that will assist the National Academies of Sciences, Engineering, and Medicine in making each published report as sound as possible and to ensure that it meets the institutional standards for quality, objectivity, evidence, and responsiveness to the study charge. The review comments and draft manuscript remain confidential to protect the integrity of the deliberative process.

We thank the following individuals for their review of this report:

AMY ANDO, The Ohio State University
TIM COLLINS, University of Utah
KRISTI PULLEN FEDINICK, George Washington University
ADRIENNE HOLLIS, National Wildlife Federation
DIANA LIVERMAN (NAS), University of Arizona
MIKE RATCLIFFE, U.S. Census Bureau
ANA DIEZ ROUX, Drexel University
MITCHELL SMALL, Carnegie Mellon University
LAUREN ZEISE, California Environmental Protection Agency Office of Environmental Health Hazard Assessment

Although the reviewers listed above provided many constructive comments and suggestions, they were not asked to endorse the conclusions or recommendations of this report, nor did they see the final draft before its release. The review of this report was overseen by **ALICIA CARRIQUIRY (NAM)**, Iowa State University, and **DAVID DZOMBAK (NAE)**, Carnegie Mellon University. They were responsible for making certain that an independent examination of this report was carried out in accordance with the standards of the National Academies and that all review comments were carefully considered. Responsibility for the final content rests entirely with the authoring committee and the National Academies.

Acknowledgments

Many individuals assisted this committee by providing important input. The committee would like to thank the following people who gave presentations and participated in discussions throughout the course of the study.

MICHELLE ALVARADO, Resiliency Law Center
LOKA ASHWOOD, University of Kentucky
CHITRA BALAKRISHNAN, Urban Institute
LUCAS BROWN, Council on Environmental Quality
SHARUNDA BUCHANAN, Center for Disease Control and Prevention
WEIHSUEH CHIU, Texas A&M
WINN CONSTANTINI, Drexel University
NATASHA DEJARNETT, Environmental Justice Data and Evaluation
KE JACK DING, GTI Energy
DAVID FOLCH, Northern Arizona University
LARRY LAMBERT, 7th District, Delaware General Assembly, House of Representatives
MATTHEW LEE, Environmental Protection Agency
TAI LUNG, Environmental Protection Agency
CECILIA MARTINEZ, Bezos Earth Fund
NAYAMIN MARTINEZ, Central California Environmental Justice Network
BENJAMIN MCKENZIE, Center for Disease Control and Prevention
MICHAELA SAISANA, European Commission's Competence Centre on Composite Indicators and Scoreboards (COIN), Joint Research Centre
ALEXIS SHULMAN, Drexel University
MARISA SOTOLONGO, Northeastern University
MATHY STANISLAUS, Drexel University

VIOLA WAGHIYI, Alaska Community Action on Toxics
SACOBY WILSON, University of Maryland, College Park
KRISTEN WOOD, Office of Policy Development and Coordination
NEZAHUALCOYOTL XIUHTECUTLI, Farmworker Association of Florida

The committee would like to acknowledge **SHELLEY HOOVER** at Princeton University whose doctoral work helped shape the committee's scan of EJ tools and workshop exercise to explore the Climate and Economic Justice Screening Tool results at the community level.

Contents

Preface

Environmental injustice is a pervasive, persistent, and largely unaddressed problem in the United States. Its roots are an amalgamation of longstanding public and private policies and norms that have resulted in a differential concentration of environmental hazards and vulnerabilities across communities. Decades of environmental justice research and activism have shown that the communities most disadvantaged by society exist at the intersection of high levels of hazard exposure, racial and ethnic composition, and poverty. Redressing damage suffered in disadvantaged communities requires intentional actions to mitigate the harms caused by societal marginalization, pollution overburden, and chronic underinvestment. Such mitigative actions are particularly imperative given the profound climate crisis facing the United States and the rest of the world. Disadvantaged communities are particularly vulnerable to ravages of extreme weather induced by global heating.

The Biden Administration's Justice40 Initiative seeks to rectify these vulnerabilities and build greater resilience by ensuring that 40 percent of benefits from certain federal investments flow to these communities. A fundamental challenge is to identify which communities are disadvantaged, and thus priorities for investment. This report, sponsored by the Bezos Earth Fund, supports these efforts by evaluating the Council on Environmental Quality's (CEQ's) Climate and Economic Justice Screening Tool (CEJST), a geospatial mapping tool for identifying "Justice40 communities," and suggesting data strategies to maximize the tool's effectiveness. CEJST parallels efforts at the state level to identify the most impacted places and prioritize corrective investments. The primary audiences for this report are the CEQ, federal agencies that will use the current or future versions of CEJST to support investment decision making, and others who may use the tool to evaluate policies and seek funding to increase resilience in American communities.

The committee took their responsibilities seriously and listened carefully to experts, practitioners, and activists in a series of meetings and workshops. A major theme of the report is the importance of integrating the lived experiences and perspectives of communities into multiple aspects of tool development. It is particularly vital to gain this understanding from people and their representatives who are the most overburdened by pollution and adversely affected by underinvestment. Our workshop, Representing Lived Experience in Climate and Economic Justice Screening Tool, helped cement the principle of centering community perspectives, and we thank all participants for taking the time to share their knowledge.

Several experts presented to the committee about environmental and demographic data and indicators. They include David Folch of Northern Arizona University, Kristin Wood from the Department of Transportation, Weihsueh Chiu from Texas A&M University, and Michaela Saisana from the Joint Research Centre of the European Commission. We thank them for taking the time to share their knowledge and perspectives, as they positively informed the committee discussions. In particular, Dr. Saisana's presentation and publications of the Joint Research Centre helped establish a second central theme of the report: the need for a systematic approach when constructing composite indicators used in policy decision making. Creating a reliable and valid composite indicator requires more than identifying and combining relevant quantitative measures. It should also be based on a clearly defined and vetted conceptual framework that is thoroughly coherent with the selection of indicators and their integration. The combination of expanding spatial data availability, growing policy interest at the intersection of physical and social environments, and increasing need for publicly accessible decision-making tools suggests that the role of composite indicators in public policy is likely to rise. One of our desired outcomes of this report is to foster in the United States the careful and systematic approaches to indicator construction that we see in European policy and practice.

The committee formed for this study represents a diverse group of people, disciplinary backgrounds, and professional communities of practice. It has been a great privilege working with them to advance understanding of the state of knowledge and paths forward for environmental justice tools. The work was intellectually rich, collegial, and equally shared, resulting in a truly consensus report. We thank the members of the committee for the commitment, thoughtfulness, professionalism, and spirit they brought to this important task. We also thank the hardworking NASEM staff, particularly Samantha Magsino and Anthony DePinto, who are the unsung heroes in this study. They kept the committee on task and moving forward and played no small part in helping to transform our ideas into the actionable knowledge in this report.

Personal Notes

Harvey Miller: We are at a hinge point in the history of humanity, and the choices we make now will reverberate for generations. One does not often have the opportunity and privilege to participate in an activity that addresses the profound and consequential questions at the heart of this consensus study. I sincerely hope that this report helps to move our nation forward toward a future with environmental justice for all. My personal thanks to my co-chair, Eric Tate, the other members of the committee, and the National Academies staff, all of whom made this process as smooth and productive as possible.

Eric Tate: I am deeply appreciative of the opportunity for meaningful public service afforded by co-chairing this study. Reflecting on connections between Justice40 and the establishment of this committee reminded me of a passage from the final public speech of Frederick Douglass. Offering a roadmap to realizing the principles of liberty and equality, he called for America to "*recognize the fact that the rights of the humblest citizen are as worthy of protection as are those of the highest, and your problem will be solved.*" My hope is that this report plays a constructive role in more closely aligning our national ideals of equal protection for all, with our scientific practices for modeling environmental injustice and our public policies for dismantling it.

Harvey Miller, *Co-Chair*
Eric Tate, *Co-Chair*
Committee on Utilizing Advanced Environmental Health and
Geospatial Data and Technologies to Inform Community Investment

Summary

President Biden's Executive Order (EO) 14008 (Tackling the Climate Crisis at Home and Abroad) established the Justice40 Initiative, which sets a goal that disadvantaged communities reap 40 percent of federal investment benefits in the areas of clean energy and energy efficiency, clean transit, affordable and sustainable housing, training and workforce development, legacy pollution, and clean water infrastructure. EO 14008 directed the White House Council on Environmental Quality (CEQ) to create a geospatial tool to identify the communities across the United States and its territories eligible for Justice40 investment benefits. The Climate and Economic Justice Screening Tool (CEJST)[1] was developed in 2022 in response.

Mapping and geographical information systems have been crucial for analyzing the environmental burdens of marginalized communities since the 1980s, and several federal and state geospatial tools have emerged to address a variety of environmental justice (EJ) concerns. These include the Environmental Protection Agency's EJScreen[2] and California's CalEnviroScreen.[3] CEJST is the first tool of this type developed at the federal level to identify disadvantaged communities in terms of climate, energy, sustainable housing, employment, and pollution burden for the purpose of guiding federal investment. As with any novel initiative, it requires breaking new ground in terms of research methodologies and data use. It calls for data of sufficient granularity and scientific validity to compare communities across states and regions, and in terms of their rural and urban contexts. Federal agencies have been instructed to use CEJST to identify communities that meet Justice40 goals in their programming. States, territories, Tribal governments, and other organizations also look to this national-level tool for guidance.

[1] See https://screeningtool.geoplatform.gov/ (accessed October 3, 2023).

[2] See https://www.epa.gov/ejscreen (accessed March 27, 2024).

[3] See https://oehha.ca.gov/calenviroscreen/report/calenviroscreen-40 (accessed March 27, 2024).

Definitions of terms associated with identifying disadvantaged communities and environmental and economic justice vary. A community is a group of people who share common experiences and interactions. In the case of CEJST, communities also share geographic proximity based on U.S. Census Bureau-defined census tract boundaries. Community disadvantage results from a complex interplay of factors that inhibit or prevent people in some communities from achieving the positive life outcomes that are commonly expected in a society. Community disadvantage is also a consequence of structural factors and historical processes that create conditions undermining resilience to shocks and disruptions (e.g., those associated with climate change, economic transitions, or other social and environmental pressures). Cumulative impacts (also called cumulative burdens) are the combined total burden from stressors, their interactions, and the environment that affects the health, well-being, and quality of life of an individual, community, or population. The concept of burden is included in nearly all definitions of disadvantaged communities and indicates an activity or agent with negative consequences for human health and well-being. Measuring community disadvantage in a tool such as CEJST requires careful conceptualization and rigorous application of model construction if the results are to reflect the real world and support effective policy. The challenge of reducing a multidimensional concept to a single composite indicator is not specific to CEJST or EJ tools.

STUDY CHARGE AND APPROACH

Under the sponsorship of the Bezos Earth Fund, The National Academies of Sciences, Engineering, and Medicine convened an ad hoc multidisciplinary committee of 11 experts to consider the different types of environmental health and geospatial data and data integration approaches used in existing environmental screening tools that could be applied to identify disadvantaged communities. The committee was not asked to review CEJST, but rather to conduct a scan of existing EJ tools including CEJST to identify types of data (e.g., environmental, socioeconomic) needed for CEQ tools; to evaluate data availability, quality, resolutions, and gaps; and to discuss data integration approaches. The committee was asked to provide recommendations that could be incorporated into an overall data strategy for CEQ's tool(s). The committee concluded early in its deliberations that a good data strategy appropriate for CEQ and CEJST would also be appropriate for other EJ tool developers and tools. Concepts and recommendations in this report are therefore generalized and applicable to the development or management of any EJ tool. Therefore, this chapter synthesizes the main concepts of this report into actionable recommendations intended for EJ tool developers at CEQ and elsewhere.

On the basis of a review of existing EJ tools and the literature, through discussions with state and federal EJ tool developers and experts from the European Commission's Competence Centre on Composite Indicators and Scoreboards at the Joint Research Centre, and through a public workshop focused on community engagement,[4] the committee developed its own conceptual framework for composite indicator and

[4] Proceedings in Brief from the workshop are available at https://nap.nationalacademies.org/catalog/27158/representing-lived-experience-in-the-climate-and-economic-justice-screening-tool (accessed March 14, 2024).

tool construction. While the statement of task requested recommendations regarding elements of a data strategy for CEQ tools, recommendations in this report are rooted in the fundamentals of sound indicator development and are applicable to EJ tools generally.

SYNTHESIS OF REPORT CONCEPTS

At the heart of many geospatial EJ tools is a composite indicator that represents a multidimensional concept or condition related to the EJ questions of concern—for example, community disadvantage in CEJST to guide investment under the Justice40 Initiative. Quantitative indicators are proxies for abstract concepts and are represented by one or more existing datasets. Calculating a composite indicator brings together measurements from multiple dimensions of the concept (e.g., different aspects of community disadvantage) to determine a single value intended to reflect the condition being measured. Sound composite indicators are developed with a clearly defined purpose and intended audience and reflect real-world conditions. The validity of a tool rests on a foundation of scientific and methodological rigor, meaningful and sustained participation and input from community and other interested and affected parties, transparency, and acceptance by institutional actors (e.g., government agencies), communities, and other affected parties.

There are published, systematic methodologies for developing composite indicators, for evaluating decisions related to their construction and internal robustness, and for external validation. In these methodologies, composite indicator construction is an interrelated system of steps or components for identifying the concept to be measured; selecting indicators and data, analyzing, normalizing, weighting, and aggregating the data; evaluating the results for coherence, internal robustness, and validity, and presenting the resulting information. Effective community engagement—described by the Centers for Disease Control and Prevention as the collaborative process of working with people connected by geographic proximity or interests to address issues of health and wellbeing—is necessary to validate every component of composite indicator construction to determine if each of those components represents real-world conditions and lived experience. The committee developed a conceptual framework (Figure S.1) to demonstrate the interrelationships of the different components of tool construction and used this framework to organize recommendations around an EJ tool data strategy. The outermost ring represents the objectives for constructing any tool, the innermost ring represents activities related to composite indicator construction, and the middle ring represents the communication activities necessary to move inside out.

An EJ tool that measures cumulative impacts reflects the combined effects of environmental and socioeconomic burdens. A binary approach such as that applied in CEJST does not discern communities facing single and multiple burdens. However, systematically identifying all the burdens faced by communities and how those burdens combine and interact is challenging. A certain amount of error is inevitable, but tool validation techniques can be applied to create a tool that is stable, accepted, scientifically sound, and based on cumulative impact scoring approaches. Considering appropriate indicators and measures, including measures of economic burden beyond the federal

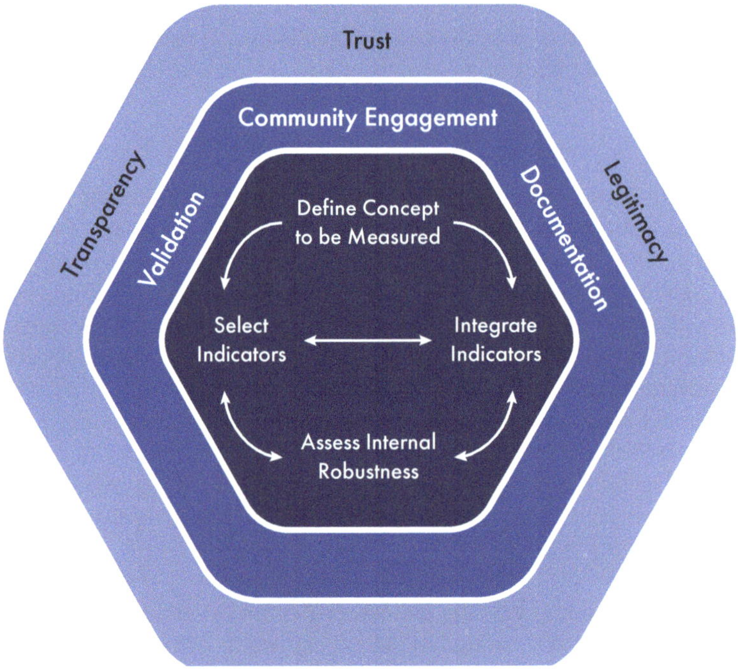

FIGURE S.1 Conceptual framework of environmental justice geospatial tool development. The arrows in the innermost ring indicate the direction of influence that each aspect of composite indicator construction has on another (i.e., defining the concept to be measured will influence the selection and integration of indicators, selection of indicators influences integration of indicators and vice versa, assessing internal robustness influences selection and integration of indicators and vice versa).

poverty level, and explicitly incorporating measures of race and racism into an EJ tool will result in a tool that better reflects lived experience.

DATA STRATEGY RECOMMENDATIONS

Transparency implies that a tool's goals, processes, data, and uncertainties are recognized and understood by tool users and the people interested in tool results. The committee formulated a set of science-based recommendations to inform a data strategy for EJ tool development based on current research in geospatial tool development and EJ. These recommendations are not intended to advocate for changes in legal policy, but rather represent the committee's conclusions regarding how data may be used to achieve model results that reflect reality. The legitimacy of a tool is related to acceptance of the tool and its results by communities. Therefore, a good data strategy considers how transparency, trust, and legitimacy can be created and reinforced in all components of tool construction (i.e., those depicted in Figure S.1). As the guiding forces of the tool development process, the recommendations outlined in this report inherently are meant to promote trust, transparency, legitimacy in the tool.

As of this writing, 35 different state and federal EJ tools have been released, and half of those have been released since 2021. Several reviews of EJ tools have found that definitions of community and disadvantage are often inappropriate and not reflective of community self-determinations or lived experiences; that indicators and measures of burden incorporated into tools are often incomplete or out of date (e.g., collected many years previously and not reflective of current conditions); that consideration of race and ethnicity is warranted; that the ways in which multiple burdens interact (cumulative impacts) are important, and that community input and engagement are vital to a relevant tool. However, many good indicator construction and analysis practices have been employed in EJ tool development. The recommendations provided here are based on the sound practices identified by the committee that could help tool developers such as CEQ improve their tool development and data strategies.

Community Engagement

The committee's conceptual framework (Figure S.1) emphasizes the importance of community engagement in tool construction. Choosing appropriate indicators, datasets, and integration approaches requires more than statistical robustness to achieve valid results. Community engagement validates the choices made in tool development as well as tool results and allows developers to understand the types of errors that are likely, why and where they occur, and how they might be overcome. Community engagement helps developers understand how uncertainties in tool results might impact the policy decisions the tool is meant to inform.

> **Recommendation 1: Create and sustain community partnerships that provide forums and opportunities to identify local environmental justice issues, identify the indicators and datasets for measuring them, and determine whether tool results reflect community lived experiences.**

There is a spectrum of possible engagement approaches based on the desired level of involvement by community members and interested and affected parties. For community engagement to be meaningful, it will be collaborative and sustained, and allow communities to feel involved in governmental decisions with local implications. Many EJ issues are local in scope, and close community engagement helps bring local issues into context, not only for understanding burdens across communities, but also for finding targeted solutions that address unique needs. However, community in-depth engagement cannot be accomplished meaningfully or sustained with every community represented by a tool, and an engagement program will necessarily be designed with the aid of experts in community engagement and with advisory committees to appropriately identify representative communities, to design tool feedback methodologies, and to validate decisions made during indicator construction. EJ tools such as CEJST often use indicators developed for different purposes for other tools or datasets. Ensuring that processes are established to meaningfully validate those derived indicators with communities will support the transparency, trust, and legitimacy of the tool.

Documentation

Community engagement, validation, and documentation (middle ring of Figure S.1) are all dependent on communication between the tool developer and communities, tool users, and a variety of interested and affected parties. Documentation is the means for a tool developer to describe tool components and explain the rationale behind decisions related to indicator and data selection, data integration and analysis approaches chosen, and about all aspects of robustness and validation analyses.

> **Recommendation 2: Provide thorough and accessible documentation of indicator design processes and decisions, including descriptions and rationale for all major indicator construction components.**

The current documentation of CEJST methodology and data includes descriptions of the burden categories and datasets, instructions on how to use the tool, and access to the code via a public repository. Less clear are the processes and rationale for decisions regarding indicator selection and weighting, data normalization, data treatments, thresholds, indicator aggregation, assessing statistical and conceptual coherence, epistemic uncertainty analysis, external validation via community engagement, and the design of the user interface. Thorough documentation of all tool components and approaches is vital to ensure proper tool use and to help decision makers understand where and how the tool may be accurate, what kinds of uncertainties should be expected, and when tool results need to be supplemented with other types of information. Good documentation makes the strengths and weaknesses of the tool clear to a variety of technical and nontechnical users or community members and provides guidance regarding the best use of the tool for decision making.

Validation

Composite indicator construction requires a certain amount of compromise. In alignment with the executive order that mandated the creation of CEJST, CEQ uses census tracts to define communities knowing that census tract boundaries do not always align with community boundaries and that large disparities in community health and well-being within a census tract may exist. However, the choice also takes advantage of national data sets available at that scale. Given that such compromises are inevitable during the development of any tool, and given that no single definitive measure might be available to validate for the purpose of a tool (e.g., "overburden" or "disadvantage" for CEJST), validation methodologies need to be applied throughout the construction of a tool to determine how well the tool relates to real-world conditions.

Effective validation spans how well the indicators measure what they are supposed to (construct validity), the degree of alignment among indicators (concurrent validity), and the indicators' representativeness of the underlying concept (content validity). Methodological components and processes can be applied during tool construction to ensure that a tool and its findings are rooted in the realities and lived experiences of communities. Validation may take the form of a combination of technical, statistical, and community engagement activities. Ground truthing, for example, includes a comparison of measured versus modeled information through a variety of approaches, including:

- **Convergent validation:** comparing tool components or results with those of similar tools (e.g., correlation analysis of tools results).
- **Community validation:** an iterative process conducted via collaborative engagement with communities to compare how well the tool reflects lived experiences. Although challenging given the scope of many tools, consistent engagement throughout tool development allows developers to test decisions, approaches, and tool results against community member narratives, while empowering communities to accept or refute definitions being assigned to them.
- **Mixed methods:** allowing collection and analysis of both qualitative and quantitative datasets to better understand multiple perspectives on any issue. Mixed methods challenge the "traditional" scientific mindset focused on quantitative data, but their use will result in the incorporation of lived-experience information into data analyses.

Supplemental analysis using independent external datasets outside a tool development process can be used to check indicator data sources for gaps or inaccuracies and to compare, for example, the spatial correlations between the results of different tools. CEQ might conduct supplemental analysis to, for example, compare the distribution of race/ethnicity indicators and CEJST outputs to understand the impact of the tool's current formulation on underrepresentation of racial disparities. The analysis can result in a greater understanding of sociodemographic composition, identification of the determinants of health in communities identified, and generation of localized narratives to better understand lived experience. Because communities and burdens are dynamic, repeated validation of indicators and results is necessary.

Recommendation 3: Validate tool development and evaluative processes throughout the construction of an environmental justice tool using approaches such as ground truthing, convergent validation, and community validation to ensure that the tool's indicators and results reflect lived experiences.

Using a Structured Development Process

Composite indicator construction involves carefully considered interlinked decisions. The first decisions to be made are related to the objective of the tool, identifying the concept to be measured and developing a clear definition of the concept. There are multiple structured frameworks for the construction of a composite indicator that ensure that all composite indicator construction decisions are considered explicitly, lead to the stated objective, and are then documented thoroughly and carefully. Tool developers such as CEQ can utilize one of these frameworks to improve the transparency, trust, and legitimacy of their tools.

Recommendation 4: Initiate environmental justice tool and indicator construction with the development of clear objectives and definitions for the concept(s) to be measured. Follow a structured composite indicator development process that requires explicit consideration and robustness analysis of all major decisions involved with indicator selection and integration; assessment of uncertainties; and validation and visualization of results.

A good data strategy requires an explicit, systematic structure. The lack of explicit structure in CEJST linking the concept to be defined, its dimensions, indicators, and integration strategies results in an implicit weighting scheme. If CEQ incorporates more sophisticated indicator integration methods for capturing cumulative burdens in future iterations of CEJST, the lack of an explicit conceptual structure may be problematic. The state of the art and practice in composite indicator and EJ tool construction includes:

- Defining the concept to be measured and developing a description of its multiple facets or dimensions;
- Selecting the indicators that measure each dimension;
- Analyzing, treating, normalizing, and weighting the indicators as appropriate;
- Integrating/aggregating the indicators;
- Assessing statistical and conceptual robustness and coherence and determining the impact of uncertainties; and
- Validating the results and presenting them visually (e.g., choice of category breaks and colors).

Selecting and Assessing Indicators

The selection of indicators and datasets is part of the structured approach described above and requires consideration of their technical and practical characteristics and how objectively and well they support the indicator and tool. Available datasets may not be of equal quality, expressed in the same units or at the same scales, or collected for the same purposes. Many indicators are based on empirical datasets, which, through analysis, may be found to be statistically sound, but may not actually represent lived experiences. Not all information will be empirical and require different techniques to assess. Given the close interconnection between concept definition, indicator selection, weighting, and ground-truthing methods, decisions related to the selection of indicators are central to a high-quality and accurate tool. CEJST uses an apparently reasonable set of indicators and datasets, but numerous other federal and national datasets exist for EJ tools that could be used (Appendix D provides examples).

Recommendation 5: Adopt systematic, transparent, and inclusive processes to identify and select indicators and datasets that consider technical criteria (validity, sensitivity, specificity, robustness, reproducibility, and scale) and practicality (measurability, availability, simplicity, affordability, credibility, and relevance). Evaluate measures in consultation with federal agencies, technical experts, and community partners.

A systematic scan of the federal and national-level indicators and datasets, perhaps in partnership with federal agencies, other data providers, or a steering committee, could identify additional or more appropriate indicators for defining community disadvantage. Once identified, correlation analyses of potential indicators would inform indicator selection and organization into categories. Demonstration of highly correlated indicators might indicate redundancy in the indicator set, possibly resulting in an unintended implicit weighting scheme if the highly correlated datasets are used. Low, negative, or statistically insignificant dataset correlation signifies poor statistical alignment with the concept to be measured. Correlation analysis of an indicator set could provide an empirical rationale and methodological transparency for a targeted revision of an indicator set.

Selecting Economic Indicators

Metrics of income do not necessarily measure wealth, and the wealth gap between high-income and low-income households is larger than the income gap. A single, uniform low-income measure in an EJ tool such as CEJST may not accurately reflect lived experiences, even after doubling the standard poverty level and accounting for the cost of living.

Recommendation 6: Choose measures of economic burden beyond the federal poverty level that reflect lived experiences, attend to other dimensions of wealth, and consider geographic variations in cost of living.

Income-based measures deserve scrutiny because of the effects of income on all aspects of a person's or household's quality of life (e.g., nutrition, health care, and education). Other indicators can be used as socioeconomic measures (e.g., U.S. Department of Housing and Urban Development Public Housing/Section 8 Income limits for low-income,[5] homeownership rates, median home values, or weighted income metrics). Tool developers should work alongside communities to identify other dimensions of wealth that would more accurately reflect economic burdens, and they should conduct sensitivity analyses on these indicators and their thresholds.

[5] HUD's FY2023 methodology for determining Section 8 limits can be found here: https://www.huduser.gov/portal/datasets/il//il23/IncomeLimitsMethodology-FY23.pdf (accessed March 8, 2024).

Consideration of Racism and Race

The enduring effects of historical race-based policies on housing, transportation, and urban development continue to shape contemporary environmental inequalities. There is ample research demonstrating racism as a fundamental cause of disadvantage and social, economic, health, and environmental inequalities in the United States. Racism itself is a key factor leading to unequal exposures and outcomes for specific populations, and research demonstrates that race and ethnicity—more so than economic indicators—are reliable predictors of disparity.

Recommendation 7: Use indicators that measure the impacts of racism in policies and practices that have led to the disparities observed today. If indicators of racism are not used, explicitly factor race and ethnicity as indicators when measuring community disadvantage.

The best available research suggests that incorporating racism as an indicator in an EJ tool can strengthen and add legitimacy to the tool; however, doing so may pose legal policy challenges. If a tool developer chooses not to explicitly factor race or ethnicity as an indicator to measure community disadvantage as the best available research would suggest (e.g., developers of CEJST and some other EJ tools), then other approaches are necessary to acknowledge the history of racism and land use policies that have led to EJ disparities observed in communities populated by peoples of color. There are readily available disaggregated data on race and ethnicity that could be used in a national-level EJ tool (e.g., U.S. Census data on race and ethnicity). Those data could be used until appropriate indicators for racism are chosen or developed. Tool developers can work with those who have developed other tools, representatives of communities of color, and technical experts to identify existing empirical data (see Recommendation 5) and consider whether and how well the metrics, quantitative data, and qualitative data reflect community lived experiences. To measure residential segregation, CEJST includes an indicator of historic underinvestment based on redlining data. While those data are important for understanding the multidimensional nature of structural racism, the data are incomplete and unavailable nationally.

Supplemental analysis to compare distributions of race or ethnicity indicators and CEJST outputs could help CEQ tool developers better understand how well CEJST captures community disadvantage in its current formulation. The results could reveal how the input and output indicators are distributed by racial and ethnic composition. Such analyses can improve the understanding of the degree of racial and ethnic disparities in the designation of disadvantaged communities and can address how well CEJST identifies disadvantage without the inclusion of race or ethnicity indicators. Publication of the analysis results would show CEQ's responsiveness to numerous public comments received on this issue and increase trust in the tool development process and tool results.

Integrating Indicators and Cumulative Impact Scoring

Measuring and redressing the cumulative impacts of environmental stressors is a stated objective of EO 14008, the CEQ, and EJ advocates. This is important because the interplay of multiple concurrent stressors interacting with sociodemographic, environmental, and public health factors leads to the possibility of the total burdens on a community being greater than the sum of the individual stressors.

Recommendation 8: Designate communities as disadvantaged based on cumulative impact scoring approaches that are informed by the state of science; the knowledge, needs, and experiences of agencies, tool developers, and users; and validation efforts conducted in partnership with affected communities. Choose an approach to represent cumulative impacts, such as threshold approaches (e.g., summing thresholds or categories exceeded) or aggregation-based approaches for composite indicator construction (e.g., compensatory, noncompensatory, or a combination of approaches).

Cumulative impact scoring is an established practice applied in state-level EJ tools such as CalEnviroScreen and Maryland EJ Screen.[6] Such scoring enables clearer comparison of communities and prioritization of investment based on the severity of burden. The current binary approach used in CEJST to designate communities as either disadvantaged or not disadvantaged is based on meeting the threshold of a burden indicator and its associated socioeconomic indicator. This type of approach does not allow the tool to discern communities facing single or multiple burdens. A future CEQ data strategy could incorporate the use of an aggregated approach that allows combining indicators mathematically into a single number, thus capturing a representation of cumulative impacts.

Decisions related to indicator selection, weighting, and aggregation to capture cumulative burdens are intertwined and should be made in an iterative and engaged manner, reflecting scientific knowledge, tenets of indicator construction, perspectives from interested and affected parties, and lived experiences. Explicit weighting of indicators is necessary, given their major impact on composite indicator results when aggregating.

Assessing Internal Robustness

When constructing composite indicators that are used for high-consequence resource allocation and project prioritization, it is crucial to understand the degree to which modeling decisions affect the robustness of the outputs. Numerous modeling decisions—each of which includes multiple plausible options based on scientific knowledge, available data, and community preferences—can independently and conjointly influence which communities a tool identifies as disadvantaged. Uncertainty

[6] See https://mde.maryland.gov/Environmental_Justice/Pages/EJ-Screening-Tool.aspx (accessed March 26, 2024).

and sensitivity analyses are methodologies that can illuminate the degree and drivers of instability in model outputs. Uncertainty analyses quantify the variability in model outputs based on changes in model inputs. Sensitivity analyses apportion variability in model outputs to different input parameters or model structures. Both types of analyses can be conducted as a local analysis (i.e., one parameter evaluated at a time) and global analysis (i.e., multiple parameters and their interactions assessed simultaneously using Monte Carlo simulation). Interactive methods exist for understanding the implications of modeling decisions, visualizing the impacts of different decisions on the results, and visualizing the composite indicator decomposed into subgroups and individual indicators in both chart and map form.

Recommendation 9: Perform and document uncertainty and sensitivity analyses to evaluate how decisions made during tool development affect tool results. Decisions to be assessed may relate to, for example, the selection of indicators and indicator thresholds; model structure; processes related to the normalization, weighting, and aggregation of indicators; and the criteria used for the final designation or classification of communities.

Uncertainty and sensitivity analyses are core best practices for quality assurance in composite indicator construction and should be a part of a data strategy for any EJ tool, including CEQ tools such as CEJST. A global uncertainty analysis of CEJST will improve understanding of the precision with which communities are designated as disadvantaged when the model is subjected to alternative construction choices. The modeling decisions creating the most uncertainty can be identified in subsequent global sensitivity analyses. Uncertainty can then be diminished through subsequent research, targeted data collection, and improved modeling. The ultimate goals are reducing statistical fragility and increasing the transparency of the modeling process. Global uncertainty and sensitivity analyses can also provide empirical results that support response to public queries about the certainty of overall and geographically-specific designations of community disadvantage.

CLOSING THOUGHTS

A good data strategy to create a trusted, transparent, and legitimate EJ tool is one that uses a structured composite indicator development process that facilitates decision making on the independent and interrelated design factors in the construction process. Decisions related to clear identification of the concept to be measured; selection of indicators and data; analysis, normalization, weighting, and aggregating the data; evaluation of results for coherence, internal robustness, and validity; and the visual presentation of results need to be made considering how those decisions affect each other. For example, to create a legitimate tool that represents lived experience, decisions regarding effective community engagement need to be made in consideration of how every component of composite indicator construction can be validated to determine if real-world conditions are being represented. Collaborative partnerships with community members, government entities, other tool developers, advisory groups, and technical experts are a means to identify appropriate indicators and datasets, check modeling decisions, validate indicators and results, and refine model approaches. Documenting community engagement, response to input, tool approaches, rationale for decisions, uncertainties in tool data and results, and how and when tool users need to supplement tool results with additional information will create needed transparency.

Careful analysis of indicators and datasets to choose—in partnerships with communities, governments, and technical experts—appropriate representative measures and correlation among datasets. Considering measures of economic burden beyond the federal poverty level and explicitly incorporating measures of race and racism into an EJ tool will result in a tool that also better reflects lived experience. Data strategies based on cumulative impact scoring approaches are the state of the science and advanced practice and can reflect the combined effects of environmental and socioeconomic burdens over time in a manner that reflects reality. Implementing the recommendations in this report will allow the construction of valid geospatial tools for EJ that can inform better targeting of community investment.

1

Introduction

President Biden's Executive Order (E.O.) 14008 (Tackling the Climate Crisis at Home and Abroad) established the Justice40 Initiative and directed the White House Council on Environmental Quality (CEQ) to develop a whole-of-government approach to environmental justice (EJ) (EOP, 2021). Justice40 sets a goal that disadvantaged communities reap 40 percent of federal investment benefits in specific sectors: energy, housing, health and resilient communities and infrastructure, economic and workforce development, transportation, and water. To advance this initiative, E.O. 14008 charged CEQ with creating a geospatial tool, the Climate and Economic Justice Screening Tool (CEJST),[1] that will be used to identify the communities across the United States and its territories eligible for Justice40 investment benefits.

CEJST represents the first time a tool of this kind has been developed at the federal level to identify the disadvantaged communities in terms of climate, energy, sustainable housing, employment, and pollution burden for the purpose of federal investment. As with any novel initiative, it requires breaking new ground in terms of research methodologies and data use. It calls for data of sufficient granularity and scientific validity to compare communities across states and regions, and in terms of their rural and urban context. Figure 1.1 illustrates an example of CEJST output, identifying a specific census tract in South Carolina as disadvantaged. CEJST highlights this tract based on meeting criteria for its climate change and health indicators, and socioeconomic status. These criteria are discussed in more detail throughout the report.

This report summarizes the conclusions and recommendations of a National Academies of Sciences, Engineering, and Medicine (National Academies) committee regarding the directions of a future data strategy for CEQ tools. This chapter describes the statement of task provided to the committee, and a brief history of the use of geospatial

[1] See https://screeningtool.geoplatform.gov/ (accessed October 3, 2023).

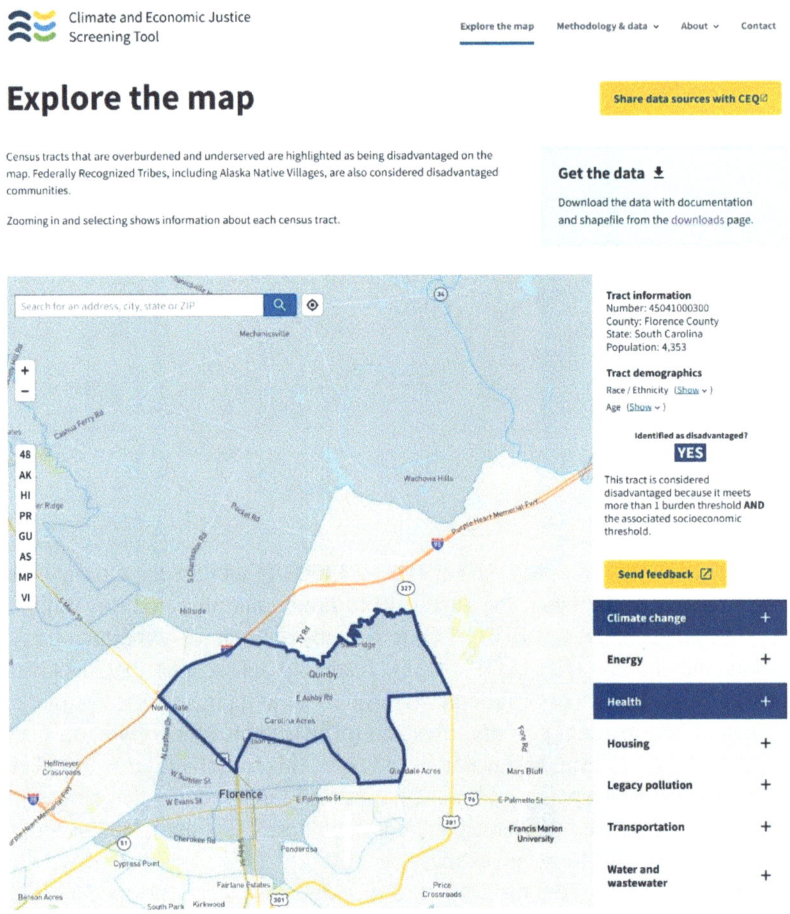

FIGURE 1.1 Screenshot of the Climate and Economic Justice Screening Tool showing a census tract in Florence County, South Carolina designated as disadvantaged due to climate change, health indicators, and socioeconomic status. SOURCE: Climate and Economic Justice Screening Tool (2024).

tools for addressing EJ issues. The committee and committee processes are described below, as is the report organization. Whereas the committee was explicitly asked to provide recommendations regarding CEQ tools, the good practices described in this report are applicable to the developers of other geospatial tools.

STATEMENT OF TASK

Under the sponsorship of the Bezos Earth Fund, the National Academies convened an ad hoc multidisciplinary committee of 11 experts to consider how environmental health and geospatial data and approaches were built into various environmental screening tools to identify disadvantaged communities. The committee was asked to consider how data at a variety of scales and resolution may be integrated and analyzed and to make recommendations for an overall data strategy for CEQ in the development of future versions of CEJST or other tools. The statement of task provided to the committee is found in Box 1.1.

HISTORY

Since the emergence of a national EJ movement in the 1980s, mapping and the use of GIS (geographic information systems) have been important for analyzing and communicating unequal and disproportionate environmental burdens for communities of color and other historically marginalized communities (Commission for Racial Justice, 1987; GAO, 1983; Kumar, 2002). Following issuance in 1994 of E.O. 12898 "Federal Actions to Address Environmental Justice in Minority Populations and Low-Income Populations" (EOP, 1994), federal agencies, especially the Environmental Protection Agency (EPA), state governments, and nongovernmental institutions sought to define "environmental justice" and "disproportionate burden" and to explore ways to map

BOX 1.1
Statement of Task

A committee of the National Academies of Sciences, Engineering, and Medicine will analyze how environmental health and geospatial data and environmental screening tools can inform CEQ's Climate and Economic Justice Screening Tool by conducting a data assessment to assist CEQ in considering the disparities it has prioritized. The committee's assessment will build on the following tasks:

- Scan of existing screening tools for types of data and approaches used to identify disadvantaged communities (e.g., CEQ-funded Climate and Economic Justice Screening Tool, Environmental Protection Agency Environmental Justice Screen)
- Identification of the types of data (e.g., environmental, socioeconomic, energy, transportation) needed for CEQ's screening tool(s)
- Evaluation of current data availability, quality, and spatial and temporal resolutions, as well as key data gaps
- Discussion of approaches to process, integrate, and analyze these data (e.g., weighting, consideration of additive effects)

The committee will provide recommendations to be incorporated in an overall data strategy for CEQ's tool(s).

and identify both (University of California Hastings College of Law, 2010). By 2001, each of the 10 EPA regional offices used some form of EJ survey tool internally to map potential EJ communities for enforcement prioritization of environmental regulations or enhanced scrutiny, based primarily on demographic indicators for the likelihood of experiencing environmental injustices (e.g., the number or percentage of racial minority residents, indicators of income or poverty, and population density) (Kumar, 2002). In 2015, EPA released EJScreen, the first publicly available interactive mapping tool that used a nationally consistent dataset and approach for combining environmental and demographic socioeconomic indicators at the Census Block Group level.[2] EJScreen did not provide thresholds for identifying or prioritizing communities for action, but it did become one of a few examples or templates for the development of EJ mapping and screening tools for other federal agencies, states, and local communities.

Publicly accessible state-level EJ mapping or screening tools were developed in parallel with those of the federal government but took diverging paths on how to define and map EJ or disadvantaged communities (Payne-Sturges et al., 2012). In 2008, Massachusetts released its Environmental Justice Viewer,[3] an interactive, online map that displayed "environmental justice populations" defined by demographic criteria thresholds by Census Block Group—percent minority, percent lower income, percent limited English speaking, or percent foreign born—reflecting the EJ policy issued by the state's Office of Environmental Affairs in 2002. Some states followed the Massachusetts model, defining and mapping "environmental justice" or "overburdened" communities primarily based on demographic thresholds (e.g., Connecticut [2022][4], Illinois [2018],[5] Maryland [2017],[6] Pennsylvania [2022],[7] and Rhode Island [2022]).[8] A different model was released as California's CalEnviroScreen in 2013 after more than a decade of development (Lee, 2021). The latter defined and mapped "cumulative impact" scores for every ZIP code in the state (later changed to census tracts)—a cumulative summary of environmental burdens, exposures, health status, and social vulnerabilities. CalEnviroScreen's approach, and specifically its mapping of cumulative impacts, has been influential for community and academic researchers, state governments (Michigan, New Jersey, New York, and Washington State), and the federal government (Centers

[2] See U.S. Environmental Protection Agency's (EPA)'s website for "How Was EJScreen Developed?" https://www.epa.gov/ejscreen/how-was-ejscreen-developed (accessed December 22, 2023).

[3] See updated Environmental Justice Map, https://www.mass.gov/info-details/massgis-data-2020-environmental-justice-populations (accessed December 22, 2023).

[4] See Connecticut's definition of EJ communities at https://portal.ct.gov/DEEP/Environmental-Justice/05-Learn-More-About-Environmental-Justice-Communities (accessed February 9, 2024).

[5] See Illinois's definition of EJ at https://epa.illinois.gov/topics/environmental-justice/ej-policy.html (accessed February 9, 2024).

[6] See Maryland's EJ definition at https://mde.maryland.gov/Environmental_Justice/Pages/Landing%20Page.aspx (accessed February 9, 2024).

[7] Pennsylvania's Office of Environmental Justice includes their definition of EJ at https://www.dep.pa.gov/publicparticipation/officeofenvironmentaljustice/Pages/default.aspx (accessed February 9, 2024).

[8] See Rhode Island's definition of EJ at https://dem.ri.gov/environmental-protection-bureau/initiatives/environmental-justice (accessed February 9, 2024).

for Disease Control/Agency for Toxic Substances and Disease Registry [CDC/ATSDR], Department of Energy, and Department of Transportation). Some have described "cumulative impact" mapping as the "next generation" of EJ mapping (Lee, 2020). Cumulative impacts will be discussed in further detail in Chapter 3 of this report.

In response to Section 223 of E.O. 14008 "Tackling the Climate Crisis at Home and Abroad" (EOP, 2021), CEQ and a variety of covered federal agencies began the process of developing or launching publicly accessible mapping or screening tools that could be used to identify "disadvantaged communities" to meet the Justice40 Initiative goal that "40 percent of the overall benefits of certain Federal investments flow to disadvantaged communities that are marginalized, underserved, and overburdened by pollution."[9] In July 2023, the Office of Information and Regulatory Affairs (OIRA) released guidance to federal agencies on community engagement practices for regularity purposes.[10] While CEJST is not a regulatory tool, the OIRA guidance provides support for the federal government utilizing community engagement. CEQ launched a beta version of CEJST in February 2022 (Chemnick, 2022), identifying "disadvantaged communities" (DACs) as census tracts in which higher percentiles of environmental or health criteria are paired with markers of economic vulnerability (described in more detail in this chapter). The White House subsequently issued guidance in January 2023 directing covered federal programs to use CEJST as "the primary tool" to geographically identify DACs for the purposes of initiating implementation of the Justice40 Initiative by October 2023 (EOP, 2023).

However, the White House guidance also acknowledged and allowed that some federal agencies already use other tools and methodologies to identify DACs that predate CEJST. In addition to EPA's EJScreen (2015),[11] these earlier tools included the CDC/ATSDR Social Vulnerability Index (2016),[12] the Census Bureau's Community Resilience Estimates (2020),[13] and the Federal Emergency Management Agency (FEMA) National Risk Index (2020).[14] Other federal tools were launched at about the same time as CEJST, including CDC/ATSDR's Environmental Justice Index (2022),[15] the National Oceanic and Atmospheric Administration's Climate Mapping for Resilience and Adaptation (2022),[16] the Department of Energy's Energy Justice Mapping Tool—

---

[9] See the White House's web page for "Justice40: A Whole-of-Government Initiative" https://www.whitehouse.gov/environmentaljustice/justice40/ (accessed October 4, 2023).

[10] See the White House's Office of Information and Regulatory Affairs (OIRA), "Broadening Public Engagement in the Federal Regulatory Process," https://www.whitehouse.gov/wp-content/uploads/2023/07/Broadening-Public-Participation-and-Community-Engagement-in-the-Regulatory-Process.pdf (accessed December 15, 2023).

[11] See https://www.epa.gov/ejscreen (accessed September 18, 2023).

[12] See https://www.atsdr.cdc.gov/placeandhealth/svi/index.html (accessed September 18, 2023).

[13] See https://www.census.gov/programs-surveys/community-resilience-estimates.html (accessed February 9, 2024).

[14] See https://hazards.fema.gov/nri/ (accessed February 9, 2024).

[15] See https://www.atsdr.cdc.gov/placeandhealth/eji/index.html (accessed February 9, 2024).

[16] See https://resilience.climate.gov/ (accessed February 9, 2024).

Disadvantaged Communities Reporter (2022),[17] and the Department of Transportation's Equitable Transportation Community Explorer (2022).[18] All these federal mapping tools focus on identifying socially vulnerable or environmentally burdened communities, and all employ cumulative burden measures or composite indexes of social and environmental risk or burden. None of these tools or metrics share the same methodology. More information on each of these tools can be found in Chapter 4 of this report.

COMMITTEE COMPOSITION

An interdisciplinary group of experts was convened specifically to deliberate on the task to inform the future data strategies for CEQ's tool(s), as described in Box 1.1. Expertise on the committee included data and geographical sciences, geospatial analysis, environmental economics, environmental health, public health, EJ, and environmental science. Those with experience using or creating relevant geospatial tools at the federal and state levels were specifically sought for committee membership, as well as those with experience applying information derived from such tools in research or decisions regarding public health, water, energy, the environment, or infrastructure. Members of the committee have experience addressing EJ issues from the research, decision-making, and advocacy points of view. A range of perspectives and diversity were intentionally considered in committee composition, including diversity in where members live and work, race/ethnicity, age, and gender. Appendix A provides biographies of committee members.

INTERPRETATION OF THE STATEMENT OF TASK AND STUDY BOUNDARIES

The task given to the committee (Box 1.1) included many nuances. The sponsor of the study (Bezos Earth Fund) asked the committee to develop recommendations "to be incorporated in an overall data strategy for CEQ's tool(s)." The committee was directed by the sponsor that this study was not a review or evaluation of CEJST (thus the use of "scan" in the statement of task). Instead, the committee would consider approaches used in several geospatial EJ tools and how those approaches may inform CEJST and potential future CEQ tools. CEJST is CEQ's only tool at present, but the language in the task implies interest in strategies applicable to future tools. Based on language in the statement of task and discussions with the study sponsor, the committee established some boundaries of what would and would not be addressed in its report:

- The committee would focus on geospatial EJ tool development and the selection and integration of the information used as input for such tools;

[17] See https://curie.pnnl.gov/document/energy-justice-mapping-tool-disadvantaged-communities-reporter (accessed February 9, 2024).

[18] See https://experience.arcgis.com/experience/0920984aa80a4362b8778d779b090723 (accessed February 9, 2024).

- The committee would not comment on policy or political decisions informing any approaches taken in the development of any EJ tool; and
- The committee would not comment on any resource allocation decisions made based on the results (i.e., outputs) of any tool.

COMMITTEE INFORMATION GATHERING

The extensive collective expertise and experience in several disciplines and practices relevant to the statement of task was a major source of internal information for the committee. That expertise was augmented by research of the published scientific literature and of technical documentation from several geospatial tools, and from input from a variety of sources including invited speakers, panel discussions, and guests to multiple public meetings, webinars, and a public workshop. Many of the experts consulted were intimately familiar with EJ tools developed at different scales (e.g., state, national) or with tools developed internationally. Appendix B provides the agendas for the committee's public events.

An especially important and influential part of the committee's information gathering was its public workshop, summarized in a proceedings-in-brief (NASEM, 2023a). The committee invited speakers and participants representing community organizations, government entities, academe, and Tribal nations (a participant list is included in Appendix B), and the agenda was divided into presentations, panel discussions, a hands-on exercise in which participants could explore CEJST, and plenary discussion. Participants shared their lived experiences and observations and how those compared with results generated by CEJST. They provided input on the types of indicators (i.e., the individual concepts being measured) and datasets for inclusion that could be useful in the tool, how data might be analyzed, the visual presentation and the way tool results were communicated, on aggregating indicators and determining thresholds of disadvantage, on analyzing the data against localized tools, investment tracking, the importance of community involvement, and other topics.

The committee's workshop emphasized the importance of incorporating community voices in an EJ tool's data strategy—one that integrates community input with quantitative data, that uses community input to validate choices regarding how community and disadvantage are defined and how disadvantage could be measured, and that uses community input to validate tool results against the lived experience of the community. These concepts formed the foundation of the committee's conceptual framework—its vision—for a data strategy and approach to EJ tool development (see Chapter 3).

REPORT ORGANIZATION

The committee developed an approach to its work based on some early conclusions: (1) a valid and useful geospatial EJ tool development requires a transparent structured process; (2) community voices are integral to valid EJ tool construction; and (3) a data strategy appropriate for CEQ and CEJST would also be appropriate for other EJ tool developers and tools. To increase the usefulness of this report to CEQ—and to expand the utility of the report recommendations to others—the concepts and recommendations in this report are generalized and applicable to the development or management of any EJ tool. Although there are many sections in the report that describe specific CEJST processes (as appropriate given the statement of task), the committee attempts to highlight many of those as examples for all tool developers.

The committee organized its report to first provide some background and definitions of important concepts in Chapter 2. These include the concepts of community and community disadvantage, burdens, stressors, and impacts, and cumulative impacts. Chapter 2 also explains some reasons behind disproportionate exposure to different hazards. Chapter 3 then describes different approaches and components for building geospatial tools that compile multiple individual indicators into a single tool to represent the complex multidimensional concept being measured (such as community disadvantage, as CEJST attempts to measure). The committee then offers a conceptual framework for tool development and information about community engagement as an elemental part of the framework. In Chapter 4, the committee provides information about previous surveys of EJ tools, and an overview of the approaches employed in 12 different EJ tools. The committee then describes the selection and analysis of indicators and datasets in Chapter 5, including more detail on the categories of burden used to define a disadvantaged community by CEJST. Chapter 6 describes different approaches for integrating indicators (e.g., combining multiple datasets collected for different purposes and scales) and treating the data. Chapter 7 discusses different methods to validate quantitative and qualitative data. Finally, the committee synthesizes its recommendations for a data strategy in Chapter 8.

2

Background and Definitions

The White House Council on Environmental Quality's (CEQ's) Climate and Economic Justice Screening Tool (CEJST)[1] was designed to screen for communities that qualify for extra consideration for investment under the Justice40 Initiative.[2] Because definitions of the terms associated with identifying disadvantaged communities and economic justice vary depending on the context, this chapter provides definitions of terms as used in this report from climate and economic justice perspectives. The definitions of community and community disadvantage are described, including the types of exposures that lead to that disadvantage. This is followed by some reasons for disproportionate exposure, and definitions for burdens, stressors, and impacts. Because a community is rarely disadvantaged as the result of a single stressor, the concept of cumulative impacts is also introduced here. The chapter concludes with a discussion of how the magnitudes of individual stressors, the presence of multiple stressors, and the interaction across stressors might be considered contributors to disadvantage. A discussion of how different measures may be incorporated into geospatial environmental justice (EJ) tools is provided in Chapter 5.

COMMUNITY AND COMMUNITY DISADVANTAGE

Community is defined as a group of people who share common experiences and interactions. While traditionally associated with specific places (i.e., geographic communities)—such as a neighborhood, town, city, or region—more general definitions decouple community from geography. Communities can exist at multiple scales—from local to global—and places can exist without community (Bradshaw, 2008).

---

[1] See https://screeningtool.geoplatform.gov/en/ (accessed December 15, 2023).

[2] Information on Justice40 can be found on the White House's website, https://www.whitehouse.gov/environmentaljustice/justice40/ (accessed December 15, 2023).

Communities can also be more loosely coupled with geography. Examples include people who move from place to place, such as migrant and seasonal farmworkers, and people who identify as lesbian, gay, bisexual, transgender, queer (or questioning their gender), intersex, asexual, (or their allies) (LGBTQIA+) who share experiences and interactions with others who live in disparate locations.

Residential location often defines the proximity of a geographic community due to the importance of homes in anchoring social networks, activities, and exposure to environments. A geographic community often corresponds to what is generally referred to as a neighborhood in a city or town, although the geographic extent can vary depending on the location and context: neighborhoods tend to be compact in dense urban settings and more geographically extensive in rural areas. While the Justice40 Initiative recognizes both nongeographic and geographic communities, geospatial tools such as CEJST

BOX 2.1
What Is Redlining?

Redlining refers to the discriminatory practice of conditioning access to mortgage lending and insurance on the racial composition of neighborhoods. To address housing price collapses and a foreclosure crisis in the wake of the Great Depression, the Roosevelt Administration transformed housing finance from short-term loans with balloon payments to fully amortized longer-term loans. The Federal Housing Administration (FHA) introduced mortgage insurance and associated underwriting guidelines. The Federal National Mortgage Insurance Agency created a secondary loan market, and the Federal Home Loan Bank Board (FHLBB) chartered and oversaw federal savings and loan associations.

The Home Owners' Loan Corporation (HOLC), created in 1933 as an agency operating under the FHLBB, was charged with assessing the relative riskiness of lending in neighborhoods in more than 200 U.S. cities (Aaronson, Hartley, and Mazumder, 2021). Between 1935 and 1940, HOLC developed a series of neighborhood descriptions with color-coded maps summarizing mortgage lending risk. The HOLC maps gave grades of A, B, C, and D to neighborhoods reflecting increasing estimated risk; the neighborhoods on the corresponding maps were colored green, blue, yellow, and red, respectively (Fishback et al., 2020). Black households were disproportionately concentrated in "D" or "redlined" neighborhoods and considered hazardous for mortgage lending. HOLC shared these with the FHA, which had already woven racial segregation into the provision of mortgage insurance (Mallach, 2024). There is also (disputed) evidence of the sharing of these maps with private lenders, although that was not official policy (Aaronson et al., 2021). The Mapping Inequality project at the University of Richmond maintains an online collection of georeferenced HOLC maps across the United States (https://dsl.richmond.edu/panorama/redlining/), which allows users to interactively explore these historic residential security maps for over 200 cities across the country. Figure 2.1.1 shows a 1936 HOLC map for Columbus, Ohio.

may only identify geographically proximate communities. Justice40 also considers geographic areas within Tribal jurisdictions to be communities of interest (EOP, 2023).

Because of their physical proximity, those in a geographic community may be both positively and negatively influenced by common forces. Neighborhood effects refer to these shared influences on individual social, economic, and health outcomes; these stem from the interactions and exposures among individuals and with the physical and built environments (Dietz, 2002; Diez Roux and Mair, 2010). Geographic communities may also be influenced by the policies and situations that establish, sustain, and otherwise impact the social determinants of health. In the case of these political determinants of health, both action and inaction are considered to have impacts and outcomes. Examples of political determinants include public and private policies, such as redlining (see Box

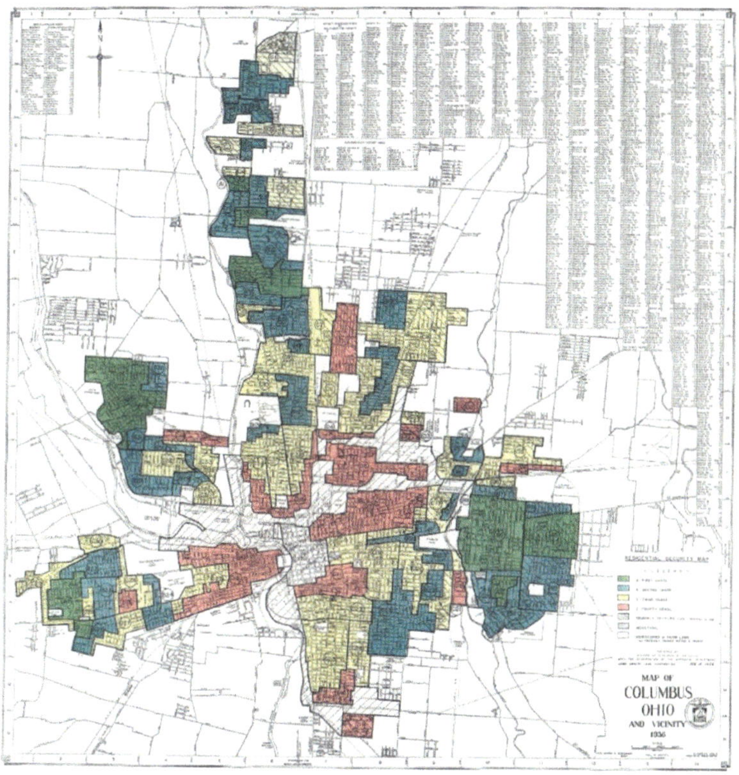

FIGURE 2.1.1 A HOLC map for Columbus Ohio (1936). "A" neighborhoods (indicated in green on the map) were rated most desirable for mortgages. In contrast, "D" neighborhoods (indicated in red) were rated highest risk for mortgages. Black people were disproportionately concentrated in the "redlined" neighborhoods in Columbus and many other U.S. cities. SOURCE: City of Columbus Historical Data (1936).

2.1), mortgage insurance, racially restrictive covenants, and exclusionary zoning, which have led to disinvestment and disadvantage (Dawes, 2020).

Defining community disadvantage is akin to the "wicked problems" that often face policy and planning decisions (Rittel and Webber, 1973); it is multifaceted and value-laden with intricate interdependencies that make it hard to describe comprehensively. The precise definition may shift depending on the problem context and lens brought to bear on the issue. Nevertheless, there are scientific principles and empirical evidence that provide a foundation for operational and actionable definitions of community disadvantage and related concepts.

Community disadvantage results from a complex interplay of factors that inhibit or prevent people in some communities from achieving the positive life outcomes that are commonly expected in a society. These factors include the social, economic, built, and natural environments in which they live (Price-Robertson, 2011). Traditional conceptualizations and measures of community disadvantage focus solely on economic disadvantage (e.g., poverty and unemployment), pointing directly to redistributive and workforce development policies. Multifaceted approaches, including social capital, financial capital, resources and social inclusion/exclusion, capture more of the complex interplay of factors (Price-Robertson, 2011). Factors that influence community resilience and disadvantaged status are rooted in world and national history. These factors go back centuries, European colonization, the transatlantic slave trade, capitalism and patriarchy took hold in what we now call the United States, as Europeans began colonizing and trafficking enslaved individuals (Berberoglu, 1994; Tarlow, 2024).

The social capital approach (e.g., Putnam, 2000) focuses on the social norms and networks in a community and the neighborhood and place effects that encourage or inhibit strong norms and networks. The social inclusion/exclusion approach (e.g., the UK Indexes of Multiple Deprivation; see Deas et al., 2003) focuses on individuals in a community and the barriers they face. Building on the work of economist Amartya Sen and philosopher Martha Nussbaum, the capabilities approach views disadvantage as restrictions on the ability to take actions and participate in activities that have meaning and value to people. The latter approach is comprehensive across the human experience, including lifespan, health and wellness, bodily autonomy, and integrity and control over one's environment (Price-Robertson, 2011).

Community disadvantage is also a consequence of structural factors and historical processes that create conditions undermining resilience to shocks and disruptions, such as those associated with climate change, economic transitions, or other social and environmental pressures. One major factor is marginalization; it occurs when people are excluded from mainstream social, economic, educational, or cultural life based on personal attributes or history. Examples include, but are not limited to, groups excluded due to racial identity, gender identity, sexual orientation, age, disability status, language, or immigration status (Sevelius et al., 2020). Marginalized people are frequently segregated into specific geographic communities due to sorting based on ability to pay for housing and as a result of discriminatory practices such as racially restricted covenants and exclusionary zoning (Galster and Sharkey, 2017; Reardon and Bischoff, 2011; Rothstein, 2017; Shertzer, Twinam, and Walsh, 2022). Marginalization can also occur due to economic and social changes resulting from market forces, often shaped by policy decisions; examples include rural areas losing population due to continuing urbanization and the loss of traditional economic bases in regions such as the U.S. Midwest and Appalachia. Another cause of marginalization is through displacement; examples include gentrification (Finio, 2022) and the forced movement and concentration of Native Americans (Farrell et al., 2021). Marginalized groups can form communities of safety from the intentional rejection of what is considered mainstream society. (Betts and Hinsz, 2013).

Disadvantaged communities can face lower levels of private and public capital investment, with detrimental impacts on personal and community resources and the built environment. In urban areas, the historic practice of redlining—the discriminatory practice of basing mortgage eligibility on racial identity—enforced segregation and deprived people of the ability to build personal wealth through homeownership and to transfer this wealth to subsequent generations (Aaronson, Hartley, and Mazumder, 2017; Faber, 2020; Fishback et al., 2021; Rothstein, 2017; Woods, 2012, 2013). Compounded with consequent poor-quality housing stock and infrastructure, these effects are profound and persistent: historically redlined neighborhoods are correlated with heat stress (Li et al., 2022; Schinasi et al., 2022; Wilson, 2020), air pollution (Bose, Madrigano, and Hansel, 2022; Cushing et al., 2023a; Lane et al., 2022; Schuyler and Wenzel, 2022), traffic violence (Taylor et al., 2023), lack of healthy food (Li and Yuan, 2022; Shaker et al., 2023), and a range of health disparities (Li et al., 2022) that persist to the present day.

DISPROPORTIONATE EXPOSURE TO ENVIRONMENTAL HAZARDS

Disadvantaged communities are disproportionately exposed to technological and natural environmental hazards. These hazards may include highways and railroads, extractive and industrial processes, natural hazards, and waste disposal sites (Di Fonzo, Fabri, and Pasetto, 2022; Donaghy et al., 2023; Johnston and Cushing, 2020; Johnston et al., 2020; Karas, 2015). The potential effects include increased exposure to harmful chemical or biological agents, poor air quality, noise, and disruptive events that undermine resilience. Marginalized communities are often located in places with such undesirable exposures, on floodplains, near industrial areas, or in neighborhoods that lack tree canopy and green space (Volin et al., 2020). Different reasons for disproportionate exposures can be placed into economic, sociopolitical, and racially discriminatory categories, described in the following paragraphs.

Economic Explanations

Economic explanations of disproportionate exposure to environmental hazards focus on market dynamics. The spatial concentration of environmentally hazardous facilities in areas that are predominantly composed of people of color and various socioeconomically disadvantaged communities is a consequence of market forces that encourage industries to look for inexpensive land and labor. Areas providing inexpensive land and labor often coincide with neighborhoods where people of color and people with low income are more likely to reside. Additionally, after industrial facilities and other pollution sources have been sited, their negative impacts (e.g., air pollution, community targeting, and stigmatization) may decrease property values. Housing becomes more affordable in those areas for people of lower income. Market forces can thus result in greater in-migration of socially disadvantaged people in polluted communities and out-migration of socially advantaged people who can afford to move elsewhere (Saha and Mohai, 2005). Recent studies of the economic forces contributing to disproportionate exposure focus on how those forces affect residential location decisions (often referred to as "sorting") and siting decisions by firms, as well as how these effects are shaped by discrimination (in markets as well as in policy and enforcement of regulations) (Banzhaf, Ma, and Timmins, 2019a,b; Cain et al., 2024). Some studies also recognize that markets themselves are inherently social and political constructs resulting from economic, social, and political power as well as initial resource allocations that can lead to persistent inequalities and stratification (Chelwa, Hamilton, and Stewart, 2022).

Sociopolitical Explanations

Sociopolitical explanations of disproportionate exposure to environmental hazards focus on the power relations and politics that drive unequal environmental outcomes. Specifically, communities with lower levels of educational attainment, voter participation, home ownership, social capital, and political clout are frequently targeted by industries seeking less local resistance to development (Bullard and Wright, 1987). Communities where opposition to industries will be minimal or less effective are also likely to contain higher proportions of people of color and low-income residents, including people with limited access to resources and political representation. The success of white "not-in-my-backyard" attitudes—often referred to as "NIMBYism"—has led to more hazardous waste sites, polluting industries, and other undesirable land uses in socially disadvantaged communities with the fewest number of resources to oppose such siting (Mohai and Saha, 2015).

Racial Discrimination Explanations

Scholars from multiple disciplines have long recognized racism as a fundamental cause of disadvantage and other social, economic, health, and environmental inequalities in the United States (Bonilla-Silva, 1997; Bullard, 2001; Callahan et al., 2021). Earlier scholarship and legal theory focused on interpersonal discrimination and questions of racist intent, but the focus has now shifted to understanding the role of historical and current institutional actions and structures that lead to racially unequal and discriminatory outcomes—referred to as systemic, institutional, or structural racism. While scholars increasingly differentiate these terms,[3] this report uses the generic term

[3] For elaboration on various theories of racism, see Dean and Thorpe (2022).

racism to refer to the phenomenon of structural racism. Following the definition offered by Bailey and others (2017), structural racism refers to "the totality of ways in which societies foster racial discrimination through mutually reinforcing systems of housing, education, employment, earnings, benefits, credit, media, health care, and criminal justice. These patterns and practices, in turn, reinforce discriminatory beliefs, values, and distribution of resources."

Key to this conceptualization of racism is the understanding that systematic differences between racial and ethnic groups in socioeconomic status, wealth, education, political power, and health are a consequence of historical, social, institutional, or political circumstances and are not reflective of innate biological or cultural differences ascribable to race or ethnicity itself. Indeed, leading scholars across the disciplines of anthropology, sociology, public health, health care, and population genetics have been at pains to educate their peers in the scientific community and the public that race and racial categories are social and political constructs, and they are not coherent or credible biological categories. Differences between races are a consequence of racism, not race itself (Adkins-Jackson et al., 2022; Bailey et al., 2017; Boyd et al., 2020; Braveman et al., 2022; Lett et al., 2022; NASEM, 2023b; Payne-Sturges, Gee, and Cory-Slechta, 2021; Smedley and Smedley, 2005; Yudell et al., 2016).

Empirical Evidence of Racism as a Relevant Factor in Unequal Exposures and Outcomes

A large body of research demonstrates the association of race or ethnicity with disproportionate exposure to environmental stressors. Since its emergence in the early 1980s, EJ scholarship, in particular, has highlighted the central role of racism in explaining geographic and population-based patterns of unequal exposures to environmental stressors, including hazardous waste (e.g., Bullard et al., 2007; Mohai and Saha, 2007), air pollution (e.g., Bravo et al., 2022; Kodros et al., 2022; Lane et al., 2022), water pollution (Konisky, Reenock, and Conley, 2021; Martinez-Morata et al., 2022), toxic metals (e.g., Martinez-Morata et al., 2022; O'Shea et al., 2021), and noise (e.g., Collins, Nadybal, and Grineski, 2020; Trudeau, King, and Guastavino, 2023). Conversely, the same communities are shown to suffer disproportionately from the absence of various environmental and public amenities, such as access to recreational and open space (e.g., Fernandez, Harris, and Rose, 2021; Sims et al., 2022), access to affordable and healthy sources of food, adequate tree canopy coverage (e.g., Locke et al., 2021; Schwarz et al., 2015), functional public infrastructure (e.g., Kim, M. et al., 2023; Luna and Nicholas, 2022), and consistent or equal enforcement of environmental laws and regulations (e.g., Bae and Kang, 2022; Bae, Kang, and Lynch, 2023).

While some scholars have raised questions about the relative importance of race versus class or socioeconomic status (SES) in explaining these patterns, numerous investigations have repeatedly demonstrated that race is an important quantitative predictor of unequal exposure and outcomes, independently and sometimes more significantly than income or SES. Indeed, one common finding is that minoritized racial or ethnic groups within the same SES systematically experience greater exposures or burdens relative to their white counterparts. The implication is that associations of exposure with race are not explained simply as a function of underlying SES differences, and SES is not a substitute for racial differences (see Bullard et al., 2007; Mohai and Saha, 2007; Liu et al., 2021; Tessum et al., 2021). At the same time, race and ethnicity are often highly correlated with various measures of SES, especially income and educational attainment, suggesting that race and ethnicity interact with SES in complex ways to produce these unequal exposures and outcomes.

Race is not equivalent to racism (Adkins-Jackson et al., 2022). While indicators of race and ethnicity are consistently associated with various forms of social, environmental, and health inequities, these indicators do not reveal the processes by which these inequities arise. Structural racism has emerged as a powerful theoretical framework for understanding and measuring the phenomenon of racism and its impacts, particularly in the population health sciences. Research on structural racism has drawn renewed attention to the utility of long-established sociological theories that racism is not a singular or separable phenomenon but rather intersects with other systems—economic, political, institutional—to produce and maintain racial inequalities in power, status, access to resources, and health (Bonilla-Silva, 1997; Du Bois, 1889; Massey, 1990). Health science scholars, schools of medicine and public health, and health organizations now explicitly recognize racism as a key social determinant of health and a fundamental cause of health inequities—e.g., asthma, cardiovascular diseases, cancer, diabetes, kidney disease, low-birth-weight pregnancies—many of which are also connected to differential exposures to environmental stressors (e.g., lead, air pollutants, hazardous waste, contaminated water) (Ahmed, Scretching, and Lane, 2023; Commission on Social Determinants of Health, 2008; Dean and Thorpe, 2022; Dennis et al., 2021; Paradies et al., 2015).

The interlocking systems of structural racism and their cascading effects on social and health inequities can be understood in consideration of the exclusion of Black people from housing and intergenerational wealth accumulation (Brown, 2022; Chadha et al., 2020). In the aftermath of the Reconstruction Era, Jim Crow laws throughout the former confederate states mandated racial segregation in all public facilities—including schools, public transit, and government buildings—and companion laws excluded Black people from voting and thus political representation (Daniels, 2020; Du Bois, 1935). Laws enforcing racial segregation and second-class citizenship remained in place through the mid-20th century, institutionalizing economic, educational, and other social disadvantages (Abrams, 1955; Jackson, 1987). It is important to understand that racist policies and practices were not confined to the South (Luxenberg, 2019; Purnell, Theoharis, and Woodard, 2019).

New Deal federal policies following the Great Depression institutionalized other forms of racial discrimination nationally. For example, the Social Security Act of 1935 created a system of employment-based old-age insurance and unemployment compensation, which has become a cornerstone of the country's social safety net and dramatically reduced poverty among the elderly. However, to secure passage of the Act from southern Democrats, agricultural workers and domestic servants were excluded from the program, occupations held largely by Black men and women. Social Security provided mostly white recipients with an opportunity to protect and pass on wealth to their children. By contrast, those who were excluded were not afforded this opportunity, forcing them to depend on their children in retirement, further diminishing the opportunity to pass on intergenerational wealth (Bailey et al., 2017; Omi and Winant, 2014). Other federal policies were more racially explicit.

As described in Box 2.1, discriminatory redlining practices had disproportionately negative impacts on communities of color. While the New Deal and immediate post-World War II periods witnessed a national expansion in homeownership and household wealth as a result of programs and policies to support homeownership, residents of redlined neighborhoods, especially Black residents, were effectively excluded from this opportunity for building intergenerational wealth, or even from moving out of segregated, inner-city neighborhoods (Aaronson, Hartley, and Mazumder, 2021; Faber, 2020; Rothstein, 2017). Although official redlining maps were ruled unconstitutional by the U.S. Supreme Court in the 1940s, the footprint of redlining has endured to the present day. Researchers have found that the locations of formerly redlined neighborhoods show strong correlations with a range of social, environmental, and health inequalities for the minoritized groups that live there (Aaronson et al., 2021; Berberian et al., 2023; Blatt et al., 2024; Bompoti, Coelho, and Pawlowski, 2024; Hoffman, Shandas, and Pendleton, 2020; Kephart, 2022; Lane et al., 2022). These persistent harms are compounded by persistent discrimination in the housing and rental markets (Howell and Korver-Glenn, 2021; Langowski et al., 2020).

Other federal policies aimed at supporting housing and wealth creation were similarly denied to Black people. Veterans returning from World War II in 1945 were entitled to G.I. Bill benefits, which provided free college and low-cost home and business loans to veterans and has been credited with lifting millions of veterans and their families into the middle class. Although Black veterans were technically entitled to these same benefits, local practices of housing and educational racial discrimination throughout the country meant that most were never able to use these benefits to secure quality housing, education, and the socioeconomic benefits that such advantages foster (Agbai, 2022; Lawrence, 2022; Meschede et al., 2022; Turner and Bound, 2002). Racial discrimination in housing was not outlawed nationally until 1968, and banks were not required to practice equitable lending until 1977.

Although housing discrimination based on race is illegal, research shows that it is still widespread and operates through both institutional and social mechanisms. Housing discrimination's most visible manifestation is the persistence of racial residential segregation across the country (Massey, 2020). Scholars have identified residential segregation as a pillar in the foundation of structural racism and a direct contributor to racialized health inequities (Bailey et al., 2017). Residential segregation contributes to poor health outcomes by concentrating people of color in neighborhoods with dilapidated housing, substandard quality of the social and built environment, greater concentration of and exposure to pollutants and toxics, limited access to high-quality educational and employment opportunities, and restricted access to health care. Health outcomes associated with residential segregation include higher rates of adverse birth outcomes (Acevedo-Garcia et al., 2003), increased exposure to air pollutants (Bravo et al., 2016; Lane et al., 2022; Smiley, 2019), less access to parks and greenspace (Kephart, 2022), shorter lifespans (Collins and Williams, 1999; Williams and Collins, 2001), increased risk of chronic disease (Acevedo-Garcia et al., 2003; Kershaw et al., 2011; Williams and Collins, 2001), and increased rates of homicide and other crime (Collins and Williams, 1999; Krivo et al., 2015).

Housing and employment have also been shown to intersect with disproportionate policing and incarceration of Black people (Chadha et al., 2020). The federal War on Crime policies that originated in the mid-1960s and peaked in the 1980s encouraged and supported more aggressive and punitive policing and was directed most heavily against communities of color through disproportionate incarceration and harsher sentencing for the same crimes committed by white people (Alexander, 2012; Bailey et al., 2017; Chadha et al., 2020; Hinton, 2016). Mass incarceration of people of color has had both direct and indirect effects on incarcerated individuals and their communities. In addition to the psychological and economic impacts on incarcerated individuals who face numerous barriers to securing housing and employment after returning to society, a toll is also taken on their families and communities. Disproportionate incarceration rates are both correlated with and predictive of a variety of community-level health inequities, such as adverse birth outcomes (Larrabee Sonderlund et al., 2022), demonstrating another system by which structural racism intersects with health and economic inequities (Bailey, Feldman, and Bassett, 2021; Dennis et al., 2021). Greater appreciation of historical and sociological understandings of racism, in conjunction with the adoption of the social determinants of health framework, is providing population and environmental health scholars with a growing array of conceptual tools to operationalize, measure, and document racism and its consequences.

Measuring Racism

There are numerous methods for operationalizing and measuring structural racism. In a systematic review of public health literature, Groos and others (2018) identified 20 research articles assessing structural racism across the domains of residential neighborhood/housing, perceived racism in social institutions, SES, criminal justice, immigration and border enforcement, political participation, and workplace environment. These studies found that structural racism was associated with mental and physical impacts, including stress, anxiety, poor psychological well-being, colorectal cancer survival, myocardial infarction, mean arterial blood pressure, episodic memory function, behavioral changes, poor adherence to hypertensive treatment, and delayed HIV testing across the population.

In a comprehensive and systematic review, Ahmed, Scretching, and Lane (2023) identified over 1,700 relevant peer-reviewed research articles and reviewed 54 of these in depth to understand the use of different designs, measures, and measurement indexes when studying structural racism as a social determinant of health. Among the 58 different measurable health outcomes in these studies, they found that infant health outcomes (e.g., preterm birth, low birth weight, infant mortality) and quality of life (e.g., dementia, disability patterns, years of life lost) were primarily affected by structural racism, more than by other social determinants of health, such as educational attainment, employment, income, or access to health care. An array of chronic health conditions was also associated with structural racism, including cardiovascular disease, acute respiratory syndrome, body mass index, and late-stage diagnosis of cancer. Ahmed, Scretching, and Lane (2023) identified 73 measurement scales or indexes of structural racism. The most common scales of measurement were the Concentration of Extremes,[4] the Dissimilarity Index (see Hammonds and Herzig, 2009; Kramer and Hogue, 2009; Pursch et al., 2020), the Everyday Discrimination Scale (see Commission on Social Determinants of Health, 2008; Dennis et al., 2021; Hardy-Fanta et al., 2006; Lukachko, Hatzenbuehler, and Keyes, 2014), the Experience of Discrimination Scale (see Alson et al., 2021; Commission on Social Determinants of Health, 2008; Dennis et al., 2021; Dougherty et al., 2020; Lukachko, Hatzenbuehler, and Keyes, 2014), the Five Segregation Scale (see Levac, Colquhoun, and O'Brien, 2010; Tester, McNicoll, and Tran, 2012), the Index of Race Related Stress (see Arksey and O'Malley, 2005; Chambers et al., 2020; Hammonds and Herzig, 2009; Hankerson, Suite, and Bailey, 2015), the Isolation Index (see Hammonds and Herzig, 2009; Hansen, 2015; Harnois et al., 2019), and the Perceived Racism Scale (Commission on Social Determinants of Health, 2008) Hankerson, Suite, and Bailey, 2015).

Reviews of the scientific literature have observed that measures of segregation, especially Black-white segregation, have been the most common indicators used to describe and model structural racism. This is not surprising given the ease of data availability and the well-documented association of residential segregation with environmental and health inequities (see, e.g., Bravo et al., 2016; Jones et al., 2014; Kodros et al., 2022; Kramer and Hogue, 2009; Morello-Frosch and Jesdale, 2006; Morello-Frosch and Lopez, 2006; Rice et al., 2014; Woo et al., 2019; Yitshak-Sade et al., 2020. More recently, scholars have argued that structural racism needs to be measured using an index method that better reflects its multidimensional nature (Adkins-Jackson et al., 2022; Dean and Thorpe, 2022; Furtado et al., 2023). However, there is currently no scientific consensus on the best way to measure structural racism (Ahmed, Scretching, and Lane, 2023). But rather than seeking consensus on one best way, Furtado et al. (2023) and Wien, Miller, and Kramer (2023) suggest that the appropriate measure of racism should be selected based on the context and specific question at hand, as well as the specific population of concern. Indeed, one of the common shortcomings identified in most measures of structural racism, including multidimensional indexes, is the reliance on Black-white indicators of racism and the implicit assumption that racialized experiences are the same for all minoritized racial and ethnic groups. That assumption is belied by the available evidence. For example, research by multiple independent research teams has documented how exposure to contaminated drinking water and unequal enforcement of Clean Water Act regulations across the country are most strongly associated with populations of Hispanic/Latino and American Indian/Alaskan Native residents across the South and Southwest, but not with populations of non-Hispanic Black residents

[4] Also described as the Index of the Concentration of Extremes. See Feldman and Bassett (2021); Krieger et al. (1993); and Nyika and Murray-Orr (2017).

(Bae and Kang, 2022; Bae and Lynch, 2023; Konisky, Reenock, and Conley, 2021; Martinez-Morata et al., 2022).

In addition to problematic assumptions about a uniform racialized experience across racial and ethnic groups, intraethnic differences are also important. Marquez-Velarde (2020) disaggregated asthma prevalence data for Mexican Americans, separating foreign-born and U.S.-born, as well as Black Mexican Americans and white Mexican Americans, and found that Black Mexican Americans had a significant disadvantage in relation to both their white Mexican American counterparts and white non-Mexican Americans, suggesting that ethnic experiences are modified by race. Lett and others (2022) have raised similar concerns about treating "Hispanic" as a single ethnic category separable from race and lumping together "multiple, heterogeneous populations spanning dozens of countries and racial groups as well as highly variable socioeconomic, political, and cultural contexts." They draw the example of research on "Hispanic health," which would simultaneously refer to "groups of impoverished Guatemalan migrant workers of largely Indigenous origins as categorically equivalent to wealthier, whiter, Cuban U.S. citizens of significantly more European descent" (Lett et al., 2022, p. 160). Such an undifferentiated approach would provide limited utility in identifying health inequities and might possibly hide or erase important differences.

Similar critiques could be raised about other large and usually undifferentiated racial categories, such as "Asian," which include peoples of vastly different cultural and historical experiences and economic strata, from China to Japan to India. Finally, but no less importantly, are the experiences of Indigenous populations, such as Native Americans, Alaskan Natives, and Native Pacific Islanders, all of whom are only infrequently disaggregated from "minority" or "people of color" (Balakrishnan et al., 2022). Although Indigenous populations have experiences that parallel those of other non-white, minoritized groups (e.g., higher rates of unemployment and poverty, disproportionate exposure to environmental burdens), the settler colonial structures that created these disparities are unique to them, and their concerns are land based and connected to issues of political sovereignty (Dennis et al., 2021; Wispelwey et al., 2023). Numerous researchers of structural racism argue that credible and useful measures of race and ethnicity, as well as racism, must take these differences into account by critically assessing how racial and ethnic categories are being conceptualized and by disaggregating racial and ethnic data in order to reflect these different experiences and their relation to disparities (Adkins-Jackson et al., 2022; Braveman et al., 2022; Casey et al., 2023; Dennis et al., 2021; Furtado et al., 2023; Lett et al., 2022; Payne-Sturges, Gee, and Cory-Slechta, 2021).

Research on health and environmental inequities has demonstrated that measures of racism are essential to identifying and understanding inequity. Within the health sciences, structural racism is identified as a fundamental cause of health inequities and a key element of the social determinants of health. Scholars advise that structural racism is best measured through a multidimensional index that reflects a theoretical understanding of the phenomenon of racism for a given domain or set of domains, institutional context, and population of concern. Appropriate measures of racism require proper conceptualization of and data on race and ethnicity. Properly collecting data on race and ethnicity, and disaggregating racial and ethnic categories, is essential in order to monitor the state of racial or ethnic disparities, to properly identify differences in population experiences of racism and inequity, and to avoid perpetuating or exacerbating structural racism through the erasure of real differences between and within population groups (Adkins-Jackson et al., 2022; Braveman et al., 2022; Kauh, Read, and Scheitler, 2021; Polonik et al., 2023; Wang et al., 2022).

Please see Chapter 5 for further discussion of how to apply measures of racism in geospatial EJ tools.

Social Vulnerability

Disproportionate exposure stemming from economic, sociopolitical, and racial discrimination processes is not the only major mode of societal marginalization pertinent to environmental injustice. Some of the same populations experiencing persistent exposure and exclusion are also less resistant to its effects due to marginalization-induced social vulnerability. Social vulnerability arises due to combinations of political, economic, social, and institutional processes that reduce the capacity of individuals and groups to anticipate, cope with, resist, and recover from hazard impacts (Wisner et al., 2004). EJ research and activism have traditionally focused on marginalization at intersections with anthropogenic hazards, whereas social vulnerability scholarship and modeling have done so in the context of disasters and natural hazards (Ryder, 2017).

The groups facing persistently heightened adverse hazard impacts and outcomes are multiple and intersectional (Elliott and Pais, 2006; Kuran et al., 2020; see also Box 2.2 on intersectionality), extending beyond the common focus on socioeconomic status (Fothergill and Peek, 2004) to also include renters (Lee and Van Zandt, 2019), populations at extremes of age (Ngo, 2001; Peek and Stough, 2010), with limited English language proficiency (Santos-Hernández and Morrow, 2013), living with disability (Wisner, Gaillard, and Kelman, 2012; Chakraborty, Grineski, and Collins, 2019), racial and ethnic minorities (Fothergill, Maestas, and Darlington, 1999), federally subsidized housing residents (Chakraborty et al., 2021), and undocumented immigrants (Méndez, Flores-Haro, and Zucker, 2020). Starting in the 2000s, researchers began quantifying, modeling, and mapping social vulnerability by collecting demographic indicators reflective of sensitive populations and aggregating them into indexes (Clark et al., 1998; Cutter, Boruff, and Shirley, 2003). The aggregated index values reflect both multidimensionality and accumulation of social vulnerabilities across indicators. Social vulnerability indexes quantified processes that had been observed but rarely measured, enabling exploration of its spatial distribution across places (Emrich and Cutter, 2011; Nayak et al., 2018), temporal variation (Cutter and Finch, 2008), and intersection with an array of natural hazards (Schumann et al., 2024; Wu, Yarnal, and Fisher, 2002) and disaster impacts (Finch, Emrich, and Cutter, 2010).

The Social Vulnerability Index (SoVI) developed at the University of South Carolina (Cutter, 2024) was the first social vulnerability index to find widespread use in natural hazards research. SoVI applies principal components analysis to reduce more than two dozen indicators to a smaller number of statistical factors, which are aggregated to create the index (Cutter, Boruff, and Shirley, 2012). The U.S. Centers for Disease Control and Prevention later developed an index also called the Social Vulnerability Index (SVI), which is based on an aggregation of roughly 15 indicators (see Chapter 4; Flanagan et al., 2018). The SoVI and SVI remain the dominant indexes of social vulnerability in the United States and have found utility as screening tools in planning and resource allocation in pre-disaster mitigation (Community Disaster Resilience Zones Act of 2022, Pub. L. No, 117-255; FEMA, 2022) and post-disaster assistance (Blackwood and Cutter, 2023; West Virginia Development Office, 2020). The major questions raised about social vulnerability indexes surround their fit for policy application (Hinkel, 2011), lack of stakeholder participation in their development (Preston, Yuen, and Westaway, 2011), stigmatization of people and places as inherently deficient (Marino and Faas, 2020), correlation with disaster outcomes (Bakkensen et al., 2017; Rufat et al., 2019), degree of statistical robustness (Schmidtlein et al., 2008; Spielman et al., 2020; Tate, 2012), and obfuscation of underlying vulnerability drivers (Greco et al., 2019). Such critiques can help guide the development of the newer field of composite indicators for EJ tools.

BURDENS, STRESSORS, AND IMPACTS

The concept of burden is included in nearly all definitions of disadvantaged communities. For example, the California Public Utilities Commission defines disadvantaged communities as "areas … which most suffer from a combination of economic, health, and environmental burdens. These burdens include poverty, high unemployment, air and water pollution, presence of hazardous wastes as well as high incidence of asthma and heart disease."[5] Burden is a widely used concept that indicates an activity or agent with negative consequences for human health and well-being. In other cases, burden refers to the impacts themselves (i.e., outcomes experienced by individuals), which stem from a combination of exposures and barriers. Under the latter use of the term, for example, excessive heat can be an exposure for an entire population of a given location but might only be a burden for those without the ability to escape the heat (e.g., those without access to air-conditioned space).

An alternative terminology that emphasizes the distinction between causes and effects is stressors and impacts. Following Tulve and others (2016), the U.S. Environmental Protection Agency's Office of Research and Development (ORD) defines stressors as "any physical, chemical, social, or biological entity that can induce a change (either positive or negative) in health, well-being, and quality of life (either now or into the future)" (EPA, 2022a). More broadly interpreted, stressors can encompass both exposures and preexisting conditions that can lead to (cause) impacts (or burdens) either by themselves or when combined. For example, asthma or heart disease would lead to negative health impacts when suffered alone and even more so when combined with

[5] See https://www.cpuc.ca.gov/industries-and-topics/electrical-energy/infrastructure/disadvantaged-communities#:~:text=Disadvantaged%20communities%20refers%20to%20the,of%20asthma%20and%20heart%20disease (accessed September 15, 2023).

other stressors such as exposure to high levels of particulate matter. Similarly, poverty negatively impacts well-being both by itself and even more so when combined with environmental exposures. On the other hand, given the disproportionate impact on certain age groups (e.g., children or the elderly) of exposures such as lead or excessive heat, age might be a stressor when combined with these exposures (other stressors) even though it would not be viewed as a stressor by itself. All of these would fall under a broad interpretation of the term stressor.

The magnitude of a stressor is the quantity, size, or degree of the stressor. This dimension can also capture the difference between temporary exposure and long-term (e.g., lifetime) exposure, as well as the potential for increased future exposure (due, for example, to climate change). For example, being exposed to a given stressor for a longer period can lead to a larger total (cumulative) burden, especially when impacts exhibit threshold effects or nonlinearities. Data reported as continuous values—either in raw form as absolute values or in relative measures such as percentiles, rankings/ratings, or continuous indexes—provide information about the magnitude (or relative magnitude) of a given stressor, allowing for variation across communities in the measured level of burden. By contrast, data reported simply based on exceeding a threshold (in the form of a binary variable of "above" or "below") allow for only a very crude representation of impact that does not distinguish among different levels of impact in the "above" or "below" categories. A community that is only very slightly above the threshold is viewed as being identical to one that is far above the threshold, and vice versa. The implication of this approach is an underestimation of the burden impact of the stressor for groups that have relatively high or extreme magnitudes of exposure.

MULTIPLE STRESSORS

Many EJ studies examine impacts based only on single environmental stressors. Examples include investigating proximity to landfills (Commission for Racial Justice, 1987), accidental releases of hazardous substances (Chakraborty et al., 2014), exposure to air pollutants (Collins, Grineski, and Nadybal, 2022), disaster impacts (Bullard and Wright, 2009; Chakraborty, Collins, and Grineski, 2019), tree canopy coverage (Landry and Chakraborty, 2009), extreme heat exposure (Mitchell and Chakraborty, 2019), occupational exposure (Arcury, Quandt, and Russell, 2002; Carrillo and Ipsen, 2021), and accessibility to healthy food (Alkon, 2018). This single-stressor approach is also prevalent in quantitative risk assessments that examine the health effects of environmental exposure or toxins. A principal critique of this approach is that it fails to reflect concurrent and compounding impacts faced by many communities (Sadd et al., 2011). Capturing the combined impacts of multiple stressors requires transitioning beyond a single-media, single-location (e.g., residential, workplace), and single point-in-time analysis; this is difficult but crucial (Huang and London, 2016).

Capturing the combined impacts of multiple stressors requires considering the magnitude of individual stressors, the presence of multiple stressors, and interactions among the stressors. Typical approaches to accounting for these factors include count, intersection, or aggregation methods. The following discussion provides an overview;

BOX 2.2
Intersectionality

Exposure and resilience to environmental burdens differ for individuals and population groups along a variety of social dimensions, including race, ethnicity, class, gender, disability status, age, immigration status, indigeneity, sexual orientation, or identity, and other demographic or identity characteristics (Goldsmith, Raditz, and Méndez, 2022; Méndez, Flores-Haro, and Zucker, 2020). Within the broad scholarship on environmental justice, race and class have dominated discussions about the underlying demographic factors that explain or predict environmental inequalities, sometimes as competing theories about causation. Indeed, the racial and class characteristics of communities are both predictors of a wide variety of environmental disparities, independently and in tandem. However, other socio-demographic characteristics beyond race and class can aid in understanding differences in risk or vulnerability.

Significant gender power imbalances due to custom, culture, or law mean that environmental vulnerabilities and adaptive capacities are mediated by gender, often placing women and girls at a systematic disadvantage and greater risk (Denton, 2002). For example, gender-based differences in mortality from natural hazards have been described, with women exhibiting a lower life expectancy than men (Neumayer and Plümper, 2007). Some risk exposures are unique to women, such as exposure to toxins in female beauty products (Collins et al., 2023; McDonald et al., 2022; Storr, 2021) or complications during pregnancy (Giarratano et al., 2015), while other risks are not gender-specific but nevertheless have differential impacts. Risks such as climate change are likely to exacerbate existing gender-based inequalities (Huyer et al., 2020; Vinyeta, Whyte, and Lynn, 2015).

Disability status is another dimension of both marginalization and vulnerability. In addition to heightened physical susceptibility to environmental risks, disabled individuals face numerous societal barriers to inclusion and participation in decision making which result in reduced access to education, health services, employment, and resulting poverty and lower levels of information and resources (Kosanic et al., 2022). Recent studies also indicate how people with disabilities are disproportionately exposed to various environmental hazards and pollution sources, compared to the non-disabled population (Chakraborty, 2019, 2020, 2022). The evidence suggests that individuals with disabilities often experience a 'multiple jeopardy' defined by the convergence of disability status with other social disadvantages such as racial/ethnic minority, poverty, and elderly status, to amplify their vulnerability to environmental risks (Chakraborty, 2020).

Finally, age is a significant factor affecting vulnerability and exposure. Children are more sensitive than adults to toxins, such as lead, and the long-term developmental impacts are more consequential (McFarland, Hauer, and Reuben, 2022; Lu, Levin, and Schwartz, 2022; Schneider, 2023). At the other end of the age spectrum, older adults and the elderly are often more vulnerable to disasters or extreme weather events. Chronic health conditions, reduced sensory or cognitive capacity, coupled with commonly lower socioeconomic status can mean lower resilience and greater need for support or assistance for older adults and the elderly (Phraknoi et al., 2023).

The intersection of these identities or socio-demographic characteristics is also important. While race and class have often been examined in isolation, their intersection (i.e., lower income, non-white communities) identifies an important subpopulation that faces unique challenges. The same logic can be extended to other dimensions of identity. The intersection of marginalized identities means that these individuals or populations may be oppressed or differentially affected by a combination of interconnected societal structures—racism, sexism, ableism, ageism, etc. (Crenshaw, 1991; Ryder, 2017). While women are at heightened risk of exposure to toxins in beauty products due to societal pressures and expectations of femininity, as well as historic lack of governmental regulation of the cosmetics industry, women of color who have lower income are more likely to consume more products with higher levels of toxicity as a result of both social expectations of femininity and the compounding pressures to conform to white standards of beauty (e.g., using harsh chemicals to lighten their skin or straighten their hair), and less awareness or access to safer alternatives (Collins et al., 2023; McDonald et al., 2022; Storr, 2021). Disability can further exacerbate vulnerabilities of race/ethnicity, income, gender, and age (Chakraborty, 2020; Jampel, 2018). While all women are at heightened risk of sexual violence, the risk of experiencing sexual violence is heightened during disasters or disruption, and even more so for women with disabilities who are displaced or forced to reside in shelters. Children with disabilities are more likely to be left homeless after a disaster (Kosanic et al., 2022). The intersectionality of identity and its impacts on exposure and vulnerability is an important factor in understanding how risk and burdens vary both between and within population groups.

Chapter 4 discusses these approaches in practice through a scan of existing EJ tools and their methods.

Count methods simply count the presence or number of stressors, typically based on threshold data and the number of thresholds exceeded. The intersection approach is an extension of the count method for capturing interactions among stressors: a given stressor creates a burden for a community only if a second stressor is also present. Thus, under this approach, total impact is measured by meeting thresholds for both stressors. For example, to be designated as disadvantaged, a community would have to meet a threshold for both income and an environmental indicator or composite measure. This is a binary conceptualization: the community meets the thresholds for multiple stressors or does not.

By contrast, aggregation methods use mathematical or statistical methods to combine the individual stressors into a single number. Aggregation methods capture variations in stressor magnitudes into the summary measure rather than simply presence versus absence, as in count and intersection methods. In addition, aggregation methods capture the magnitude of interactions among stressors. Typical aggregation approaches include addition and multiplication. An additive approach conceptualizes interactions as being linear: the combined impact is the sum of the individual stressors; they accumulate but do not directly interact to create synergistic impacts. The multiplicative approach is nonlinear and captures direct, synergistic impacts among stressors: the whole impact is greater than the sum of the individual stressors. It is also possible to use a hybrid approach in which some stressors interact additively and others in a multiplicative manner. Statistical methods include correlation, regression, and principal components/factor analysis. The individual stressors can be weighted equally or unequally to reflect their relative importance or impacts (Gan et al., 2017).

Simplicity and legibility are important considerations for analytical approaches to capturing multiple stressors, especially when conducting stakeholder and community engagement (NRC, 2009; Solomon et al., 2016). August and colleagues (2012) provide an example of an analytical framework for measuring the impacts of multiple stressors in a clear and transparent manner, in this case, pollution; Figure 2.2 provides a graphic illustration. Individual indicators are averaged into groupings of related indicators such as exposures, public health and environmental effects of the exposures, sensitive populations, and socioeconomic factors. These in turn are summed into the higher-level components of pollution burden or population characteristics. Note that these higher-level components are combined in a multiplicative manner, meaning that, for example, lower pollution burdens cannot compensate for less vulnerable population characteristics such as lower sensitivity and/or higher SES. As such, this framework is both linear and synergistic, reflecting one model of interrelationships among social and environmental burdens, and how the population characteristics modify their effects (Solomon et al., 2016). This hierarchical arrangement is intended to support clear thinking and understanding about how the indicators should be combined; but it is not intended as a causal model. As discussed in Chapter 3, this is a common approach to constructing an indicator for multidimensional concepts.

The interplay of multiple stressors interacting with sociodemographic, environmental, occupational, and public health factors leads to the possibility of the total impacts being greater than the sum of the individual stressors. Because of these synergistic effects, capturing cumulative impacts is crucial for a comprehensive assessment of community disadvantage.

There is a lingering knowledge gap in the understanding of the interactions among stressors, especially causal relationships, but this understanding can inform how to capture the cumulative impacts facing a community (EPA, 2022a). What is not well understood, for example, is if the impact of multiple stressors on populations is akin to the sum of their individual effects, or if the total burden occurs through synergism in which stressors and populations interact and amplify one another and result in greater cumulative impacts. In addition, understanding is limited regarding the effects of long-term exposure to stressors where effects accumulate over time. Although additional empirical research is necessary to fill these gaps, it is challenging to develop completely specified process models of these interactions, especially for nonchemical stressors. Possible paths forward include data-driven models that leverage novel data sources and mixed-method approaches (discussed further in Chapter 7; Schäfer et al., 2023, Shamasunder et al., 2022).

CUMULATIVE IMPACTS

As noted, the concept of cumulative impacts includes both a temporal component (i.e., the accumulation of impacts over time through persistent exposure) and a contemporaneous component due to the presence of multiple stressors. To date, the focus of EJ work on cumulative impacts has been on the latter component, attempting to incorporate into EJ tools the understanding that communities that face multiple stressors are more burdened than those facing a single stressor (e.g., Popovich et al., 2024). However, efforts such as Justice40 are typically predicated on concerns about "historic" underinvestment that has led to disproportionate burdens that have accumulated over time. In addition, burdens faced by communities can change over time, potentially either increasing or decreasing, due to a variety of factors such as investment/disinvestment or environmental, political, or social crises or changes (including, for example, disastrous environmental accidents or epidemics). Thus, although much of the discussion of cumulative impacts in the EJ community as well as the discussion in this report focuses on the presence of multiple stressors, a dynamic perspective on the accumulation of and changes in impacts over time is critical for fully capturing the cumulative burden faced by communities.

Because of its complexity, defining and measuring cumulative impacts have evolved over time. Leaders in the EJ community and federal and state policy makers have developed definitions of cumulative impacts to guide scientific investigations, policies, and agency guidance. The earliest identified definition of cumulative impacts comes from the California Environmental Quality Act of 1970:

> *...two or more individual effects which, when considered together, are considerable or which compound or increase other environmental impacts .... The cumulative impact from several projects is the change in the environment which results from the*

> *incremental impact of the project when added to other closely related past, present, and reasonably foreseeable probable future projects. Cumulative impacts can result from individually minor but collectively significant projects taking place over a period of time* (CalEPA, 2004; EPA, 2022a).

Federal government definitions of cumulative impacts date back to 1978 when the CEQ published implementing regulations for the 1969 National Environmental Policy Act (NEPA), which defined cumulative impacts as:

> *...the impact on the environment which results from the incremental impact of the action when added to other past, present, and reasonably foreseeable future actions regardless of what Agency (federal or non-federal) or person undertakes such other actions. Cumulative impacts can result from individually minor but collectively significant actions taking place over a period of time* (CEQ, 1978).

Some states—including California, Hawaii, Illinois, Maryland, Massachusetts, Minnesota, New Jersey, New York, Vermont, and Washington—have introduced or enacted legislation that includes cumulative impacts. Several of the states listed above—as well as Michigan, New Mexico, and Oregon—have reports or mapping tools that include cumulative impacts. Scientific definitions of cumulative impacts span disciplines such as public health, toxicology, environmental science, geography, and planning (Tishman Environment and Design Center, 2022). The collaborative nonprofit, Coming Clean, Inc., and its strategic partner, the Environmental Justice Health Alliance for Chemical Policy Reform, define cumulative impacts as the combination of nonchemical and chemical stressors on health, quality of life, and well-being. This definition was created by communities, researchers, health care workers, lawyers, and other EJ advocates (Coming Clean, Inc., n.d.).[6]

A web tool by the Tishman Environment and Design Center at The New School provides a timeline, summaries, and links to federal and state government reports, tools, and legislation and to scientific journal articles that define, explicitly or implicitly, the concept of cumulative impacts and the indicators and thresholds used to measure cumulative impacts. Table 2.1 provides example definitions of cumulative impacts. These definitions do not represent a comprehensive list but rather are selected to illustrate features of cumulative impacts that will be critical to this report.

The definitions in Table 2.1 commonly emphasize multiple environmental stressors across natural, built, and social environments and encompass where people live, work, play, and pray (Bullard, 2001). The Collaborative definition is rooted in the lived experiences of the communities that compose the Environmental Justice Health Alliance for Chemical Policy Reform. Chemical stressors damage organisms and ecosystems from the release of pollutants into the environment through activities such as waste treatment and disposal, manufacturing, natural resource extraction, energy production, transportation, and agriculture. Chemical stressors have historically been the focus of one kind of

[6] Coming Clean, Inc., "Cumulative Impacts and Mandatory Emissions Reductions Team" https://comingcleaninc.org/projects/ci-mer (accessed December 18, 2023).

TABLE 2.1 Example Government Agency Definitions of Cumulative Impacts and Effects

Entity	Definition	Source
Federal government definitions		
Council on Environmental Quality (CEQ)	Exposure to one or more chemical, biological, physical, or radiological agents across one or more media (e.g., air, water, soil) from one or more sources over time in one or more locations that have the potential for deleterious effects on the environment and/or human health.	CEQ (1997)
National Environmental Justice Advisory Council	The interaction among multiple stressors within a population or community where individuals differ based on susceptibility, exposure, preparedness, and resilience (ability to recover)	NEJAC (2004)
U.S. House of Representatives	Any exposure to a single or multiple public health or environmental risk in the past, present, and reasonably foreseeable future occurring in a specific geographical area, taking into account sensitive populations and other factors that may heighten vulnerability, including socioeconomic characteristics.	H.R.2021 - Environmental Justice for All Act (U.S. Congress, House, 2021)
U.S. Senate	Similar to H.R.2021	S.2630—Environmental Justice Act of 2021 (U.S. Congress, Senate, 2021)
U.S. Environmental Protection Agency	The totality of exposures to combinations of chemical and non-chemical stressors and their effects on health, well-being, and quality of life outcomes throughout a person's lifetime, encompassing both direct and indirect effects through impacts on resources and the environment, accounting for the context of individuals, geographically defined communities, or definable population groups. Cumulative impacts reflect the potential state of vulnerability or resilience of a community.	EPA (2022)
State government definitions		
California Environmental Protection Agency	Cumulative impacts are: exposures, public health or environmental effects from the combined emissions and discharges, in a geographic area, including environmental pollution from all sources, whether single or multi-media, routinely, accidentally, or otherwise released. Impacts will take into account sensitive populations and socioeconomic factors, where applicable and to the extent data are available. Formula for estimating relative cumulative impact among communities: Cumulative Impacts = [Exposures + Public health effects + Environmental effects] × [Sensitive populations + Socioeconomic factors]	CalEPA (2010) (definition adopted in February 2005)

continued

TABLE 2.1 Continued

Entity	Definition	Source
State of California	Two or more individual effects which, when considered together, are considerable and compound or increase other environmental impacts. These effects may be from single or multiple sources ("projects") in the past, present, and reasonably foreseeable probable futures.	California Environmental Quality Act (CEQA), as codified in Cal. Code Regs. Tit. 14, § 15355—Cumulative Impacts
State of Vermont Senate	The cumulative impacts of environmental harms disproportionately and adversely impact the health of BIPOC and communities with low income, with climate change functioning as a threat multiplier. These disproportionate adverse impacts are exacerbated by a lack of access to affordable energy, healthy food, green spaces, and other environmental benefits.	S.148—An Act Relating to Environmental Justice in Vermont (Vermont State Legislature, 2022)
State of Washington Senate	"Environmental harm" means the individual or cumulative environmental health impacts and risks to communities caused by historic, current, or projected (a) exposure to pollution, conventional or toxic pollutants, environmental hazards, or other contamination in the air, water, and land (b) adverse environmental effects, including exposure to contamination, hazardous substances, or pollution that increase the risk of adverse environmental health outcomes or create vulnerabilities to the impacts of climate change (c) loss or impairment of ecosystem functions or traditional food resources or loss of access to gather cultural resources or harvest traditional foods (d) health and economic impacts from climate change.	SB5141—The Healthy Environment for All Act (Washington State Legislature, 2021)
Scientific definitions		
K. Sexton and S. H. Linder, 2011 article in *American Journal of Public Health*	Combined threats from multiple environmental stressors. Uses the concept of *allostatic load* to characterize the cumulative biological toll taken on the human body through physiologic responses to life's everyday stress-provoking demands. Notes that exposure to multiple environmental agents, including biologic, chemical, physical, radiologic, and psycho-social stressors, can result in outcomes that are either antagonistic (less than additive) or synergistic (more than additive)	Sexton and Linder (2011)

TABLE 2.1 Continued

Entity	Definition	Source
G. M. Solomon, R. Morello-Frosch , L. Zeise, and J. B. Faust, 2016 article in *Annual Review of Public Health*	Identifies four key concepts that underlie composite indicator: i) Health disparities are linked to social and environmental factors for many diseases; ii) Inequalities in exposures to environmental hazards are significant; iii) Intrinsic biological and physiological factors can modify the effects of environmental factors; iv) Extrinsic social vulnerability factors at the individual and community levels may amplify the effects of environmental hazards	Solomon et al. (2016)
Collaborative (Community and advocates) definition		
Coming Clean Inc. and the Environmental Justice Health Alliance for Chemical Policy Reform	The combination of non-chemical and chemical stressors on health, quality of life, and well-being.	Coming Clean, Inc. (n.d.)

SOURCE: Based on Tishman Environment and Design Center, 2022.

cumulative risk assessment due to greater data availability and alignment with regulatory frameworks (Lewis et al., 2011). Nonchemical stressors include lifestyle health hazards, weather extremes, poverty, racial discrimination (policy and institutionalized practices), and crime (Bullard, 2001; Morello-Frosch et al., 2011).

Another common theme in Table 2.1 is environmental stressors accruing to vulnerable populations, labeled as "definable" by the EPA and "sensitive" by CalEPA (2010). The separation of stressors, exposures, and aggravating effects (such as public health and environmental conditions) from population characteristics enables examination of the individual drivers of cumulative impacts (Alexeeff et al., 2012; August et al., 2012). The framework illustrated in Figure 2.1 supports this type of analysis.

Sensitive populations are demographic groups most likely to face disproportionate effects from environmental stressors. The academic literature has largely bifurcated these populations in relation to intrinsic and extrinsic factors (Liévanos, 2018). Intrinsic factors are biological and physiological characteristics that amplify how environmental exposures translate into adverse health outcomes. Examples are extremes of age, genetics, epigenetics (i.e., how behaviors and environment cause functional changes in genes), biological sex (including pregnancy and birth outcomes), and chronically impaired health (Morello-Frosch et al., 2011). Extrinsic factors are social characteristics that are associated with differential outcomes from stressors, including race, ethnicity, gender identity, disability, indigeneity, and immigration status. Socioeconomic factors are resources that influence capacity to reduce exposure to environmental hazards. These resources include socioeconomic status elements of income, wealth, education, and occupation. EJ scholarship is replete with empirical examinations of how discrimination, marginalization, classism, and exclusion in decision making related to sensitive

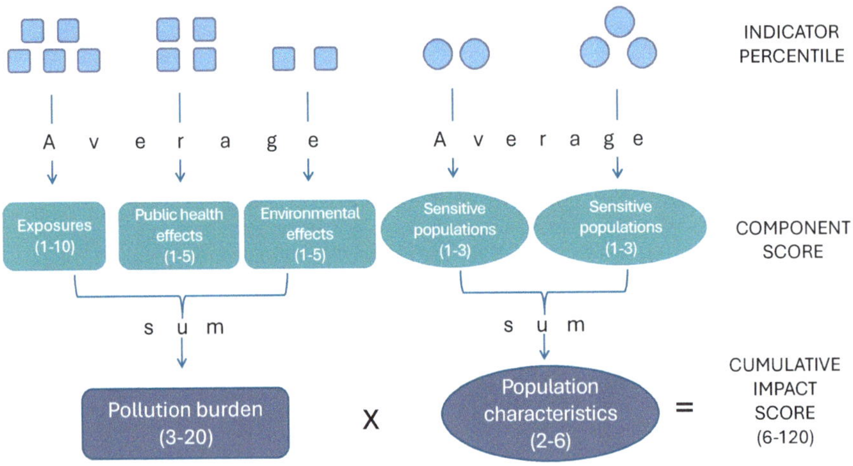

FIGURE 2.1 Example analytical framework for measuring the impacts of multiple stressors; in this example, chemical stressors (pollution). Note the careful arrangement of indicators (measurable variables) into higher-level components, reflecting the burden and the differential impacts based on population characteristics and the different ways both indicators and components can be combined to estimate the overall impacts. SOURCE: Redrawn from August et al., 2012.

populations and socioeconomic factors lead to differential and disproportionate environmental outcomes, as the discussion earlier in this chapter suggests.

Table 2.1 also illustrates in the EPA and CalEPA definitions that geography is a central organizing principle and provides an analytical framework for understanding how pollution effects accumulate; pollution impacts are aggregated in geographic communities. Such geographic emphasis is supported by numerous EJ studies demonstrating how combinations of de jure (in law) discrimination and de facto (in practice) discrimination, combined with other socioeconomic processes and institutional practices, generate and reproduce spatially concentrated community disinvestment (Rothstein, 2017) and population disparities in exposures and effects (MCRC, 2017).

Cumulative impacts are complex in their effects and those who are affected, but also in their interactions. The framing of allostatic load may be well suited to understanding such interactions. The concept of allostatic load refers to the cumulative biological risk due to wear and tear on the human body from multiple and repeated stresses over time (Gustafsson et al., 2014). Research on allostatic load seeks to understand how human health effects result from interactions among chemical, social, and psychological stressors (Sexton and Linder, 2011), with a growing focus on the role of neighborhood deprivation in generating health disparities (Gudi-Mindermann et al., 2023). Findings indicate elevated rates of death and disease associated with higher allostatic load, yet there is an underdeveloped understanding of the pathways between social stressor exposures and disease outcomes (Ribeiro et al., 2018).

An important element in identifying and measuring cumulative impacts is the continuous involvement of concerned or affected communities throughout the process. The National Environmental Justice Advisory Council and the White House Environmental Justice Advisory Council have specifically recommended that cumulative impact assessment should incorporate participatory engagement with communities in all phases of the process, from planning and interpretation through implementation (NEJAC, 2004, 2011; WHEJAC, 2021). This community-engaged approach is also central to recommendations by the EPA's Office of Research and Development on cumulative impacts research (EPA, 2022a).

For the purposes of this report, cumulative impacts are defined as the total burden—adverse, neutral, or beneficial—from stressors, their interactions, and the environment that affects the health, well-being, and quality of life of an individual, community, or population at a given point in time and that accumulates over time (EPA, 2022a). Figure 2.2 illustrates a conceptual framework of the principal factors and their interrelationships that generate cumulative impacts. These impacts are generated by combinations

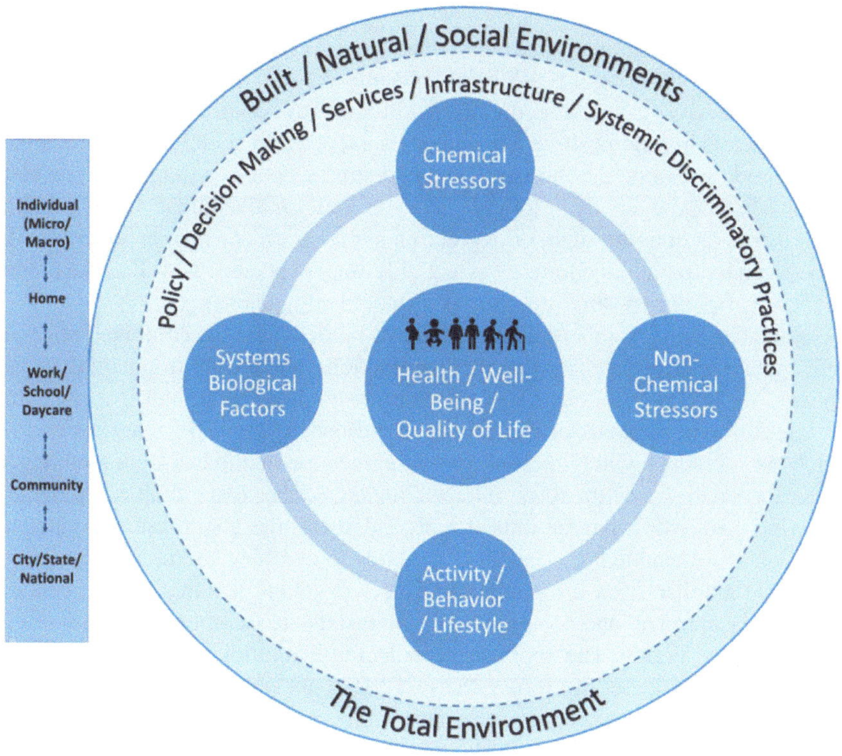

FIGURE 2.2 Stressors from the total environment (built, natural, social) interact with systems biology (genetic and epigenetic factors) and activity, behavior, and lifestyles to create cumulative impacts. SOURCE: EPA (2022a); adapted from Tulve et al., 2016.

of chemical stressors, nonchemical stressors, biological factors (e.g., age, sex, health, genetics) that affect susceptibility to stressors, and the human activities, behaviors, and lifestyles that affect both exposure to stressors and the mitigation or exacerbation of their effects. The framework for cumulative impacts in Figure 2.2 is holistic, highlighting that (a) the factors that influence human health, well-being, and quality of life span multiple dimensions, and (b) the factors operate interactively rather than independently. This multiple and interactive framing of cumulative impacts is directly pertinent to the development of EJ tools in how disadvantage is conceptualized, which indicators are selected (Chapter 5), and how indicators are integrated (Chapter 6). The context for these interactions is the total environment: the built, natural, and social environments created from policies and decisions, services, and infrastructure investments that are often shaped by systematic discriminatory practices. Stressors in the past, present, and anticipated in the future can aggregate and accumulate over time; they also operate at multiple scales, with impacts affecting individuals, populations, and geographic communities (EPA, 2022a).

CHAPTER HIGHLIGHTS

Communities are defined as groups of people who share common experiences and interactions and, in this report, are geographically defined. Disadvantaged communities experience adverse social, economic, built, and environmental factors that create barriers to achieving the positive life outcomes expected in society. Those factors can be considered "stressors" or "burdens" and encompass both exposures and preexisting conditions that independently or jointly lead to (cause) negative impacts. The magnitude of a stressor is its quantity, size, or degree. This dimension can also capture the difference between temporary and long-term (e.g., lifetime) exposure, as well as the potential for increased future exposure (due, for example, to climate change).

Cumulative impacts (also called cumulative burdens) are the combined positive or negative total burden from stressors, their interactions, and the environment that affects the health, well-being, and quality of life of an individual, community, or population. There are different methods of capturing the presence of multiple stressors and their interactions, including count, intersection, or aggregation methods. Count methods only consider the existence of stressors. Intersection approaches expand on count methods, considering a stressor a burden only if a second stressor is also present. Aggregation methods use mathematical or statistical methods to combine the individual stressors into a single number. Measuring community disadvantage requires careful conceptualization, measurement, and construction if the results are to reflect the real world and support effective policy. The problem of reducing a multidimensional concept to a single composite indicator in a tool such as CEJST is not specific to EJ tools. However, there are many issues that create challenges when creating an EJ tool for measuring community disadvantage. Research demonstrates an association of race or ethnicity with disproportionate exposure to environmental stressors, and historical and current structural racism has led to persistent disparities and outcomes for communities of color. There are several indexes that can be used to measure structural racism, including geographically based methods to quantify the magnitude of its impacts, but there is not consensus on the best way to do so. Care needs to be taken to avoid treating people of color as a monolithic group.

The next chapter of this report reviews good practices for composite indicator construction, with special reference to supporting EJ screening and analysis tools. Understanding these principles will set the stage for the scan of existing EJ tools presented in Chapter 4, the description of burden indicators in Chapter 5, and discussions of integration analysis and validation in Chapters 6 and 7.

3

A Conceptual Framework for Building Composite Indicators

The White House Council on Environmental Quality's (CEQ's) Climate and Economic Justice Screening Tool (CEJST)[1] is an example of a geospatial tool that calculates a composite indicator. A composite indicator represents a complex, multidimensional concept or condition and assesses the existence or intensity of that concept for real-world entities such as countries, states, cities, regions, companies, or universities. While an indicator relies on data measurements that are chosen to represent the indicator, a composite indicator is a combination of those measurements. The calculation of a composite indicator is useful when the concept being described is better represented through a combination of different factors. This is accomplished by bringing together measurements across the multiple dimensions of the concept so that a single number and small sets of numbers reflect the condition with the greatest validity possible. For example, as discussed in Chapter 2, community disadvantage is a multidimensional concept. CEJST attempts to identify which census tracts meet this condition by combining data measures across its "categories of burden"[2] to indicate if the census tract is considered disadvantaged.

This chapter provides a foundation for a structured iterative process for constructing tools that calculate composite indicators, such as CEJST and other environmental justice (EJ) tools. Such a structured process can be applied in the development of EJ composite indicators, regardless of the application, because it provides a logical workflow and mechanisms to validate construction decisions and tool results. The chapter first describes the purpose of EJ tools and how decisions regarding construction can affect tool utility. This is followed by a discussion of the conceptual foundation for composite indicators, an overview of the process for constructing composite indicators,

[1] See https://screeningtool.geoplatform.gov/en/#3/33.47/-97.5 (accessed September 15, 2023).

[2] Note that the literature uses the concept of "impact" where CEJST refers to "burden." When referring to CEJST, this report will use CEJST's terminology.

and the questions that must be addressed in the process. Following this, the essential role of community engagement in tool construction is presented. The chapter concludes with the description of a conceptual framework for EJ tool development that emphasizes the importance of trust, transparency, and legitimacy for tools such as CEJST.

TOOL PURPOSE

Geographically based EJ tools are typically intended to identify disadvantaged communities or measure the extent to which those communities are burdened or disadvantaged (Balakrishnan et al., 2022). Understanding how this information might be used and by whom is critical to the development and evaluation of EJ tools (EPA, 2022a). A tool might have multiple users—including local, state, and federal regulators, community advocates, grant applicants, academic researchers, and individual members of the public—all using the tool for different purposes. The tools might be used for educational purposes (e.g., to provide information and raise awareness about EJ concerns in specific communities and support advocacy for those communities) or as inputs into decisions related to, for example, local land use and facility siting, state or federal permitting, investment, funding prioritization, and regulatory compliance. One tool, California's CalEnviroScreen,[3] is used to direct investment of proceeds from the state's greenhouse gas Cap-and-Trade Program toward disadvantaged communities.[4] It is also central to multiple municipal and state programs and planning, including incorporation of EJ into the general plans of California municipalities, CalEPA's Environmental Justice Enforcement Task Force, the California Air Resources Board's Community Air Protection Program, and to identify vulnerabilities for tracking progress related to implementing the human right to water (Lee, 2020). Analogously, Washington State uses its Environmental Health Disparities (EHD) Map tool[5] to prioritize state grants and other resources. Washington state law directs utilities to focus their efforts on reducing pollution and increasing the benefits of clean energy to "Highly Impacted Communities," defined as census tracts in the highest two ranks of cumulative impacts according to the EHD map. Washington State's Department of Health has developed educational materials based on the EHD to bring public and environmental health tools and data into high school science classrooms in partnership with the Puget Sound Educational Service District.[6]

As the above examples illustrate, many EJ tools (including CEJST) are intended to provide input into a decision-making process. Often, their results are intended to provide an initial "screening" of eligibility or of high-priority communities. Such tools are sometimes termed "screening tools." The tool might be used to identify, for example, communities that would be eligible for funding (or, in the case of CEJST,

[3] See https://oehha.ca.gov/calenviroscreen (accessed September 15, 2023).

[4] See California Climate Investments to Benefit Disadvantaged Communities at https://calepa.ca.gov/envjustice/ghginvest/ (accessed December 21, 2023).

[5] See https://doh.wa.gov/data-and-statistical-reports/washington-tracking-network-wtn/success-stories (accessed September 27, 2023).

[6] See https://doh.wa.gov/data-and-statistical-reports/washington-tracking-network-wtn/success-stories (accessed December 28, 2023).

extra consideration for funding), but the tool would not be the sole determinant of which communities, among those that are eligible, would receive funding. Similarly, an EJ tool could be used to rule out certain communities as potential sites for new facilities that could exacerbate EJ issues impacting communities that are already disadvantaged. The tool, however, would not be used to determine ultimate siting decisions.

In the case of CEJST, the tool is intended to screen for eligibility for dispersal of different kinds of funds under the Justice40 Initiative for very different programs. Constructing indicators for this type of use requires a carefully considered structured development process that is informed by a clear definition of the tool's purpose. The tool and its indicators need to be designed in the context of the decisions it is intended to inform and in consideration of unintended consequences from the tool's use such as changes in property valuations, green gentrification, or displacement (e.g., Siddiqi et al., 2023). Results obtained from an initial screening using the tool would need to be coupled with information (e.g., community concerns, history of enforcement/violations, risk assessments, political priorities, and other policy considerations) from other sources before final allocation decisions are made, especially if the tool is general in nature and intended to support decisions in many sectors. For example, federal agencies are required to use CEJST to screen whether investment in a particular community would qualify for meeting the Justice40 goal, but no guidance is provided at present regarding the allocation of funds among those designated as qualified (EOP, 2023).

BINARY VERSUS NONBINARY TOOLS

Tools designed to determine eligibility need to ultimately lead to a categorization of a community as either disadvantaged or not disadvantaged. The Urban Institute refers to these as "binary" tools (Balakrishnan et al., 2022). Binary tools provide an "is or is not" measure of disadvantage. In contrast, "nonbinary" tools provide a continuous measure of the extent to which a community is burdened, or they reflect differences in burden levels across communities (Balakrishnan et al., 2022). The output of CEJST is binary; it is designed to identify specific communities as "disadvantaged" for the purpose of qualifying for further investment under the Biden Administration's Justice40 Initiative. While it categorizes communities as disadvantaged or not, it does not distinguish between communities that might be deemed "severely disadvantaged" and those that are "somewhat disadvantaged." In contrast, the EPA's Environmental Justice Screening and Mapping tool (EJScreen)[7] shows continuous measures (i.e., percentiles) of various environmental or socioeconomic stressors for communities across the country, but it does not categorize or label "environmental justice" communities or "disadvantaged" communities. California's CalEnviroScreen and SB 535 Disadvantaged Communities[8] have hybrid functions. CalEnviroScreen provides an index or score for community disadvantage/environmental burden, which is used to show a continuous metric of

[7] See https://www.epa.gov/ejscreen (accessed September 15, 2023).

[8] See https://oehha.ca.gov/calenviroscreen/sb535 for more information on California Senate Bill 535 and how CalEPA implements it through their tool, CalEnviroScreen (accessed February 9, 2024).

cumulative impacts across the state. Since CalEPA, following California SB 535, is responsible for identifying "disadvantaged communities" to prioritize funding, a threshold using CalEnviroScreen's cumulative impact score was chosen for that designation. Mitigation efforts can be further focused on the most heavily burdened within designated communities.

An important question regarding the development of binary tools such as CEJST is how to set the criterion or criteria for being "in" (i.e., disadvantaged) versus "out" (i.e., not disadvantaged). Because "disadvantage" is typically a matter of degree without a universally accepted standard or threshold, setting the criteria is a subjective decision of policy makers, interested and affected parties, or tool designers. The choice of criteria has important implications, both when used for initial screening (e.g., for funding eligibility) and when used as input into final decision making (e.g., allowing a permit or not). Higher thresholds or more stringent criteria for inclusion necessarily limit the communities included, potentially allowing for more targeted prioritization, yet these stricter conditions risk excluding communities that are genuinely overburdened even though they do not meet the threshold or the specific criteria chosen. Conversely, a less stringent threshold would expand the set of included (eligible) communities, thereby generating what might be viewed as a more inclusive definition of disadvantaged but also elevating communities with comparatively fewer burdens. If, in practice, limited funds are spread across a set of eligible communities, a more inclusive definition could lead to less investment in communities bearing the greatest burdens. Moreover, even when funding will not be distributed across all eligible communities, it is possible that funds will go disproportionately to eligible communities with relatively low burdens that have the resources to successfully compete for available funds rather than to communities that have been marginalized and suffer from underinvestment and overburdening based on socioeconomic, racial, and demographic composition (McTarnaghan et al., 2022). Ultimately, choosing the criterion or criteria used by the tool for categorizing a community as disadvantaged will be somewhat subjective. The impact of these choices can be further exacerbated when thresholds on single indicators are used to designate a location as disadvantaged, as opposed to thresholds on a composite indicator. This is particularly true when the scale of analysis can lead to community disadvantage being obscured by boundaries, such as census tracts, that do not reflect the underlying patterns of disadvantage. Evaluating the robustness and validity of such choices is discussed in Chapters 6 and 7, respectively.

USE OF COMPOSITE INDICATORS

Composite indicators are widely and increasingly used to reflect multidimensional concepts such as well-being (Salzman, 2003), gender inequality (World Economic Forum, 2020),[9] ecological footprint (Wackernagel and Rees, 1996), and human development (UNDP, 1990). In the case of EJ, there are numerous examples of composite

[9] See UNDP's Gender Inequality Index: https://hdr.undp.org/data-center/thematic-composite-indices/gender-inequality-index#/indicies/GII (accessed September 25, 2023).

indexes being used, as are explored in the scan of tools in this report (see Chapter 4). The advantages of using composite indicators include discriminating among competing hypotheses; structuring, understanding, and conceptualizing solutions; performance tracking in relation to goals and objectives; choosing among alternative policies; and informing interested and affected parties and the public (Failing and Gregory, 2003; Hezri and Dovers, 2006; Miller, Witlox, and Tribby, 2013). Composite indicators can have some disadvantages: since they reduce complex multidimensional phenomena to a single number or small set of numbers, they can support misleading, nonrobust policies if poorly constructed, and can invite simplistic policy conclusions, subjectivity, and the potential for inappropriate policy decisions if difficult-to-measure dimensions or features are ignored. At the same time, their ability to summarize complicated issues—while providing a "big picture" summary and transparency by making assumptions and goals explicit—makes composite indicators useful for supporting policy decisions.

Composite indicator construction can vary substantially. Variations can exist in the indicators chosen, in the methods used for normalizing and aggregating datasets, and in validating indicators and tool results. At times, subindexes—essentially forming composite indicators within the overall composite indicator which then need to be combined—will be desirable. However, there are systematic and defensible approaches to composite indicator construction that are common to many tools (Mazziotta and Pereto, 2017; OECD and JRC, 2008; Salzman, 2003). Successful composite indicators (including those employed in EJ tools) define their purpose and intended audience, and they pursue correctness or truthfulness. The validity of a tool rests on a foundation of scientific and methodological rigor, meaningful participation and input from community members and other interested and affected parties, transparency, and acceptance by institutional actors (especially government agencies or regulators), communities, and other affected parties.

CONSTRUCTING COMPOSITE INDICATORS

Building a national-scale EJ tool (such as CEJST) with far-reaching policy implications is aided by a systematic approach to ensure that the composite indicator measurements are internally coherent, transparent, easily interpreted, and externally valid, and capture relevant aspects of the concept being measured (Miller, Witlox, and Tribby, 2013). Widely used approaches for composite indicator construction already exist and are used to guide national and international policy (Mazziotta and Pareto, 2017; OECD and JRC, 2008; Salzman, 2003). Such a systematic approach provides a workflow for constructing a composite indicator and ensures that important considerations are factored into the construction. A systematic approach also provides an outline for documenting and explaining the decisions made in the creation of a composite indicator tool.

The Organisation for Economic Cooperation and Development's (OECD's) Statistics Directorate, Directorate for Science, Technology and Industry, and Econometrics and Applied Statistics Unit of the Joint Research Centre (JRC) of the European Commission developed a methodology guide for constructing composite indicators (OECD and JRC, 2008). The components of composite indicator construction laid out by OECD include conceptualizing the indicator's meaning, rigorous statistical testing of the internal consistency of indicator metrics, as well as external validation and communication. For the last 20 years, OECD and JRC guidance on composite indicator construction and evaluation has provided methodological expertise to the European Commission and its member countries, as well as to the academic community.[10] However, the process of indicator development is not strictly a technical or scientific exercise (Saisana et al., 2022).

Michaela Saisana,[11] an author of the OECD manual, described to the committee (see Appendix B for public meeting agenda) that composite indicator development is "a delicate balance between science and art." The "art" refers to subjective decision making (e.g., which variables to include or exclude, how to weight different indicators, which benchmarks to use), as well as the process of composite indicator validation. Because a composite indicator measures something that is not directly observable, developers seek to corroborate the composite indicator's truthfulness by comparing it to other well-known metrics or indicators and by soliciting input from the communities that are being measured and for whom the tool is intended. A composite indicator's utility is dependent on rigorous science and technique, clear communication of its meaning (and its limitations) to the intended audience, and, ultimately, on acceptance by interested and affected parties and policy makers (Fekete, 2012; Lee, 2020).

Saisana's insights echo the lessons learned during the development of CalEnviroScreen, an influential EJ composite indicator in the United States (Grier et al., 2022; Lee, 2020). Manuel Pastor, an EJ scholar and key architect of the Environmental Justice Screening Method (see Sadd et al., 2011) at the heart of CalEnviroScreen, recalls that the process for developing CalEnviroScreen's methodology involved extensive feedback from residents, community members, and other interested and affected parties (Sadasivam and Aldern, 2022). Arsenio Mataka, former assistant secretary for environmental justice and Tribal affairs at CalEPA, described six principles that resulted in CalEnviroScreen acceptance: the tool is (1) grounded in science, (2) informed by community experience, (3) endorsed by government, (4) universally available to everyone, (5) based on thorough public participation, and (6) able to serve as a "third-party validator" in local issues or other venues (described in Lee, 2020).

Guidance for Constructing Composite Indicators

The OECD Pocket Guide to Composite Indicators and Scoreboards (Saisana et al., 2019) outlines a 10-step process for composite indicator construction. The processes outlined in the OECD guide and other guides on indicator construction (e.g., Mazziotta and Pareto, 2017) do not focus on specific methods for accomplishing each step but rather on the importance of thinking through each step and the impact that each has on

[10] See JRC's Competence Centre on Composite Indicators and Scoreboards at https://knowledge4policy.ec.europa.eu/composite-indicators/about_en (accessed February 9, 2024).

[11] Michaela Saisana was head of the Monitoring, Indicators and Impact Evaluation Unit for the European Commission's Competence Centre on Composite Indicators and Scoreboards at the JRC in Italy at the time of this writing.

the resulting composite indicator. Although these steps are laid out linearly, the process of constructing a composite indicator is iterative and requires constant reevaluation and adjustment based on a loop of feedback. Community engagement and partnership are necessary at every step of the process. (See the next section of this chapter for more details on community engagement strategies.)

The following briefly summarizes each step of the OECD's pocket guide and can be considered a workflow for indicator construction:

1. **Define the concept to be measured.** This can be accomplished by answering questions such as: What are the objectives of the composite indicator? What is the basic definition of the concept (e.g., disadvantaged community)? What is the relationship between the definition and the objective? What are the multiple facets of the concept to be captured by indicators and are they complete, appropriate, and consistent with existing theory, empirical evidence, and lived experiences? Expert and stakeholder judgment is crucial in this step to acknowledge multiple viewpoints and achieve robustness (OECD and JRC, 2008). Thinking about the concept as a collection of facets or dimensions to be measured using a set of indicators helps with conceptualization, supports careful definitions and structure of the measurements, and helps to improve understanding. These facets can also make it easier to determine how indicators are weighted and combined (see below). The composite indicator literature refers to these concept facets as dimensions or subgroups; CEJST refers to these facets as "Categories of Burdens" (CEQ, 2022a).
2. **Select the indicators.** What are the appropriate indicators to measure the concept based on technical considerations such as validity, sensitivity, robustness, reproducibility, and scale, and practical considerations such as measurability, availability, simplicity, affordability, and credibility? Since indicator selection is closely related to concept definition, expert and stakeholder judgment is also crucial in this step (OECD and JRC, 2008).
3. **Analyze and treat the data where necessary.** Do any of the indicators exhibit significant skew, kurtosis, or outliers that might complicate interpretation or comparison? Do any of the indicator measures require significant imputation due to missing values?
4. **Bring all the indicators into a common scale (e.g., normalization).** Normalization methods can have substantial impacts on composite indicators (Carrino, 2017). Questions that could be asked include: Do all indicators exhibit the same directional meaning (e.g., do higher scores correspond to greater levels of "disadvantage")? Has a suitable normalization method been used for all indicators (e.g., percentile ranking, min-max scaling, *z*-scores)?
5. **Weight the indicators.** Questions to be asked include: what is the relative importance of each indicator and each subgroup to the concept being measured? Weighting can be accomplished using data-driven, statistical approaches or participatory approaches involving expert, stakeholder, and community perspectives (Becker et al., 2017; Greco et al., 2019). Indicator weighting is

closely intertwined with indicator selection (discussed above in Step 2) and indicator aggregation (discussed below in Step 6). Consequently, these three steps and their intermediaries are highly iterative in practice.

6. **Aggregate the indicators.** How should the indicators be combined: additively to allow compensability (i.e., compensation of a deficit in one indicator by excess in another), using a multiplicative approach which is partially noncompensatory, or in a noncompensable manner (i.e., that a deficit in one indicator cannot be compensated by a surplus in another; see Munda, 2012)? There are also hybrid approaches that allow some indicators to be compensatory and others noncompensatory to varying degrees (e.g., Blancas and Lozano-Oyola, 2022). If thresholds have been designated, do these appropriately reflect the goals of the composite indicator? Indicator weighting and aggregation are often considered in conjunction, given the interconnections between these decisions (e.g., Gan et al., 2017).
7. **Assess the statistical and conceptual coherence.** Questions to be asked include: Are the indicators organized into the appropriate subgroups? To what extent are indicators correlated with their respective categories of burden? To what extent are subgroups correlated with one another (e.g., do two or more categories of burden tell the same story)? To what extent is the composite indicator biased (e.g., statistically predisposed) toward some underlying phenomenon (e.g., population size or density, urbanicity, race/ethnicity)? It is redundant for two or more subgroups to be highly correlated; however, a high correlation between subgroups may be irrelevant or desirable if the representation of various types of burdens, regardless of statistical correlation, is important to stakeholders. Bias may be undesirable because it may mean that the composite indicator is reflective of an underlying phenomenon that is the more important driver. Bias may be irrelevant because the underlying phenomenon is inseparable or simply not important from a theoretical perspective. On the other hand, bias may be desirable because it captures a phenomenon that the tool intends to measure but which cannot be measured directly.
8. **Assess the impact of uncertainties**. Questions to ask include: What main uncertainties underlie the composite indicator? This can involve the basic concept definition, indicator selection (e.g., wrong or missing indicators), the organization of indicators into subgroups, the normalization methods, and any threshold values for indicators. How sensitive is the composite indicator to these uncertainties? What is the impact on the composite indicator when these choices are varied? Which choices have the biggest uncertainties? Sensitivity and uncertainty analyses can improve transparency and legitimacy by providing a quality assessment of the composite indicator results (Saisana, Saltelli, and Tarantola, 2005).
9. **Make sense of (validate) the data.** Questions to be asked include: To what extent does the composite indicator tell a coherent story; does it make sense? To what extent does the composite indicator mirror or harmonize with other well-known or credible indicators or variables? Although there are technical

approaches to composite indicator validation (e.g., Feldmeyer et al., 2020; Otoiu, Titan, and Dumitrescu, 2014), an important criterion is whether the result of the composite indicator reflects the lived experiences of people.

10. **Present the composite indicator visually**. Questions to ask include: How coherently does the tool present the results from the composite indicator? To what extent are tool users able to interpret the results? This step can be crucial for a geospatial mapping tool such as CEJST (see Box 3.1. for a discussion of geospatial mapping tool interface design).

There is a large literature available to guide each component of the composite indicator construction process. The widely cited *Handbook on Constructing Composite Indicators: Methodology and User Guide* (OECD and JRC, 2008) provides a comprehensive guide to the decisions involved in each step. Chapter 4 describes the composite indicator components of existing EJ tools. Chapters 5, 6, and 7 consider indicator and data selection and criteria, aggregation, and validation, respectively.

GEOSPATIAL AND ADMINISTRATIVE DATA

CEJST and some other EJ tools use U.S. census tracts to define geographic communities since they provide nationally consistent, publicly available data that encompass the variables of interest. Census tracts are spatial units in which data about individuals, households, or environmental events or conditions within the unit are represented in the form of aggregate or summary statistics (e.g., counts, percentages, concentrations). Information about individual persons, households, or specific events in a census tract is not revealed. Using aggregate spatial units to build geographic communities is a reasonable approach, but doing so can introduce measurement artifacts that affect the results but are not features of the underlying reality. Census tracts may not align with residents' definitions of neighborhoods and communities. A geographic community may have crisp boundaries in the real world: physical features such as rivers or mountain ranges, built features such as railroads, highways, major streets, and political boundaries often create barriers to interaction that are innate boundaries for communities. Census tracts may or may not follow these innate boundaries, but many environmental and social phenomena, such as air pollution or economic activity, do not. Instead, they tend to change gradually across these boundaries. Aggregating such data by crisp census tract boundaries may create a false impression of abrupt changes in concentration or activity.

Another challenge with aggregate spatial data such as census tracts is the modifiable areal unit problem (MAUP). This is the sensitivity of measurement and analysis to the choice of spatial units for collecting and analyzing data (Fotheringham and Wong, 1991). If the scale and boundaries of these units are modifiable or arbitrarily defined, changing the scale and boundaries will change the results of the measurement and analysis based on the units. Figure 3.1 illustrates the concept of MAUP as it applies to both scale and boundaries (zones). The MAUP is a result of spatial heterogeneity within the aggregate spatial unit of analysis. It is closely related to Simpson's paradox (the apparent change in trends measured in groupings of data that disappear or reverse when the groups are combined; Samuels, 1993).

BOX 3.1
Designing and Testing User Interfaces for Geospatial EJ Tools

Web-based geospatial mapping tools, such as the CEJST, combine the challenges of cartographic and user interface design. Although mapmaking involves artistry, there are fundamental scientific principles of cartographic design that need to be applied to ensure that the resulting map conveys information effectively. "A well-designed map has elegance and style," clarity and meaning, and conveys "the necessary level of precision and accuracy for the intended message" (Buckley, Hardy, and Field, 2022). Integrating digital maps with interactivity raises the design challenges by expanding the purposes of map use. An influential conceptualization by MacEachren (1994) defines three dimensions of map use (high versus low interactivity, public versus private use, and presenting knowns versus revealing knowns) with map uses and, therefore, design considerations dependent on the dimensions. Interactive cartography requires different types of user interface design and usability testing (Roth, 2013; Tsou, 2011). Spatial decision support systems (SDSS) go beyond interactive cartography to include linked nonspatial data visualizations and tools for characterizing problems, evaluating solution alternatives, and assessing their trade-offs (Jankowski, 2008; Keenan and Jankowski, 2019). Collaborative SDSS includes tools for groups of stakeholders to cooperatively explore problems, solutions, and tradeoffs (Jankowski et al. 1997).

The committee did not consider map design and user interface to be part of the scope of this report, but EJ tools such as CEJST can benefit from input from cartographic and user-interface/user-experience experts to ensure that information on disadvantaged communities is effectively conveyed and used (e.g., effective representation of category breaks and choice of colors). Because CEJST characterizes a suite of indicators that range from pollution to socioeconomic status across the country's diverse communities, there are opportunities to offer interactivity to explore these data. As described in Chapter 4, several EJ screening tools have interface functionality that allows users to view maps of individual indicators in addition to composite measures of disadvantage or burden. Other tools, such as EJSCREEN, offer infographic functionality to compare local burdens to averages at the county, state, or national level. These types of functionalities in the map interface go beyond the simple presentation of results and give users opportunities to gain insights and form narratives about their community. They can also offer practical utility: a rich, powerful user interface with data tools can be a starting point for justification of resource needs (e.g., grant applications). It is difficult for tool architects to predict which styles of cartography and which types of user interface will be most useful for their intended target audience. Therefore, careful design, testing, and workshopping of user interfaces can result in increased use and functionality of geospatial tools.

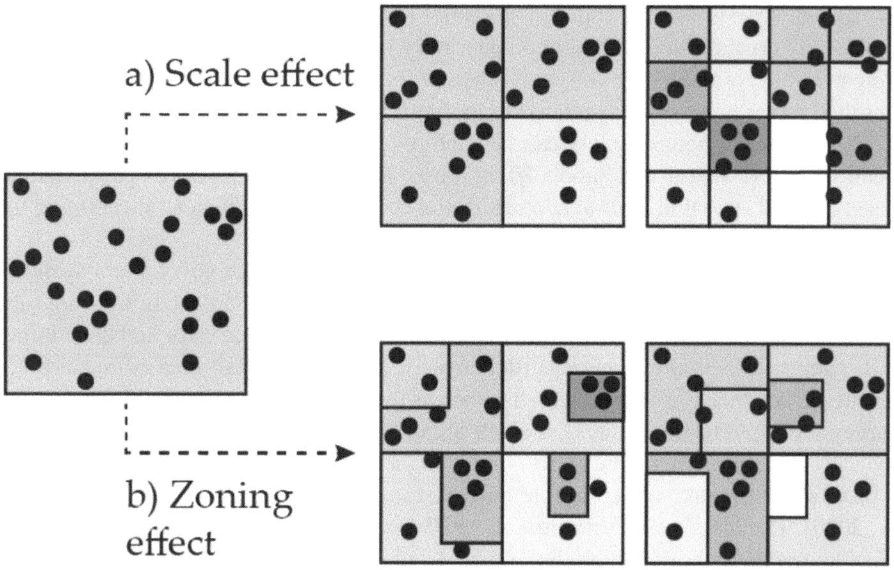

FIGURE 3.1 Illustration of scale and zoning effects in the modifiable areal unit problem. The left side of the figure shows hypothetical unaggregated data (e.g., bicycle crash locations). The top half of the right side of the figure shows aggregating the data at different spatial scales. The bottom right side of the figure shows different ways space can be partitioned at the same spatial scale. Analytical measures (e.g., bicycle crash rates per spatial unit) will vary with these different modifications of the areal units of analysis. SOURCE: Loidl et al., 2016.

There are four major strategies for managing the MAUP: (1) use datasets that are as disaggregated as possible; (2) capture the spatial nonstationarity (heterogeneity) within the aggregate spatial unit using local spatial models (see Fotheringham and Sachdeva, 2022); (3) design optimal spatial units; and (4) conduct sensitivity analysis with the spatial units (Xu, Huang, and Dong, 2018). In addition, considering spatial heterogeneity within the spatial units can help avoid committing the ecological fallacy—assuming that the generalized characteristics of an area reflect the characteristics of individuals in that area. For example, a census tract may exhibit a modest household median income. This could indicate that most households in the tract have a modest household income. However, it is also possible that there is significant income inequality within that tract. A large proportion of households in that tract could have incomes that are significantly above the median, while another proportion in that same tract is far below the median. The actual scenario is not directly discernible from statistics for an aggregate spatial unit.

Spatial heterogeneity leads to another geospatial issue when measuring geographic phenomena. If the influence of something measured by an indicator varies with location, then the indicator's influence on the composite indicator should also vary. For example, the impact of poor air quality on disadvantage may matter more in neighborhoods with poor health outcomes associated with other stressors, inadequate housing, or low income than in neighborhoods with better health, housing, and resources. This is a particularly crucial issue with noncompensatory indicator aggregation since individual indicators cannot substitute for each other as in compensatory aggregation, meaning that they retain more of their relative importance in the composite indicator (Fusco et al., 2023). Spatial heterogeneity can be captured by varying the indicator's weight by location (Fusco, Vidoli, and Sahoo, 2018). Fusco and others (2023) developed a method based on local spatial models to capture spatial heterogeneity in indicator weighting for noncompensatory aggregation.

Administrative data—data collected by public institutions or government agencies such as the Census Bureau or the Centers for Disease Control and Prevention—are subject to privacy concerns and budgetary constraints that can create error and uncertainty in its parameter estimates. This is particularly a problem with small-area estimates (e.g., census blocks, block groups, and tracts) such as those used in the American Community Survey (ACS).[12] In this context, a small area refers to a subnational area (e.g., state, county, or smaller geographic entity) in which the sample size for that subnational area is insufficient to make direct sample-based estimates with reliable precision (Logan et al., 2020). The margins of error around small-area estimates, such as median incomes of demographic cohorts, are so large that it can be difficult to determine rankings or thresholds for these areas. The larger margins of error correlate with observable geographic and demographic patterns: income estimates for center-city neighborhoods are less precise than neighborhoods farther from the city center, and neighborhoods at both extremes of the income spectrum have lower-quality estimates. These problems relate to small sample sizes, and the lack of contemporaneous population controls for small areas such as census tracts. A practical solution is to aggregate the census tracts into larger, possibly noncontiguous, units. However, doing so can dilute some of the patterns and trends in the data and can also introduce additional MAUP artifacts (Jurjevich et al., 2018; Spielman, Folch, and Nagle, 2014).

Other administrative units used by the Census Bureau with data that may be relevant for Justice40 designation. For example, Census Designated Places (CDP) could provide finer spatial resolution and alignment with specific communities such as colonias, Tribal lands, and other densely settled unincorporated places.[13] Census tract data could be augmented with other administrative units to capture population heterogeneities, but doing so may require addressing issues associated with coverage at a national scale and mixing of units with different population ranges. Issues of scale and mixing of units are also applicable when considering how community-generated data might be included.

While the decennial U.S. Census and the annual ACS represent reliable and widely used data sources for EJ tools, it is important to consider a few weaknesses. First, U.S. Census or ACS data focus exclusively on the residential or nighttime distribution of the population. These datasets cannot be used to assess daytime environmental exposure and risk burdens (Chakraborty et al., 2011), or analyze environmental injustices for

---

[12] See the U.S. Census Bureau's American Community Survey, https://www.census.gov/programs-surveys/acs, (accessed February 9, 2024).

[13] See https://www.census.gov/programs-surveys/bas/information/cdp.html (accessed June 12, 2024).

individuals in schools, workplaces, and other sites inhabited by vulnerable people (e.g., daycare centers, prisons, hospitals, and health care facilities). Second, neither the U.S. Census nor ACS includes questions on how individuals or households perceive environmental risks and respond to them and on what factors influence residential location decisions (Collins et al., 2015), thus limiting their usefulness for in-depth analysis of climate or environmental injustices. Third, multiple socially disadvantaged and vulnerable population groups (e.g., racial/ethnic minorities, young children, people with disabilities, renters, and immigrant populations) have been historically undercounted by the decennial census and ACS surveys (Stempowski, 2023). These undercounting problems have amplified since the COVID-19 pandemic, in conjunction with increased nonresponse bias in ACS estimates (Rothbaum et al., 2021).

COMMUNITY ENGAGEMENT AND PARTNERSHIPS

Composite indicator construction is intricate, multifaceted, and value laden and, therefore, cannot be reduced to a purely formulaic approach. Several of the composite indicator construction steps described above involve judgments and values, including concept definition, indicator selection, weighting, and aggregation. Technical steps—such as the application of data treatments, normalization, statistical coherence techniques, and uncertainty analysis—also affect and are affected by the judgments applied during concept definition, indicator selection, weighting, and aggregation. Finally, the tool results need to make sense and have validity in the real world; this involves human judgment beyond technical considerations. There are structured approaches to collaborative and group decision making in composite indicator construction and assessment, such as multicriteria decision analysis (El Gibari, Gómez, and Ruiz, 2019), analytical hierarchy process (Gómez-Limón, Ariazza, and Guerrero-Baena, 2020), Delphi techniques (Bana e Costa et al., 2023), and quantitative storytelling (Kuc-Czarnecka, Lo Piano, and Saltelli, 2020).

Given the consequential nature of EJ tools such as the CEJST, it is essential to incorporate as part of tool validation the perspectives and lived experiences of people and communities that may be affected by decisions informed by tool results. Understanding this requires embedding collaborative decision making within careful and thoughtful community engagement that occurs early and throughout the tool creation process, informing all steps—and the process itself—iteratively as warranted (Saisana et al., 2019). Community engagement is "the process of working collaboratively with and through groups of people affiliated by geographic proximity, special interest, or similar situations to address issues affecting the well-being of those people" (CDC/ATSDR, 1997, p. 9; quoted in McCloskey et al., 2011). Community engagement is a means to divert from traditional policy making and decision making based on executive leadership and data collected by external groups toward a policy and decision-making process that is inclusive of community input from the initial planning discussions through evaluations. As discussed below, there is no single defined process for "community engagement"; there is a spectrum of possible engagements depending on the necessary level of involvement of community members and other interested and affected parties. However, there are commonalities among the community engagement principles followed by many community organizations, universities, and state government agencies. Table 3.1 provides a high-level overview—and potential workflow—of eight aspects of community engagement as defined by Bassler and others (2008). Community engagement is a continuum and a nonlinear process, especially as collaborations mature (Schlake, 2015). Table 3.2 provides a list of additional information resources related to successful community engagement and community partnerships that can be used to establish the validity and use of the CEJST and other screening tools.

Not all participatory methodologies are effective, and some methodologies may lead to no change, continued marginalization, or even additional harm (e.g., Gaynor, 2013). Understanding who is included and excluded in community engagement and the power relations established through application of different methodologies for engagement is important in the design of community engagement programs. Trust-based and healthy community engagement and partnerships allow interested and affected parties to realize the reach of their personal power, influence, responsibility, and accountability. Below is a list of some potential benefits and returns on investment from developing and sustaining community partnerships and incorporating community engagement in each stage of the work (adapted from Bassler and others, 2008):

- Community buy-in and support for the process, program, and results;
- Increased enthusiasm and support for shared goals;
- Development of new and larger networks based on deeper understanding between interested and affected parties;
- Improved community education around important issues;
- Improved community advocacy and accountability for decision makers;
- Increased sense of empowerment and self-determination among communities;
- Increased and stronger community partnerships rooted in trust and authentic communication; and
- Potential improved community outcomes.

Community engagement also provides means to allow communities to help define themselves, empowers them with opportunities to influence how their own data or data about them might be properly and respectfully used, and to help identify, for example, any unintended consequences associated with the use of a tool (e.g., as a result of being labeled or defined a certain way).

Guiding Principles for Community Engagement

Establishing and practicing principles for community engagement and the collection of lived-experience data provide a strong foundation on which to build sustained relationships based on trust—including trust in EJ tools and their results. Principles exist (as described below) that may be adapted and used by those developing, maintaining, or evaluating EJ tools. However, specific applications need to be developed or adopted in collaboration with community members. Disadvantaged communities commonly

TABLE 3.1 Workflow for Community Engagement

Category	Principle	Details
Beginning the process	1. Define the community issue, the scope of the problem, and the framework.	Clearly define the activity's purpose. Define and determine the population(s) to engage. Define goals and ensure that the scope of the project is feasible and the goals are realistic while acknowledging that large goals can be divided into smaller goals.
	2. Define roles and level of community engagement.	Define goals and methods to obtain the goals. More in-depth information on the spectrum of community engagement will be discussed later in this chapter and can also be found in a report by Gonzalez (2019).
Items necessary for engagement	3. Identify existing tools for community engagement.	Meet in the community with which you wish to engage and involve community members. There are multiple recommended protocols for community engagement, but community engagement practices should be rooted in community-created principles such as the Jemez Principles for Democratic Organizing (Solis, 1997) and the Principles of Environmental Justice.[a]
	4. Identify stakeholders and be intentional about the community's right to self-determination.	Self-determination for the individual and community is vital to and at the core of the community process. This requires acknowledging and accepting the full humanity and authority of those traditionally not seen as "experts." While no community is a monolith, it is difficult to achieve community buy-in without making space for community influence. This also requires accepting decisions and feedback from stakeholders as valuable and without discounting their authority. For capacity-building purposes and diversity of ideas, consider intergenerational stakeholders.
Succeeding in the engagement process	5. Develop a plan to create and sustain community partnerships.	Effective community engagement requires transparent discussion and partnerships where all participants are equal. They identify opportunities, levels of contributions, and returns on investment. Develop means of community communication that are authentic and comfortable to the community. Ensure that means of engagement are as accessible as possible. Create financial plans to pay community stakeholders for their time and contributions. Pay them as you would career professionals.

continued

TABLE 3.1 Continued

Category	Principle	Details
	6. Respect community diversity and needs.	Accept that communities are not monoliths. Ensure that engagements are accessible, consider childcare, provide meals, and accept cultural and language differences.
	7. Co-create evaluation and feedback plans and determine next steps.	Accept community lived experiences when developing evaluation plans. Communities must have a say in evaluating plans and accountability. Community feedback needs to be incorporated into every step of the work, including plans for outcome dissemination.
	8. Maintain flexibility and open lines of communication.	Release the urge to control the process. Welcome the inner individual and organizational changes, even though they may be uncomfortable. Welcome changes to program development and outcome expectations. Be willing to change and adapt as society and community needs and wants change. This is vital to sustaining trust-based community partnerships. Flexibility and adaptation require open lines of communication where all stakeholders know who to contact and where their feedback will go within partnering organizations. Stakeholders must be consistently updated with the outcomes of their feedback, concerns, questions, and ideas.

[a] The Principles of Environmental Justice came out of the First National People of Color Environmental Leadership Summit in Washington, DC. https://www.ucc.org/what-we-do/justice-local-church-ministries/justice/faithful-action-ministries/environmental-justice/principles_of_environmental_justice/ (accessed December 28, 2023). SOURCE: Bassler et al., 2008.

express frustration with the lack of true community participation in decisions that will disproportionately impact their lives and well-being. Often, what is referred to as "community engagement" sustains the marginalization at the root of the issues to be addressed (Gonzalez, 2019). Unfortunately, "community engagement" is commonly little more than a survey or a single community meeting from which input received is later ignored.

Community engagement can occur at various levels along a spectrum that ranges from ignoring through informing, consulting, involving, collaborating with, and deferring to communities (see Table 3.2; Gonzalez, 2019). Partnership lies beyond engagement, involving sustained relationships and power sharing; effective engagement may lead to partnerships (Nardo et al., 2005). The urban planning literature includes the "ladders of engagement" model by Arnstein (1969), which describes a similar spectrum of community engagement. It refers to eight rungs, progressing from nonparticipation through degrees of tokenism, such as informing the public or receiving their consultation, and on to varying degrees of citizen power at the top (e.g., partnership, delegated power, and citizen control). This highly impactful model guides public engagement in planning (Slotterback and Lauria, 2019).

Many social and EJ organizations, and a growing number of nonprofits rely on the Jemez Principles for Democratic Organizing (Solis, 1997; see Box 3.2) and the Prin-

TABLE 3.2 The Spectrum of Community Engagement

Stance Toward Community	Ignore	Inform	Consult	Involve	Collaborate	Defer To
Impact	Marginalization	Placation	Tokenization	Voice	Delegated power	Community ownership
Community engagement goals	Deny access to decision-making processes	Provide the community with relevant information	Gather input from the community	Ensure community needs and assets are integrated into the process and inform planning	Ensure community capacity to play a leadership role in implementation of decisions	Foster democratic participation and equity through community-driven decision making. Bridge divide between community and governance
Message to community	*Your voice, needs, and interests do not matter*	*We will keep you informed*	*We care what you think*	*You are making us think (and therefore act) differently about the issue*	*Your leadership and expertise are critical to how we address the issue*	*It's time to unlock collective power and capacity for transformative solutions*
Activities	Closed-door meetings Misinformation Systematic	Fact sheets Open houses Presentations Billboards Videos	Public comment Focus groups Community forums Surveys	Community organizing and advocacy House meetings Interactive workshops Polling Community forums	MOUs with community-based organizations Community organizing Citizen advisory committees Open planning forums with citizen polling	Community-driven planning Consensus building Participatory action research Participatory budgeting Cooperatives
Resource allocation ratios	**100%** Systems admin	**70-90%** Systems admin **10-30%** Promotions and publicity	**60-80%** Systems admin **20-40%** Consultation activities	**50-60%** Systems admin **40-50%** Community involvement	**20-50%** Systems admin **50-70%** Community partners	**80-100%** Community partners and community-driven processes ideally generate new value and resources that can be invested in solutions

NOTE: MOU = memorandum of understanding.
SOURCE: Gonzalez, 2019.

ciples of Environmental Justice to build community partnerships. Both documents were created by individuals who have been disproportionately affected by toxic chemical exposure and economic hardship, which often occur in conjunction. The principles in the documents inform organizations on how they might hear directly from the interested and affected parties with whom they seek partnerships.

There is solid evidence that community engagement can improve health and social outcomes, particularly for disadvantaged communities, although which model of engagement is most effective can vary (Cyril et al., 2015; Haldane et al., 2019; Milton et al., 2012). Honesty and agreement are necessary parts of the choice of an appropriate model of community engagement for issues being addressed. These agreements need to be honored throughout the lifetime of the tool and need to be in place before development begins. It is also possible for engagement to move through the spectrum in a nonlinear pattern as stakeholders and priorities change. However, the goal is to move toward the highest level of community participation possible.

ACHIEVING TRUST, TRANSPARENCY, AND LEGITIMACY

The 10-step guide to indicator construction described earlier in this chapter is a robust framework that outlines the important elements of indicator construction. Validation is a part of that framework, but what is less explicit in it is that EJ tools such as CEJST need to embody trust, transparency, and legitimacy that can only be achieved through iterative community engagement and validation of tool indicators, data, processes, and results. The committee has reimagined the 10-step guide in a conceptual framework to explicitly incorporate those concepts (see Figure 3.2). The framework consists of three rings:

- The innermost ring represents composite indicator construction. This includes processes that require multiple decisions, including defining the concept to be measured, selecting indicators, and determining how those indicators will be integrated. Importantly, these decisions need to be made using an iterative process in which decisions are evaluated, modified, and reevaluated for internal robustness.
- The middle ring represents the substantive and iterative community engagement, validation of indicators and tool results, and documentation of decisions and approaches that are all integral to the tool development process.
- The outermost ring recognizes that the entirety of the environmental tool development process needs to promote the transparency and legitimacy of the tool, leading to trust in its design, results, and use.

The conceptual framework in Figure 3.2 has guided the preparation of this report and its recommendations. Following a description of existing environmental screening tools in Chapter 4, Chapter 5 provides more information regarding the selection of indicators, Chapter 6 discusses the integration of indicators and the assessment of composite indicator robustness, and Chapter 7 addresses tool validation (and presen-

BOX 3.2
The Jemez Principles

In 1996, a diverse group of environmental justice advocates met in Jemez, New Mexico, and drafted the Jemez Principles for Democratic Organizing (Solis, 1997). In the decades since, the principles have been adopted by nonprofit and private organizations around the world. The principles, along with the Environmental Justice Principles[a] developed during the First National People of Color Environmental Leadership Summit in 1991, are used as best practices when working with communities, particularly marginalized communities disproportionately affected by environmental and social injustices. Table 3.2.1 below is an overview of the Jemez Principles and provides example applications for EJ tools such as CEJST.

TABLE 3.2.1 The Jemez Principles

Principle	Summary
1. Be inclusive	Ensure all stakeholders, especially those from disproportionately impacted and marginalized communities, are partners through the entire process, from inception to evaluations, also ensuring that these groups have power and decision-making authority.
2. Emphasize bottom-up organizing	Communities do not need to be organized. Communities organize themselves. Communities do not need to be saved. They need resources and power. Communities lead.
3. Let people speak for themselves	Communities are the experts in their lived experiences, and communities can and must speak for themselves. Pass the microphone.
4. Work together in solidarity and mutuality	Act with intentionality to ensure the intersectionality of issues and concerns, building trust between groups. Share resources and strategies. Collaborate to reflect the real-life continuum of issues, including economics to environmental, breaking down false silos.
5. Build just relationships among ourselves; treat each other with justice and respect	Treat everyone justly and with respect. Intentionally develop and sustain just relationships; committing to this process that will not be linear, or quick. Make space for the acceptance and utilization of different skill sets.
6. Commit to self-transformation	Commit to working in the community, decentering individualism, and centering community. Commit yourself to authenticity, self-improvement, and living a life of “justice,” “community,” and “peace.”

[a] See https://comingcleaninc.org/assets/media/documents/ej4all-Principles2.pdf (accessed February 27, 2024).

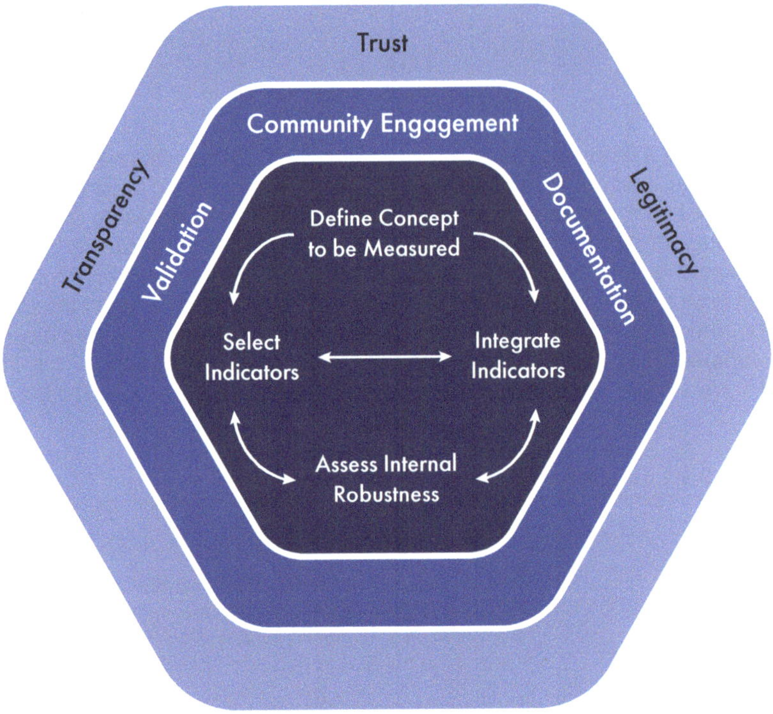

FIGURE 3.2 Conceptual diagram depicting the committee's framework to guide development of environmental justice tools. The arrows in the innermost ring indicate the direction of influence that each aspect of composite indicator construction has on another (i.e., defining the concept to be measured will influence the selection and integration of indicators, selection of indicators influences integration of indicators and vice versa, assessing internal robustness influences selection and integration of indicators and vice versa).

tation). Although the committee was charged with providing recommendations to be incorporated into an overall data strategy for CEQ's tool(s), the development of a data strategy in consideration of the conceptual framework resulted in recommendations that were more broadly applicable to EJ tools. Chapter 8 provides those recommendations.

CHAPTER HIGHLIGHTS

A composite indicator is used to represent a multidimensional concept or condition—for example, community disadvantage. Calculating a composite indicator brings together measurements from multiple dimensions of the concept (e.g., different aspects of community disadvantage) to determine a single value intended to reflect the condition being measured. Sound composite indicators are developed with a clearly defined purpose and intended audience and reflect real-world conditions. The validity of a tool rests on a foundation of scientific and methodological rigor, meaningful participation and input from community members and other interested and affected parties, transparency, and acceptance by institutional actors (e.g., government agencies), communities, and other affected parties. Composite indicators can be constructed for a variety of purposes, for example, to screen for initial results, to directly inform decisions, or for educational purposes. Results derived from a tool can be binary (e.g., "is" or "is not") or may provide a continuous measure or scale. CEJST is a binary tool intended to designate a community as disadvantaged or not based on selected criteria. Given that there is no quantitative definition or threshold for community disadvantage, there is a certain amount of subjectivity in creating a composite indicator. Composite indicator construction is multifaceted and value laden and is more than simply the result of formulas or statistics.

There are published, systematic methodologies for developing transparent, trustworthy, and legitimate composite indicators, and for evaluating their construction decisions, internal robustness, and external validity. Such methodologies consider a composite indicator as an interrelated system of steps or components for identifying the concept to be measured, selecting indicators and data, analyzing, normalizing, weighting, and aggregating the data, evaluating the results for coherence and validity, and presenting the resulting information. The methodologies provide iterative processes for making decisions about each of those components, evaluating them for coherence with the composite indicator's purpose, and validating that they reflect the real world. Community engagement is an important aspect of composite indicator and tool validation.

Effective community engagement is a means to collect community lived experience and to understand how composite indicator-related decisions can be informed by those experiences. It can also inform how composite indicator-related decisions might affect communities through the use of the resulting tool. Community engagement could be used, for example, to inform composite indicator constructors regarding the use of aggregated data, such as U.S. Census data, to define communities. A census tract may generalize data over too broad an area and may not reflect the lived experiences of a community within the tract. Community engagement could be used to identify when census tracts may or may not be appropriately used or what might be used instead. Effective community engagement is part of a comprehensive composite indicator construction process that not only validates approaches used in the composite indicator but can also result in trust in the tool and improved outcomes. Recognizing the importance of community engagement in composite indicator construction, the committee developed a vision for EJ tool development (Figure 3.2). The conceptual diagram has as its foundation a systematic methodology for composite indicator construction but explicitly recognizes the value of communication through community engagement, validation, and documentation. These components serve to build a tool that embodies trust, transparency, and legitimacy.

4

Environmental Justice Tools

An increasing number of publicly accessible environmental justice (EJ)-related mapping or screening tools have been created by federal agencies, states, regional planning agencies, and municipalities, as well as private or nongovernmental institutions in the last 15 years. This committee identified over 35 separate EJ mapping or screening tools used across the United States as of summer 2023, half of which have been released since 2021. Some of these tools were created in response to EJ-related policies, while others were created independently of any specific legislation or regulation but rather offered the opportunity to guide new policy, as well as to support information dissemination and awareness of EJ issues. As more public institutions seek to develop or refine these kinds of tools, there is a benefit to understanding and comparing the various approaches taken and the lessons learned.

This chapter summarizes the committee's scan of EJ tools that utilize social and environmental indicators to identify geographic areas that may be subject to varying levels of environmental or social inequity, burden, or underinvestment. The chapter first summarizes previous surveys of EJ tools and then provides a more detailed comparison made by the committee of a subset of tools. The chapter provides practical descriptions of how different tools format indicator data and address cumulative impacts. This subset illustrates the range of approaches or methodologies currently used in EJ mapping or screening. They were selected based on descriptions provided by tool creators to the committee during public meetings (see Appendix B for open session meeting agendas), from information in the earlier tool surveys, and based on committee member expertise and familiarity with the tools. Most of the tools are used in the United States and created or maintained by governmental (e.g., federal, state, municipal, or regional planning agencies) and nongovernmental institutions (e.g., universities and nonprofits) for public use. Descriptions of each of the tools' indicators are provided, as well as the format(s) for indicator data, the use of composite indicators to aggregate data and measure cumulative burden, and the use of thresholds to identify disadvantaged communities. The chapter then provides a brief comparison of the features of the White House Council on Environmental Quality's (CEQ's) Climate and Economic Justice Screening Tool (CEJST) with those of other tools. More information regarding CEJST is provided in later chapters. This chapter focuses primarily on describing different tool features and construction rather than comparing the outcomes of using different tools. Those outcomes can vary significantly (Popovich et al., 2024; Reckien, 2018; Tate, 2012; Wang et al., 2023).

PREVIOUS SURVEYS OF ENVIRONMENTAL JUSTICE TOOLS

Evergreen Collaborative (2020) evaluated the U.S. Environmental Protection Agency (EPA) Environmental Justice Screening and Mapping tool (EJScreen)[1] and equity mapping tools from California, Maryland, New York, and Washington State, offering recommendations to the White House on developing a national equity map. These recommendations were made prior to the development or release of CEJST. That review summarized multiple lessons from these state-level tools and highlighted gaps in EJScreen (the only federal tool extant at the time). Conclusions from the survey included the need for a clear definition of disadvantaged communities, the importance of calculating cumulative impacts that allow comparison and prioritization, the importance of official state support and enabling policies that require the use of the tools for decision making, and the centrality of community input and engagement in defining "disadvantaged communities" and relevant indicator selection.

A later report by the University of California Los Angeles Luskin Center for Innovation (Callahan et al., 2021), published shortly before the beta release of CEJST, reviewed EJ policies and programs from California, Illinois, Maryland, New York, Washington, and Virginia. That report evaluated the strengths and weaknesses of California's CalEnviroScreen[2] to frame EJ tool recommendations to the White House to "maximize the benefits of Justice40 effectively and equitably." While the report praised CalEnviroScreen's use of cumulative impact scoring based on a wide range of environmental and social factors, it critiqued the narrow focus on pollution exposure and demographics. The Luskin Center report recommended that a federal EJ tool include five categories of disparities: disproportionate exposure to pollution, uneven distribution of climate impacts, lower levels of local resources and community capacity, disproportionate occupational impacts from the transition to clean energy, and uneven distribution of the costs and benefits of environmental policies. The Luskin Center report also recommended that, unlike CalEnviroScreen, race and ethnicity should be included as important indicators in any EJ screening process. In addition to relative measures of exposure or impact (e.g., percentile scores), the Luskin report recommended that absolute changes in pollution or other indicators should be measured and reported over time to track progress. Finally, the report recommended that federal officials follow

[1] See https://www.epa.gov/ejscreen (accessed March 2, 2024).

[2] See https://oehha.ca.gov/calenviroscreen/report/calenviroscreen-40 (accessed March 2, 2024).

CalEnviroScreen's model for community collaboration and accountability. The latter included an extensive consultation process with technical experts and interested and affected parties that included community groups with local knowledge and, equally important, extensive documentation on indicator data sources and how these indicators were selected.

Similar recommendations are in other reviews, such as those by Ravichandran and others (2021) and Arriens, Schlesinger, and Wilson (2022). Ravichandran and others (2021) reviewed EJ mapping tools for California, Florida, Maryland, Cuyahoga County in Ohio, the Houston-Galveston-Brazoria region of Texas, Washington State, and EJScreen. They identified six themes that are inadequately represented in current EJ tools: social progress, vulnerability, climate equity, economic progress, health, and resilience. They recommended 20 indicators within each theme to be incorporated into such tools.

Some surveys have offered reviews across a larger number of tools. Konisky, Gonzalez, and Leatherman (2021) analyzed 19 EJ mapping tools—18 state tools and EJScreen—as part of a project to assist the state of Indiana in the development of its own tool. To compare the tools, they described the functionality of the online interface, identified the environmental and social indicators used by each tool, and described how indicator values are presented (e.g., percentiles, raw values, index values). In addition, they did qualitative assessments, based in part on interviews with state officials, to characterize tool accessibility and comprehensiveness, as well as the role of community engagement and the usage of the tools. Their analysis revealed common characteristics (e.g., all the tools are interactive, most were created by government agencies, all include social indicators, but only two-thirds include both social and environmental indicators, and most include racial composition to identify EJ communities). Konisky, Gonzalez, and Leatherman (2021) recommended a set of questions to guide the development of EJ mapping tools: (1) Who is the audience? (2) What is the purpose of the tool? (3) How do you want the map to function? (4) How do you want to implement (or use) the tool? (5) How do you get engagement from interested and affected parties?

The most comprehensive survey of EJ mapping tools to date was done by the Urban Institute (Balakrishnan et al., 2022). They compared and evaluated 31 national, state, and local EJ tools across numerous parameters and characterized these tools based on (1) data sources for social, environmental, or health indicators; (2) use (or omission) of race/ethnicity; (3) how disadvantage is defined and measured; (4) how EJ communities are prioritized (e.g., scoring, ranking, thresholds); and (5) the policy context that resulted in the tool's development as well as ongoing linkages of the tool to specific programs or funding sources. Their results are presented as both summary statistics and key findings, and they compiled most of this information into an online, interactive table that features 39 variables and associated notes.[3] The Urban Institute found that:

[3] See Urban Institute's interactive Airtable at https://www.urban.org/research/publication/screening-environmental-justice-framework-comparing-national-state-and-local (accessed September 18, 2023).

- Tool data are often out of date and lack local context. They mostly rely on data from the American Community Survey (ACS) and EJscreen; few tools are regularly updated with new data or are explicit about plans to update.
- Disaggregated race and ethnicity data are often not included, nor is an acknowledgment of the role of environmental racism. Race and ethnicity are factored into methods to identify EJ communities in more than 80 percent of the tools; they are included as context layers in the others. A more detailed breakdown of race and ethnicity is included in only six tools. Indigenous populations were rarely included as separate population groups.
- EJ communities are identified using varying methods, depending on the intended use of the tool. Most tools combine methods to quantify burdens and prioritize among EJ communities. Definitions of EJ community for some states were developed prior to tool development, while others were developed during or after.
- Specific environmental indicators are included in most tools, but important topics are overlooked. Few tools do not include environmental indicators, relying solely on socioeconomic characteristics. Air and water quality measures are included in more than 50 percent of the tools. (The measures included vary by jurisdiction.) Extreme heat, flooding, natural hazards, toxic chemicals, and waste are less commonly included.
- Rural community data and needs are not sufficiently considered in many of the tools. The paucity of data on rural and Tribal communities results in a bias toward urban issues and areas in all the tools. Region-specific issues and established community priorities are more readily captured by state and local tools.

The Urban Institute report (Balakrishnan et al., 2022) recommends that (1) tool creators explicitly state how the tool is intended to be used and allow communities to self-identify; (2) community members be a part of tool development and that the tools account for local context as much as possible, (3) more diversity in topical areas and indicators be included in the tools and updated more regularly, and (4) tools endeavor to quantify cumulative impacts. The authors call for continued research to understand the universe of related tools to further equip communities and decision makers, to determine how well different tools capture cumulative impacts or burdens and, ultimately, "how tools can advance racial equity as a central objective."

Other institutions have sought to develop public databases to identify the growing list of EJ tools and related policies. The Tishman Environment and Design Center at the New School developed a searchable tool of definitions, indicators, thresholds, and benefits focused on the various cumulative impact policies developed across the country (Baptista et al., 2022; Tishman Environment and Design Center, 2022). Thirteen states were identified (California, Hawaii, Illinois, Massachusetts, Maryland, Michigan, Minnesota, New Jersey, New Mexico, New York, Oregon, Vermont, and Washington) that had legislation, mapping tools, or agency guidance documents that include consideration of cumulative impacts. (California, Minnesota, New York, and Washington developed or were developing geospatial mapping tools for expressing cumulative impacts.) They also analyzed the policies of 11 states to understand how cumulative impacts were defined. They found that few reports or policies mention specific thresholds or methodologies useful for determining the existence or extent of cumulative impacts. They argue that the methodologies used to measure cumulative impacts or determine thresholds for "unreasonable," "significant," or even "cumulative" harm are inherently normative and subjective, and as such, there is a need for the participation of affected parties—especially those within EJ communities (Baptista et al., 2022).

The Vermont Law School's Environmental Justice Clinic released a database documenting EJ laws, policies, mapping tools, and state-recognized definitions associated with EJ across the 50 states and territories.[4] As of November 2023, 90 state EJ policies and 27 state EJ mapping tools are included in their list. No federal policies or mapping tools are included. Other reviews have focused on the experiences of individual state or regional tools. Bara and others (2018) and Driver and others (2019) described the development and application, respectively, of Maryland's Environmental Justice Screening Tool.[5] Min and others (2019) described the development and community engagement incorporated into the development of Washington State's Environmental Health Disparities Map.[6] Bhandari and others (2020) described the development and potential application of HGBEnviroScreen for the Houston–Galveston–Brazoria Region of Texas. Faust and others (2021) reflected on the lessons learned from the development of CalEnviroScreen.

New tools are emerging at a rapid pace, and federal tools aside from EJScreen have been understudied. There is significant commentary on the recently released CEJST (Associated Press, 2023; Barnes, Luh, and Gobin, 2021; Chemnick, 2022; Costley, 2022; Fears, 2023; McTarnaghan et al., 2022; Mohnot, 2023; Pontecorvo and Sadasivam, 2022; Sadasivam, 2023; Sadasivam and Aldern, 2022; Shrestha, Rajpurohit, and Saha, 2023; Sotolongo, 2022; Zeng, 2022), but comprehensive comparisons of CEJST to other tools are still emerging (Balakrishnan et al., 2022; Spriggs, Rotman, and Trauth, 2024).

OVERVIEW OF SELECTED TOOLS

Based on the above surveys, information gathered during committee meetings, and the knowledge of members of the committee, the committee chose a subset of 12 tools from which to highlight key features of geographically based EJ tools. The committee focused on tools adopted by governments. The committee was aware of at least three dozen government-sponsored EJ screening tools extant at the time of its scan, as well as numerous tools and related algorithms developed by researchers and nongovernmental

[4] See https://ejstatebystate.org/ (accessed November 4, 2023).

[5] See https://mde.maryland.gov/Environmental_Justice/Pages/EJ-Screening-Tool.aspx#:~:text=Launch%20the%20EJ%20Screening%20Tool&text=The%20demographic%20and%20socioeconomic%20data,and%20overburdened%20communities%20in%20Maryland (accessed September 18, 2023).

[6] See https://doh.wa.gov/data-and-statistical-reports/washington-tracking-network-wtn/washington-environmental-health-disparities-map (accessed September 18, 2023).

organizations (see, e.g., Baker et al., 2023; Bhandari et al., 2020; Cutter, Boruff, and Shirley, 2003; Cutter and Morath, 2013; Indiana University, 2019; Popovich et al., 2024; Ren, Panikkar, and Galford, 2023; MEJ, 2021; Tee Lewis et al., 2023; Texas Rising, 2022). However, this scan of tools was not intended to be exhaustive, but rather illustrative of the range of approaches formally adopted by communities and policy makers. Table 4.1 lists the tools that were selected and summarizes their intended uses, audiences, and output types. This section describes some general characteristics of the tools (such as audience, outputs, level of geographic resolution, user interface, and updates). The following sections describe key features of the tools, including the burden indicators used by these tools, the format(s) used for indicator data, their approaches to aggregating data and measuring cumulative impacts, and their use of thresholds to identify disadvantaged communities. Important similarities and differences across the tools are described. To facilitate comparison, the committee created a set of tables that summarize information about each tool (e.g., indicators used, geographic resolution, data sources, purpose, and the methodology employed to rank, compare, or score the index; see Appendix C). For purposes of comparison, CEJST is also listed here and described briefly at the end of this chapter. More information about CEJST is found in later chapters.

The target audiences, explicit or implicit, vary by tool and can include the agencies that created the tools; other federal agencies; partners at Tribal, state, and local levels; communities; academics; students; public health officials; policy makers; emergency response planners; nonprofits; metropolitan planning organizations; the public; homeowners; renters; and real estate professionals. Types of outputs generated by the tools also vary, often depending on the purpose of the tool. For example, tools designed to track compliance with EJ-related regulatory requirements typically provide a binary designation of a community as disadvantaged or overburdened. Other tools provide relative measures of burden (for individual burden categories or overall) based on, for example, indices, percentiles, or ratings but do not classify communities based on those measures. Some tools may do both.

The geographic resolution of the tools differs. Census block groups, census tracts, ZIP codes, counties, and Tribal lands are all used. CalEnviroScreen and federal tools other than EJScreen are available at the census-tract level, and EJScreen and the Massachusetts DPH [Department of Public Health] Environmental Justice Tool (MA-DPH-EJT) provide data down to the census block-group level. The Centers for Disease Control and Prevention/Agency for Toxic Substances and Disease Registry (CDC/ATSDR) Social Vulnerability Index (SVI), Federal Emergency Management Agency (FEMA) National Risk Index (NRI), and Census Community Resilience Estimates (CRE) can be viewed or downloaded also at the county level, while the National Oceanic and Atmospheric Administration Climate Mapping for Resilience and Adaptation (CMRA) tool can be explored via county or Tribal land boundaries. The Indices of Multiple Deprivation (IMD), which were developed as separate "geodata packs" for each of the various countries within the United Kingdom (i.e., England, Northern Ireland, Scotland, Wales) use Lower-layer Super Output Areas (England, Wales), Data Zones (Scotland), and Super Output Areas (Northern Ireland), which are statistical unit areas that represent 500 to 3,000 people, depending on the country (similar to a census-block group in the United States).

All the tools have interactive web-based platforms. Many tool outputs are available on Esri's ArcGIS Online platform and in multiple formats (e.g., as downloadable Excel, geodatabase, or shapefile formats). All of the federal tools use American Community Survey (ACS) data,[7] and many of the federal tools use each others' data layers (visual representations of geographic data sets) For example, the Department of Transportation (DOT) Equitable Transportation Community Explorer (ETCE) uses the Environmental Justice Index (EJI) and CMRA; EJScreen and CMRA use CEJST; and the Department of Energy Energy Justice Mapping Tool (DOE EJMT) uses EJScreen and NRI, among others). Notably, the EJScreen displays a "Justice40" layer from CEJST despite a mismatch of years of underlying census data (2020 source for EJScreen and 2010 source for CEJST Version 0.1). EJScreen and MA-DPH-EJT allow users to add outside data layers and define a buffer around an area. CMRA also provides information about other resources, including actions the federal government is taking, funding sources, and dataset access. The SVI and CalEnviroScreen 4.0 tools are available in Spanish.

For tools that have been released in more than one version, updates have been expanded to include Tribal lands or U.S. territories. Tribal lands were first incorporated into the tools in SVI 2014, CEJST v1.0 of 2022 (in response to feedback on the beta version), CMRA v1.0 of 2022, and EJScreen 2.1 of 2022. Unique to EJScreen 2.1 is a colonias data layer[8] for communities along the U.S. southern border. Puerto Rico was first incorporated into SVI in 2014. All U.S. territories were first incorporated into NRI v1.19.0 in 2023), CEJST v1.0 in 2022), and EJScreen v2.1 in 2022). Some tools allow users to access historical versions of their tool, including the SVI, CalEnviroScreen, and IMD, which explicitly share standardized historical data to "allow relative rankings between iterations to be compared over time" (Ministry of Housing, Communities & Local Government, 2019). CMRA has a temporal component as well since it provides climate projections (i.e., extreme heat, drought, wildfire, flooding, and coastal inundation) for three time horizons (spanning 2015 to 2099) using two Representative Concentration Pathway scenarios (4.5 and 8.5).

Burden Categories and Indicators in Selected Tools

The indicators and variables incorporated by the reviewed tools vary, as do the groupings or themes used to categorize the indicators into burden categories. The differences are typically influenced by the specific purpose of the tool and considerations such as data availability, geographic resolution, open accessibility, and national coverage. Although the exact definition and scope of the burden categories vary across tools, the categories can be broadly classified as socioeconomic, environmental, health, and climate change vulnerability. Most (but not all) of the tools include consideration of

[7] For a detailed discussion of ACS data, see Chapter 5.

[8] Descriptions of colonias data layers can be found on EJScreen's website: https://www.epa.gov/ejscreen/ejscreen-map-descriptions#plac (accessed February 13, 2024).

TABLE 4.1 Creator, Purpose, and Output Type of 13 Selected EJ Tools

Tool	Creator	Purpose	Output Type
Climate and Economic Justice Screening Tool [a]	White House Council on Environmental Quality	• Statutory or regulatory requirement • Informational	Binary designation of disadvantage or not
EJScreen[b]	Environmental Protection Agency	• Informational	Relative rank, score, index
Environmental Justice Index [c]	Centers for Disease Control and Prevention	• Informational	Relative rank, score, index
Social Vulnerability Index [d]	Centers for Disease Control and Prevention	• Informational	Relative rank, score, index
Energy Justice Mapping Tool—Disadvantaged Communities Reporter [e]	Department of Energy	• Statutory or regulatory requirement • Informational	Binary designation of disadvantage or not
Climate Mapping for Resilience and Adaptation [f]	National Oceanic and Atmospheric Administration and Department of the Interior	• Informational	Raw data values
DOT Equitable Transportation Community Explorer [g]	Department of Transportation	• Statutory or regulatory requirement • Informational	Binary designation of disadvantaged or not
Census Community Resilience Estimates[h]	Census Bureau	• Informational	Relative rank, score, index
National Risk Index [i]	Federal Emergency Management Administration	• Informational	Relative rank, score, index
Massachusetts DPH Environmental Justice Tool [j]	Massachusetts Department of Public Health	• Statutory or regulatory requirement • Informational	Tiered designation of disadvantaged or not
California CalEnviroscreen[k]	California Environmental Protection Agency	• Statutory or regulatory requirement • Informational	Binary designation of disadvantaged or not Relative rank, score, index
Indexes of Multiple Deprivation [l]	British Department for Communities and Local Government	• Statutory or regulatory requirement • Informational	Relative rank, score, index

TABLE 4.1 Continued

Tool	Creator	Purpose	Output Type
New Jersey Environmental Justice Mapping, Assessment and Protection Tool [m]	New Jersey Department of Environmental Protection	• Statutory or regulatory requirement • Informational	Binary designation of disadvantage or not

NOTE: See Appendix C for more detail regarding all these tools.
[a] https://screeningtool.geoplatform.gov/ (accessed February 12, 2024).
[b] EJScreen: https://www.epa.gov/ejscreen (accessed February 12, 2024).
[c] EJI: https://www.atsdr.cdc.gov/placeandhealth/eji/index.html (accessed February 12, 2024).
[d] SVI: https://www.atsdr.cdc.gov/placeandhealth/svi/index.html. (accessed February 12, 2024).
[e] EJMT: https://energyjustice.egs.anl.gov/ (accessed February 12, 2024).
[f] CMRA: https://resilience.climate.gov/ (accessed February 12, 2024).
[g] ETCE: https://experience.arcgis.com/experience/0920984aa80a4362b8778d779b090723 (accessed February 12, 2024).
[h] CRE: https://experience.arcgis.com/experience/b0341fa9b237456c9a9f1758c15cde8d/. Accessed 2/12/24.
[i] NRI: https://hazards.fema.gov/nri/ (accessed February 12, 2024).
[j] MA-DPH-EJT: https://matracking.ehs.state.ma.us/Environmental-Data/ej-vulnerable-health/environmental-justice.html (accessed February 12, 2024).
[k] CalEnviroScreen: https://oehha.ca.gov/calenviroscreen (accessed February 12, 2024).
[l] IMD: https://data.cdrc.ac.uk/dataset/index-multiple-deprivation-imd (accessed February 12, 2024).
[m] EJMAP: https://experience.arcgis.com/experience/548632a2351b41b8a0443cfc3a9f4ef6 (accessed February 12, 2024).

socioeconomic and environmental burdens (see Table 4.2). Many tools, such as CDC/ATSDR SVI, EJScreen, FEMA NRI, MA-DPH-EJT, and CDC/ATSDR EJI, include indicators of race or ethnicity. A summary of the four broad categories of burden included in Table 4.2 is given below.

Socioeconomic Category Indicators

Socioeconomic burden indicators are used to measure the demographic and economic characteristics of communities, such as income, linguistic isolation, housing conditions, education, employment, and transportation. Data related to these variables are typically available at a census-tract-level scale. The indicators attempt to capture and analyze data to provide insights into the social and economic conditions of communities, enabling a better understanding of disparities and vulnerabilities.

TABLE 4.2 Indicators in 12 Selected EJ Tools and CEJST

	Indicators			
Tool	**Socio-economic**	**Environmental**	**Health**	**Climate Change Vulnerability**
Climate and Economic Justice Screening Tool (CEJST)	√	√	√	√
EPA EJScreen	√	√		
CDC/ATSDR Environmental Justice Index (EJI)	√	√	√	
CDC/ATSDR Social Vulnerability Index (SVI)	√			
DOE Energy Justice Mapping Tool (EJMT)	√	√	√	√
NOAA + DOI Climate Mapping for Resilience and Adaptation (CMRA)				√
DOT Equitable Transportation Community Explorer (ETCE)	√	√	√	√
Census Community Resilience Estimates (CRE)	√			
FEMA National Risk Index (NRI)	√			√
Massachusetts DPH Environmental Justice Tool (MA-DPH-EJT)	√		√	
CalEnviroscreen	√	√	√	
Index of Multiple Deprivation (IMD)	√	√	√	
New Jersey Environmental Justice Mapping, Assessment and Protection Tool (EJMAP)	√	√		

Environmental Category Indicators

Environmental indicators encompass factors in the physical environment that have an impact on physical and mental health. The tools reviewed include a range of environmental indicators, although there is variation among them. These indicators can generally be categorized into measures of air quality (such as particulate matter [$PM_{2.5}$], ground-level ozone levels, and air toxics cancer risk), land quality (including proximity to National Priority List sites, traffic proximity, and treatment storage and disposal facilities), and water quality (such as impaired surface water).

Health Indicators

Indicators of health burdens are intended to capture the impact of physical and mental well-being on communities. They include health indicators such as asthma, diabetes, low life expectancy, and heart disease, as well as measures of health vulnerability, including hypertension, cancer risk, and mental health. These health indicators are commonly derived from the CDC's PLACES data,[9] providing valuable insights into the health burden experienced by different communities. The inclusion of these health-related variables enables a comprehensive understanding of the health conditions and vulnerabilities within specific geographic areas.

Climate Change Burden Indicators

Indicators of climate change vulnerability include variables such as susceptibility to natural hazards (e.g., droughts, wildfires, hurricanes, flooding, heatwaves), future extreme weather risk, annualized disaster losses, and the percentage of impervious surfaces. Data sources for these variables are described in Appendix C. Tools that include climate change vulnerability indicators—such as the FEMA NRI and CMRA—can aid in the development of effective mitigation and adaptation strategies for vulnerable communities.

INDICATOR DATA FORMATS

Indicator data can be displayed and compared across geographic areas in a variety of ways. This section describes different methods for displaying and comparing data.

Raw Data

A way to display demographic or environmental indicators is in their raw data form—as originally reported (e.g., pollution concentration levels or the number/location of waste sites within a certain area). In most of the tools scanned, raw data are not prominently displayed but are available for viewing by the user. For example, raw or originally reported data values for health and environmental indicators in EJScreen are only available after the user requests that a report be generated for a specific location, or the user downloads the data for analysis using another software application. (EJScreen presents data in percentile form on its interactive map.) CDC/ATSDR's EJI similarly only makes raw data available via data download. FEMA's NRI tool only presents raw values for Expected Annual Losses. (The online map presents ranks or percentiles for its risk index, social vulnerability, and community resilience layers.) The National Oceanic and Atmospheric Administration (NOAA) Climate Mapping for Resilience and Adaptation (CMRA) tool was the only tool considered by the committee that relies on raw data for presenting data for examination or comparison. An example screenshot of the CMRA tool is presented in Figure 4.1.

[9] See https://www.cdc.gov/places/index.html (accessed February 13, 2024).

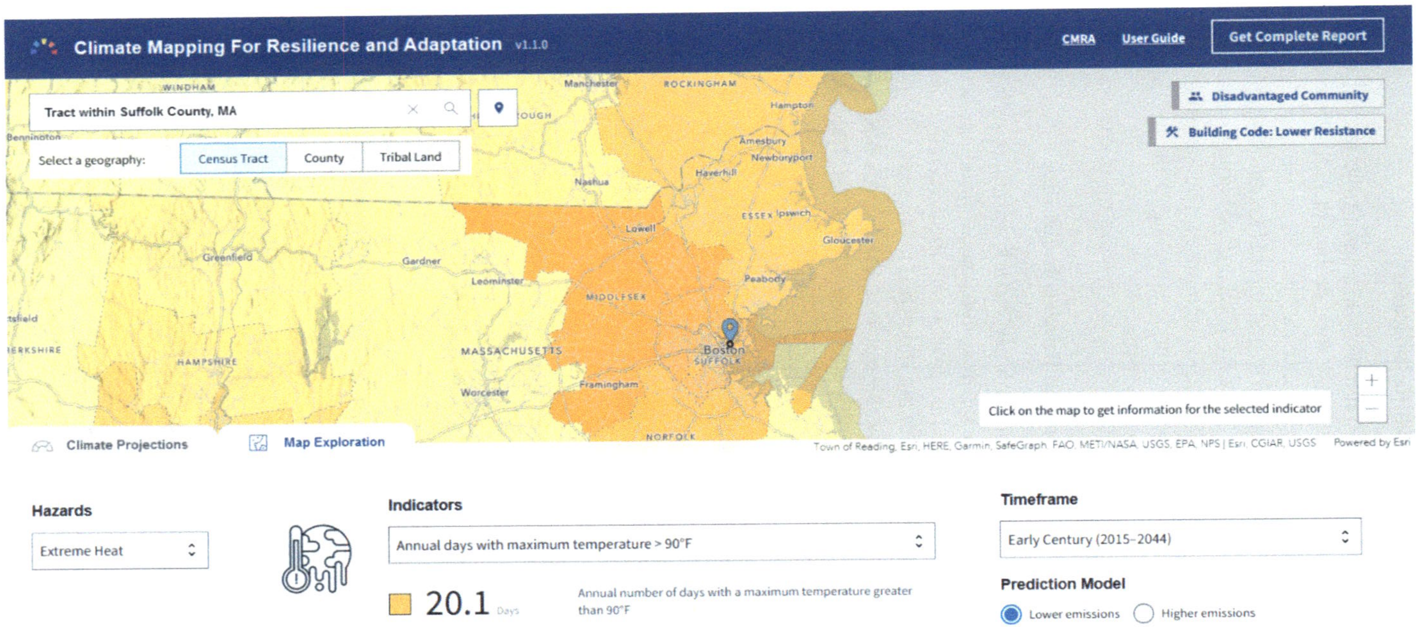

FIGURE 4.1 Climate Mapping for Resilience and Adaptation tool display illustrating presentation of raw data values. SOURCE: NOAA, 2022. Accessed November 4, 2023.

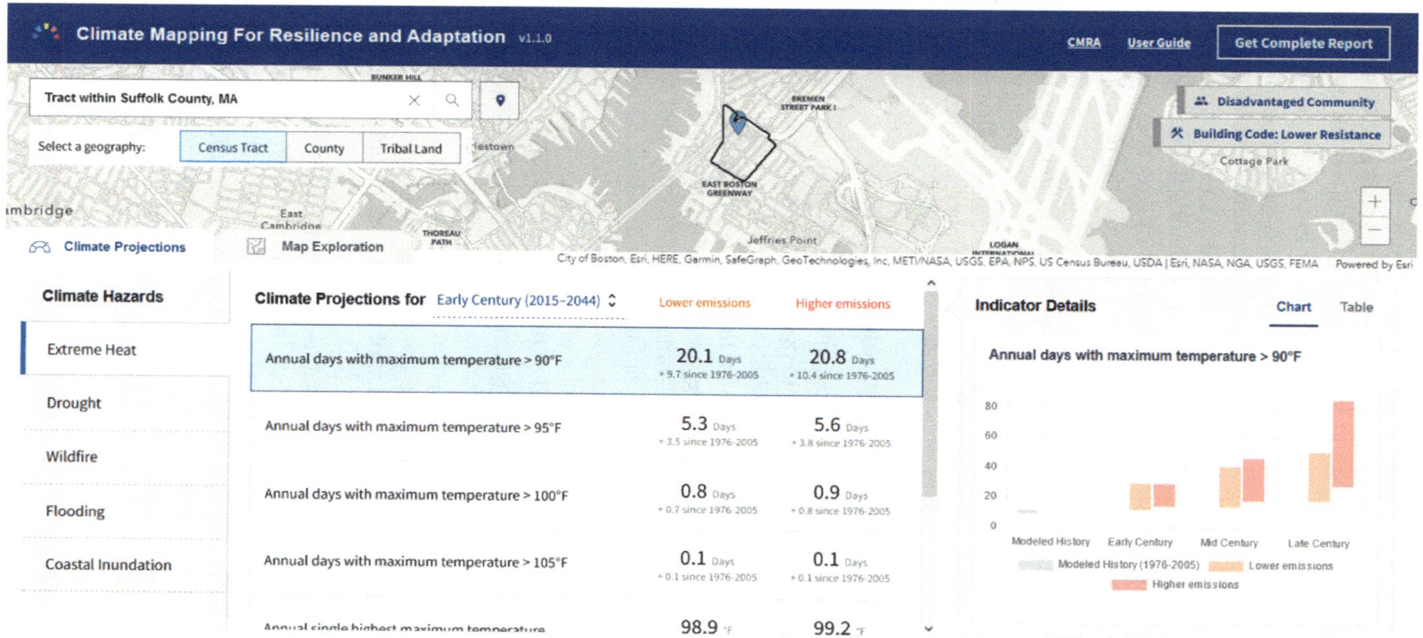

FIGURE 4.1 Continued

The CMRA tool shows current and future climate hazard information for five types of climate-related hazards across the United States: extreme heat, drought, wildfire, flooding, and coastal inundation. The Climate Projections tab shows hazards and associated indicators calculated for a selected area of interest projected for three time periods through the end of the century. In Figure 4.1, projections for extreme heat are displayed as "Annual days with maximum temperature" greater than temperature thresholds ranging from 90 to 105 degrees Fahrenheit, as well as maximum temperatures. Depending on the climate hazard, raw units include the number of days above or below specific temperature thresholds, the number of days above or below specific precipitation thresholds, average annual precipitation, and the percent of coastal areas subject to inundation.

Although some tools rely on raw data values, the primary mode of information display for most tools is one in which the indicator has been processed into a form that facilitates comparison or aggregation, either cumulatively across all indicators or within domains or subsets of indicators (e.g., health, environment, climate).

Percentiles

Most of the tools that identify EJ communities or communicate social and environmental burdens convert indicators into percentile scores. Percentile scores present indicator information for a given geographic area (i.e., census-block group, census tract) relative to all other geographic areas within the country or within a state. For example, EJScreen displays environmental, socioeconomic, and health indicators as percentiles for each census-block group relative to all other census-block groups across the country, or relative to all other block groups within its respective state (depending upon user choice). The upper panel of Figure 4.2 shows percentile scores from EJScreen for particulate matter ($PM_{2.5}$) relative to the state of Massachusetts. Census-block groups in red, within the Grove Hall neighborhood of the City of Boston, exhibit levels of $PM_{2.5}$ that are within the 95th to 100th state percentile range. This means that the pollutant levels in those block groups are higher than at least 95 percent of all other block groups in Massachusetts. However, when compared to the nation, the same location falls in the 29th national percentile (seen in the lower panel of Figure 4.2). This means that its $PM_{2.5}$ levels are higher than 29 percent of all other block groups but lower than 70 percent of all other block groups within the country.

Although most of the tools evaluated use percentiles, EJScreen is one of the few that presents percentile scores of individual indicators as the primary value for display. Most tools reviewed use percentiles as an intermediate step in the construction of more complex indicators of cumulative impacts or risk, or the percentile scores are of composite indicator values. For example, in its process of computing cumulative impact scores, CalEnviroScreen first determines percentiles for each of its 21 indicators. Each indicator is assigned to one of four component groupings (i.e., Exposures, Environmental Effects, Sensitive Populations, and Socioeconomic Factors). The indicator percentiles are averaged within their respective component groupings. These component groupings are themselves organized into one of two larger categories: Pollution Burden (Exposures and Environmental Effects) and Population Characteristics (Sensitive Populations and Socioeconomic Factors).

The Population Characteristics category score is the average of the Sensitive Population grouping score and Socioeconomic Factors grouping score. The Pollution Burden Score is computed similarly, except that the Environmental Effects grouping score is weighted half as much as the Exposures grouping score.[10] The averaged scores for the Pollution Burden and Population Characteristics categories are then scaled such that both have maximum values of 10, and the final CalEnviroScreen cumulative impact score is computed by multiplying the scaled Pollution Burden and Population Characteristics category scores. The result is values ranging from 0 to 100 for every census tract in the state, a cumulative impact score. The CalEnviroScreen tool presents these computed cumulative impact scores in its map. Users can see the original percentiles for each of the indicators after clicking on a given census tract. Other composite indicator tools reviewed here that also use percentiles as intermediate calculations include the CDC/ATSDR EJI, DOE EJMT, DOT ETCE, and New Jersey EJMAP tools.

Ratings

Some tools use ratings (sometimes referred to as "rankings") to display indicators (or constructed indexes), often based on an underlying score or percentile. Ratings indicate a simple ordinal display of relative position, usually "low" to "high." Unlike percentiles, ratings do not easily allow for quantitative assessment of the distance between rating positions or the ratings' relative frequency. Ratings are also more qualitative because they require an explicit decision by the developer about how to distinguish between the different ratings.

The CDC/ATSDR's EJI Explorer displays EJI scores as both percentiles and labeled ratings (see Figure 4.3). The latter are based on quartiles, although the quartile method of partition would not be apparent unless the user consulted the technical documentation:

- High: 75th to 100th percentile
- Moderate to High: 50th to 75th percentile
- Low to Moderate: 25th to 50th percentile
- Low: 0th to 25th percentile

FEMA's NRI also uses ratings to identify U.S. communities at greatest risk for 18 natural hazards (Figure 4.4). Three different types of results are provided for risk and their components:

[10] The rationale for weighting Environmental Effects half as much as Exposures is based on the argument that while Environmental Effects represent the presence of pollutants within a community, they do not necessarily equate to direct exposure to those pollutants. Exposure indicators are therefore assumed to represent a higher order of burden. See CalEnviroScreen 4.0 Report (August et al., 2021).

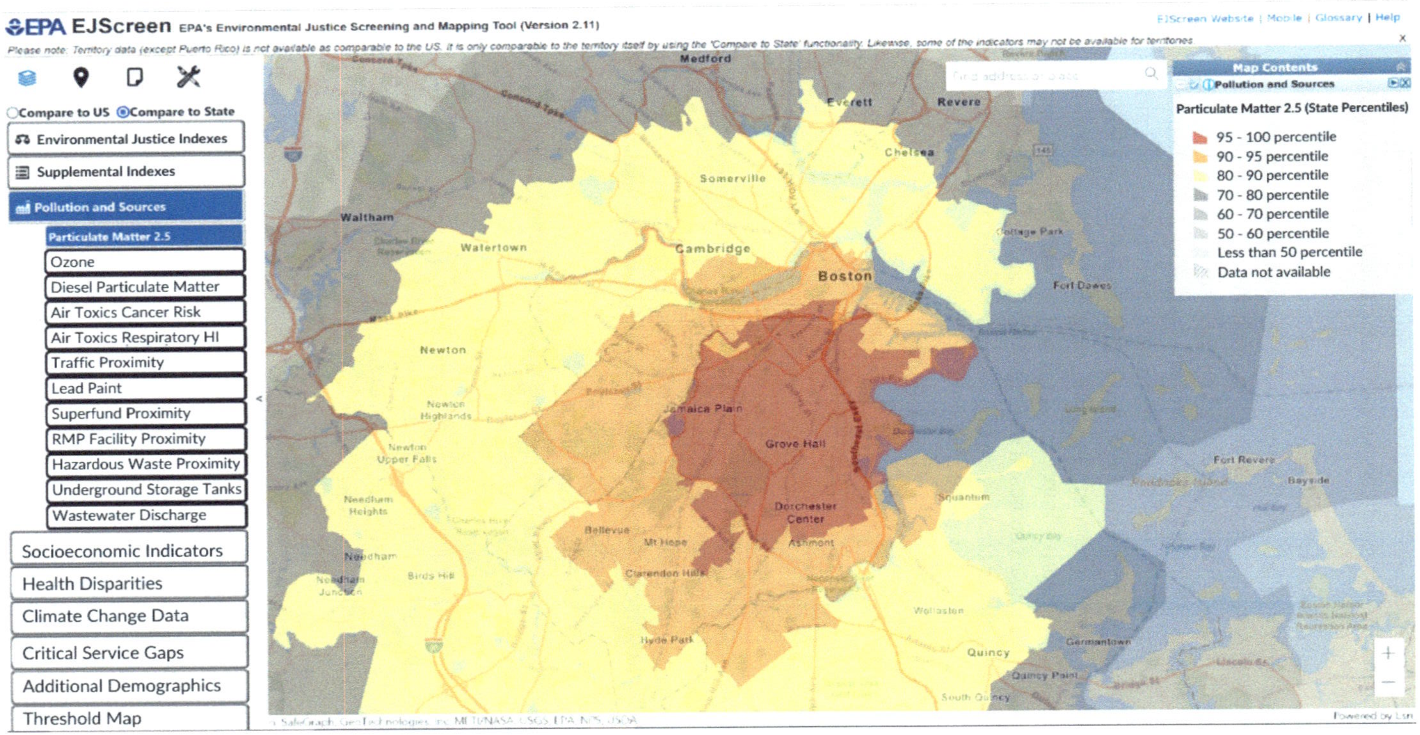

FIGURE 4.2 Screenshot from EPA's EJScreen illustrating the difference between national and state percentiles for exposure to $PM_{2.5}$. SOURCE: US EPA (2018). Accessed November 4, 2023.

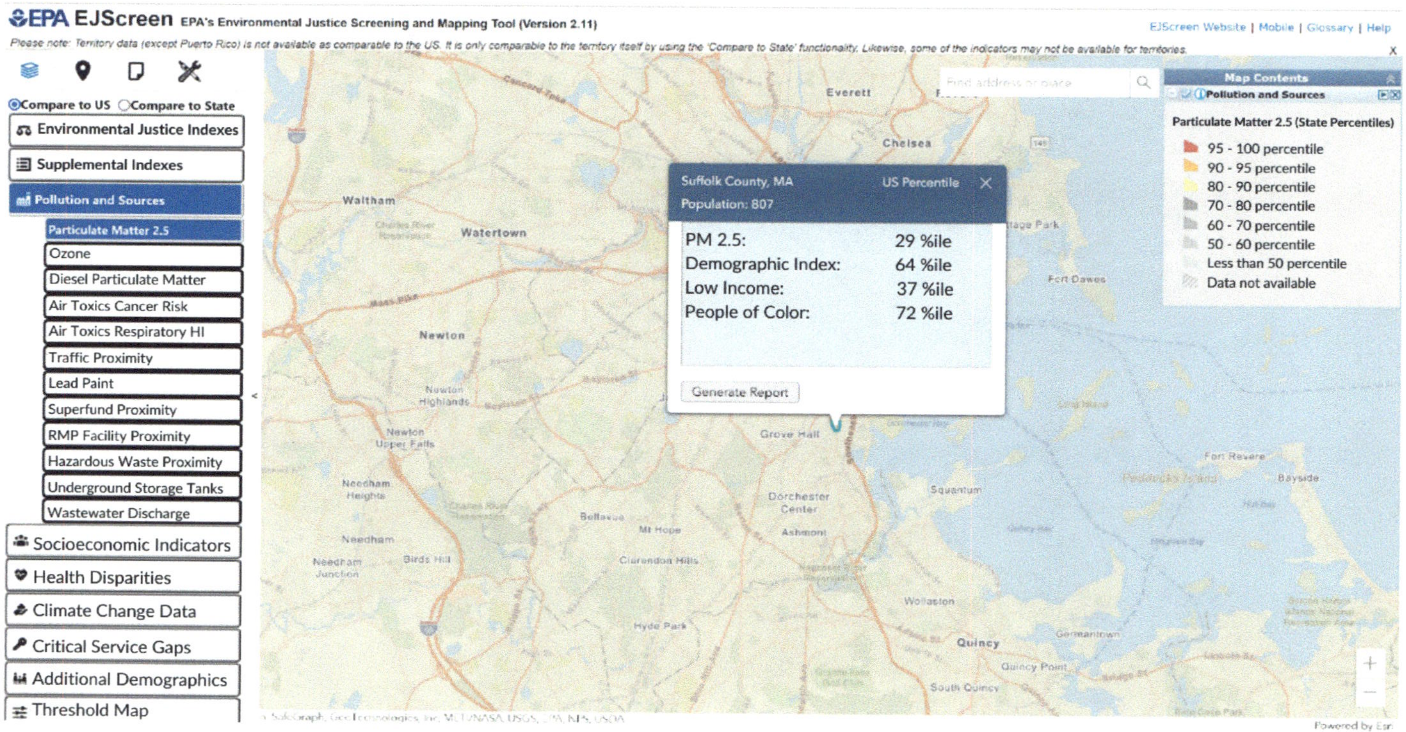

FIGURE 4.2 Continued

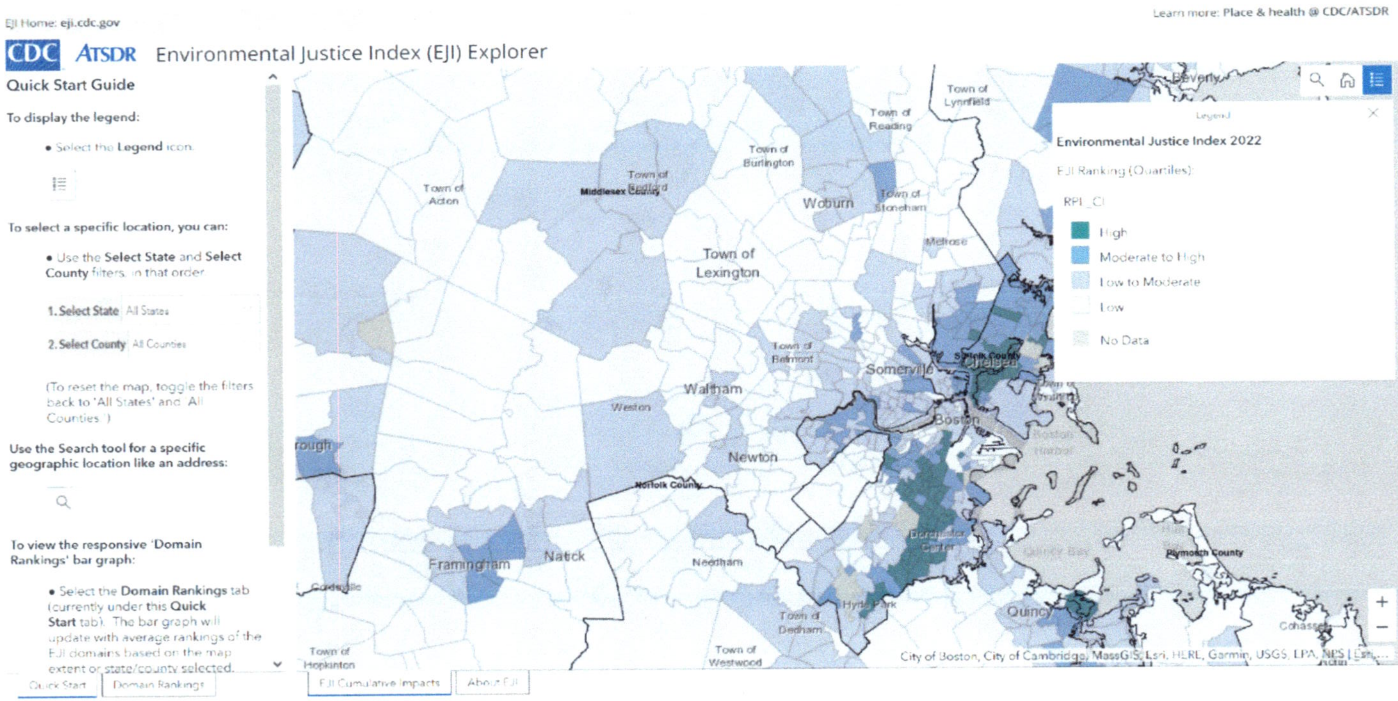

FIGURE 4.3 The Center for Disease Control and Prevention/Agency for Toxic Substances and Disease Registry Environmental Justice Index Explorer illustrating the use of qualitative ranks for comparison. SOURCE: CDC (2022). Accessed November 4, 2023.

- Values (in dollars)—for risk and expected annual loss (representing the community's average economic loss from natural hazards each year). Social vulnerability and community resilience values are index values for the community from the source data.
- Scores (percentiles)—the national percentile ranking of the community's component value compared to all other communities at the same county or census-tract level.
- Ratings—one of five qualitative categories describing the community's component value in comparison to all other communities at the same level. Rating categories range from "Very Low" to "Very High."

Social vulnerability and community resilience ratings have specific numerical boundaries, divided into quintiles based on national percentiles:

- Very High: 80th to 100th percentiles
- Relatively High: 60th to 80th percentiles
- Relatively Moderate: 40th to 60th percentiles
- Relatively Low: 20th to 40th percentiles
- Very Low: 0th to 20th percentiles

CUMULATIVE IMPACTS

Measuring cumulative impacts is an important consideration in determining those communities that are most significantly disadvantaged or burdened by multiple stressors. As mentioned in Chapter 2, this involves three major factors:

1. Magnitude of a given stressor: Those communities that have a higher level of the given stressor or indicator or exceed a threshold for that stressor by a greater amount are designated as "more" burdened.
2. Presence of multiple stressors: Those communities facing a greater number of stressors are more burdened than communities with fewer stressors.
3. Interactions across stressors: The level (or presence) of one stressor in a community can increase the burden imposed on that community by another stressor.

All the tools described in Table 4.1 incorporate one or more of these dimensions of cumulative impacts in some way, although some are more comprehensive than others. Many (but not all) use an aggregation approach to measuring cumulative impacts. As discussed in Chapter 2, aggregation can be either additive or multiplicative, combine demographic or environmental data only or together, and adopt different weights for indicators. In addition, consistent with the varying purposes of the tools, some tools incorporate some measure of cumulative impacts in providing information about the extent to which a community is burdened, while others use it in the binary designation of a community as disadvantaged. This section presents a summary of how the

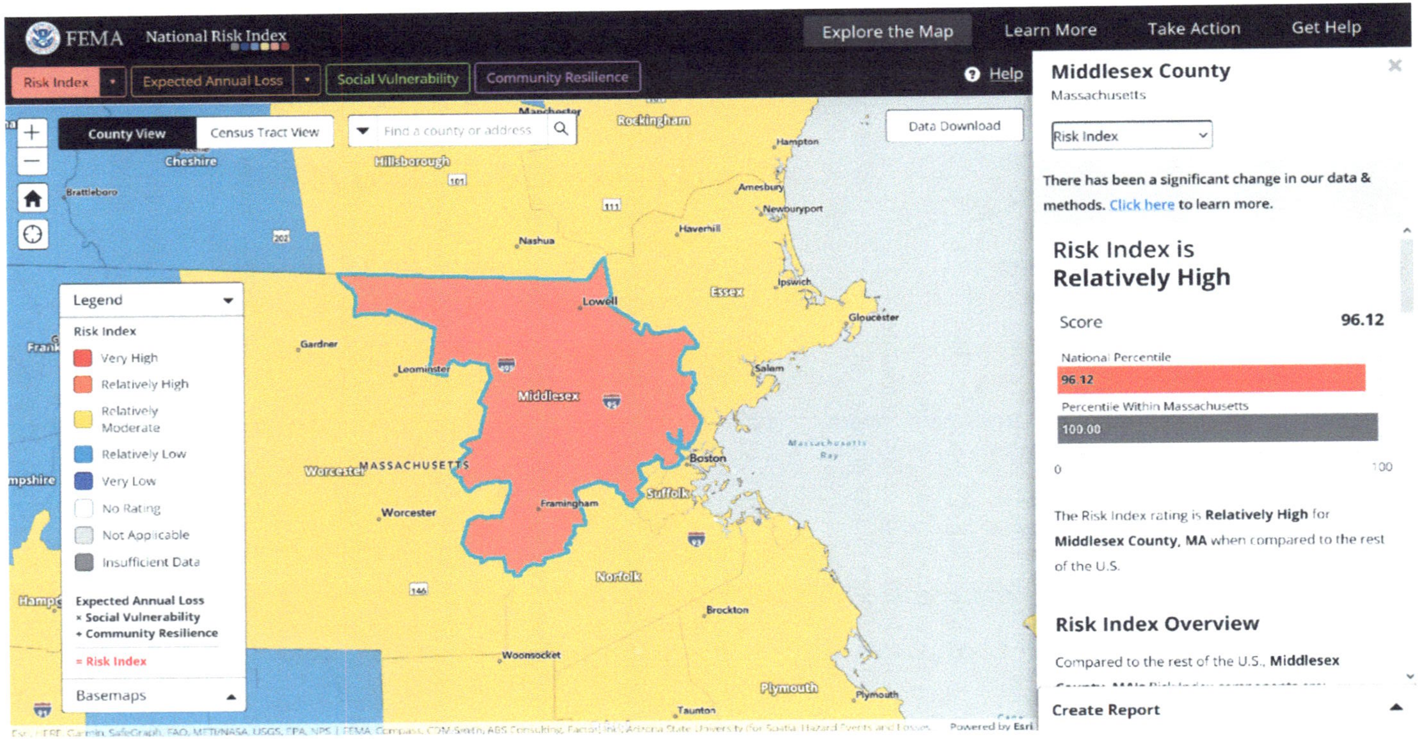

FIGURE 4.4 FEMA's National Risk Index illustrating the use of qualitative ranks for comparison. SOURCE: FEMA (2024). Accessed November 4, 2023.

three dimensions of cumulative impacts are measured by some of the tools that were reviewed.

Magnitude of a Given Stressor

Data reported in raw form or as percentiles, rankings/ratings, or continuous indexes provide information about the magnitude of a given stressor, allowing for considerable variation across communities in the measured level of burden. In contrast, data reported simply based on exceeding a threshold (e.g., a binary variable such "above" or "below") only crudely represent magnitude without distinguishing among different levels in the "above" or "below" categories. Thus, a community that is only slightly above the threshold is viewed as being identical to one that is far above the threshold, and vice versa. In practice, most tools report both continuous values and identify communities that exceed specific thresholds.

For example, the CDC/ATSDR SVI ranks each county and tract on 16 social factors and groups them into four themes: (1) Socioeconomic, (2) Housing Composition and Disability, (3) Minority Status and Language, and (4) Housing and Transportation. Counties in the top 10 percent (i.e., at the 90th percentile of values) are flagged to indicate high vulnerability (CDC/ATSDR, 2022). Similarly, the CDC/ATSDR EJI Explorer uses percentiles to rank each census tract across the country on 36 environmental, social, and health factors and groups them into three overarching modules and 10 different domains. Domains or individual indicators that meet or exceed the 75th percentile are flagged as having a "high prevalence of a chronic condition test" (CDC/ATSDR, 2023).

Multiple Stressors

The cumulative impacts arising from the presence of multiple stressors can be measured through various methods (see Chapters 2 and 3). For example, some tools simply count the presence or number of stressors, typically based on threshold data and the number of thresholds exceeded. Other tools use an additive approach that calculates the sum of indicators across various stressors, typically based on normalized continuous data of some sort, such as percentiles, to allow for comparability in the summation. They thus incorporate more variation in stressor intensity into the measure of cumulative impacts (rather than simply presence vs. absence). In addition, in calculating a summation, the individual components can be weighted equally (as is most often done) or unequally. Multiplicative approaches also involve consideration of multiple stressors but more directly allow for interaction effects of the type described above. (See further discussion below.)

The tools reviewed provided multiple examples of these approaches. For example, the score for the Health Vulnerability Module of the CDC/ATSDR EJI is calculated by summing the number of indicator "flags" for a given census tract, where a flag indicates that the tract is in the top tertile (33.33 percent) of all census tracts. This is equivalent to counting the number of flags. The New Jersey EJMT also uses a count approach. It determines the cumulative burden for a given community based on the number of stressors deemed "adverse" in that community, where the adverse designation is based on whether the stressor is above the 50th percentile.

Several tools use an additive approach for multiple stressors. For example, DOE's EJMT calculates percentiles for each of 36 burden indicators and then sums the percentiles (with equal weights) to generate an aggregate score reflecting cumulative burden. This aggregate score is then used in the determination of whether a census tract is designated as disadvantaged.

The CDC/ATSDR's EJI calculates a module score for its Environmental Burden module and its Social Vulnerability module by summing (with equal weights) the percentile ranks for each of the indicators within the module. These module scores are ranked and then added together (along with the rankings for the Health Vulnerability Module), again using equal weights, to create an overall EJI score. The DOT's ETCE calculates percentile ranks for each component, which are then summed (with equal weights) and converted to a percentile ranking that is used to determine the final disadvantage score for each tract. Where components reflect multiple stressors, they are also based on a summation of normalized indicators for those stressors. The State of California's CalEnviroScreen tool creates its "Exposure" and "Environmental Effects" components by averaging the percentiles for individual indicators within those components, with equal weights. However, when the Exposure and Environmental Effects components are combined, they are summed with unequal weights. In this summation, the Environmental Effects component receives only half the weight of the Exposure component. As a final example, CDC/ATSDR's SVI computes percentile rankings for each of 16 social factors, which are then grouped into four themes. For each of the four themes, percentiles for the variables within each theme are summed. The summed percentiles for each theme are then ordered to determine theme-specific percentile rankings. Overall tract rankings are computed by summing the sums for each theme, ordering the tracts, and then calculating overall percentile rankings.

Interactions Across Stressors

Capturing the interactions across stressors is possible by different methods. Stressors interact when the burden imposed by one stressor depends on the presence or level of another stressor. Two possible approaches to modeling the interaction of stressors are an intersection approach and a multiplicative approach. Under the intersection approach, a given stressor is deemed to create a burden for a community only if a second stressor is also present. Thus, under this approach, cumulative impact is not measured by a single composite indicator but is instead measured by meeting thresholds for both stressors. For example, to be designated as disadvantaged, a community would have to meet a threshold for both income and an environmental indicator (or possibly a composite index for that burden category). In contrast, the multiplicative approach allows for interaction based on continuous variation in the levels of two or more stressors (rather than simply their presence or absence). For most of the tools reviewed, the only interactions captured are those between environmental stressors and socioeconomic stressors (rather than between two environmental stressors). Tools that do not consider interactions typically simply include socioeconomic stressors as additional stressors along with, for example, environmental, health, and climate change stressors.

Examples of tools using an intersection approach for interactions across stressors include the Department of Energy's (DOE's) EJMT, which designates a census tract as disadvantaged if its cumulative impact index (discussed earlier) is in the top 20 percent for the state and at least 30 percent of the households qualify as low income. Similarly, the MA-DPH-EJT classifies census-block groups as Vulnerable Health EJ communities if they meet at least one EJ criterion and at least one health indicator criterion.

Several tools use a multiplicative approach for interactions across stressors. For example, although the EJScreen tool does not incorporate cumulative impacts across multiple environmental stressors (since it does not calculate overall cumulative scores), it does incorporate interaction between individual environmental stressors and a demographic index (or supplemental demographic index) through a multiplicative formula. Two communities with a similar percentile for a given environmental indicator will have different EJ indexes for that indicator if they have different demographic indexes. CalEnviroScreen calculates an overall score for each community by multiplying the combined Exposures/Environmental Effects measure (see above) by a mean percentile for socioeconomic factors and sensitive population indicators. Communities with similar combined Exposures/Environmental Effects measures will have different overall scores if they differ based on socioeconomic or health characteristics. The FEMA NRI, which measures the risk of negative impacts resulting from natural disasters, is derived by multiplying a measure of expected annual loss by a measure of social vulnerability and then dividing it by a measure of community resilience. For any given expected annual loss, the index is higher for communities that are more socially vulnerable and lower for communities that have a greater potential for resilience.

Using Thresholds to Identify Disadvantaged Communities

Many of the tools evaluated use threshold criteria to identify disadvantaged or EJ communities. In this approach, the designation of a qualifying area is based on one or more numerical thresholds. Thresholds can be applied either to individual indicators or to composite indicators. These thresholds may be based on statute, regulation, or the result of some deliberative process of the tool developers. For example, consistent with California law SB 535,[11] CalEPA uses CalEnviroScreen to identify disadvantaged communities as those in the 75th percentile or higher of cumulative impact scores. Similarly, the U.S. DOT ETCE displays continuous "Overall Disadvantage Component Scores" and separate percentile rankings for each individual indicator within a component. It categorizes a census tract as disadvantaged if the overall index score places it in the 65th percentile of all U.S. census tracts. The 65th percentile cutoff was selected to be consistent with CEJST's low-income indicator.[12]

[11] California's law SB 535: https://oehha.ca.gov/calenviroscreen/sb535 (accessed November 5, 2023).

[12] See https://experience.arcgis.com/experience/0920984aa80a4362b8778d779b090723/page/Understanding-the-Data/ (accessed November 5, 2023).

As another example, DOE's EJMT—Disadvantaged Communities Reporter identifies a community as disadvantaged if it is in a census tract that is at or above the 80th percentile of the cumulative sum of the tool's 36 burden indicators and has at least 30 percent of households classified as low income (i.e., at or below 200 percent of the federal poverty level and/or are considered low-income households as defined by the Department of Housing and Urban Development).[13]

At the state level, consistent with New Jersey's Environmental Justice Law, N.J.S.A. 13:1D-157,[14] New Jersey's EJMAP identifies Overburdened Communities (OBCs) as census-block groups in which (1) at least 35 percent of households qualify as low-income households (at or below twice the poverty threshold as determined by the U.S. Census Bureau); (2) at least 40 percent of the residents identify as minority or as members of a state-recognized Tribal community; or (3) at least 40 percent of the households have limited English proficiency. OBCs above the median (50th percentile) for a Combined Stressor Total (CST), which is based on 26 environmental or public health stressors, are designated as "subject to adverse cumulative stressors."[15]

OVERVIEW OF THE CLIMATE AND ECONOMIC JUSTICE SCREENING TOOL (CEJST)

This chapter has summarized the features of some existing EJ tools and how they may vary, illustrating the range of approaches taken in constructing EJ mapping tools. For comparison, this section provides a brief overview of key features of CEQ's CEJST, focusing on the burden categories, indicators, and the limited consideration of cumulative impacts. Subsequent chapters provide additional detail and discussion regarding CEJST.

CEJST includes eight burden categories and incorporates 30 indicators for measuring those burdens.[16] The eight categories are climate change, energy, health, housing, legacy pollution, transportation, water and wastewater, and workforce development. A community is designated as disadvantaged by CEJST if it is in a census tract that is (1) at or above the 90th percentile for one or more of the indicators and (2) meets a socioeconomic burden threshold.[17] For seven of the eight burden categories, that threshold is defined as being at or above the 65th percentile for low income (i.e., percent of households at or below 200 percent of the federal poverty level). However, for the workforce development category, the socioeconomic burden threshold is met if the percentage of people age 25 years or older whose high school education is less than a high school diploma exceeds 10 percent.

---

[13] See https://energyjustice.egs.anl.gov/ (accessed November 5, 2023)

[14] See New Jersey's Environmental Justice Law, https://pub.njleg.state.nj.us/Bills/2020/AL20/92_.PDF (accessed February 14, 2024).

[15] See https://experience.arcgis.com/experience/548632a2351b41b8a0443cfc3a9f4ef6 (accessed November 5, 2023).

[16] See Chapter 5 for more information on the burden categories and a table with their respective indicators.

[17] In addition, a census tract that is completely surrounded by disadvantaged communities and is at or above the 50th percentile for low income is also considered disadvantaged.

As already noted, most of the tools described in this chapter incorporate cumulative impacts in some way, although the extent and methods used vary across tools. Because CEJST designates a community as disadvantaged primarily on the presence or absence of at least one environmental or economic stressor (indicator) coupled with an associated socioeconomic indicator, it does not truly capture cumulative impacts. Its use of thresholds for each indicator effectively incorporates only the magnitude of stressors by distinguishing between a stressor being "high" (above the threshold) and "low" (below the threshold). There is no further differentiation based on the magnitude of a stressor. In addition, because CEJST only requires that one of the burden indicators exceed its threshold, it does not account for multiple environmental stressors (i.e., the fact that some communities are burdened with more stressors than others). The data files accompanying CEJST include information about multiple stressors as a form of cumulative impacts by reporting a count variable that measures the number of indicator thresholds that are exceeded in that community. This type of count information is used by several other tools (see earlier discussion). However, CEJST does not use this information in the designation of a community as disadvantaged. A community that exceeds only one threshold has the same designation as a community that exceeds many thresholds.

The only way in which CEJST captures some dimension of cumulative impacts is by designating a community as disadvantaged only if the census tract meets both a burden criterion (exceeding the 90th percentile for at least one burden indicator) and a socioeconomic criterion (based on income for seven of the eight burden categories and on high school attainment for the eighth category of workforce development). In this sense, it uses an intersection approach to incorporate the interaction between environmental/health/climate change stressors and socioeconomic stressors, recognizing that certain stressors impose a greater burden when coupled with socioeconomic stressors. Several of the tools reviewed also incorporate this type of interaction in some way, through either an intersection approach or a multiplicative approach. CEJST does not incorporate the possible interaction between burden categories; for example, the fact that a transportation stressor such as elevated $PM_{2.5}$ levels may impose a greater burden on a community that is also facing a health stressor such as high incidence of asthma. (Of course, it is also possible that the first stressor is contributing to the existence of the second stressor, so the two are not independent.) Most of the other tools described in this chapter do not incorporate interactions within the environment/health/climate change stressors. However, they do incorporate distinctions across communities based on the number of stressors (beyond socioeconomic stressors) they face.

CHAPTER HIGHLIGHTS

As of the writing of this report, 35 different EJ tools have been released, half of which have been released since 2021. This chapter surveys literature that has reviewed EJ tools and provided more detailed descriptions of a subset of these tools. Several of the reviews found that definitions of community and disadvantage are often inappropriate and not reflective of community self-determinations or lived experiences; that indicators and measures of burden incorporated into tools are often incomplete or out of date; that consideration of race and ethnicity is warranted; that the ways in which multiple burdens interact (cumulative impacts) are important; and that community input and engagement are vital to a relevant tool. The conclusions of these reviews are consistent with discussions about community disadvantage, the incomplete nature of many indicators, the need to address racism, and community engagement found in Chapters 2 and 3 of this report, as well as with committee member expertise and discussions with community members during the committee's open-session discussions.

In addition, the committee's review of tools shows that tools differ along a number of dimensions. Tool outputs vary by the intended use of the tool. Most tools present results as rankings (e.g., percentile comparison with other communities). Other tools may present results using ratings (e.g., "low" to "high"), a binary format (e.g., "above" or "below" some threshold), or rankings (e.g., percentile comparison with other communities). Results presented relative to some threshold (e.g., binary or threshold) represent some amount of subjective decision making in the determination of what the thresholds or rating levels should be. Binary results do not provide information about magnitude (e.g., how much above or below a threshold). CEJST includes consideration of eight burden categories plus a socioeconomic burden threshold and provides results in a binary format (i.e., disadvantaged or not disadvantaged) with no true accounting for the magnitude of stressors or cumulative impacts.

Subsequent chapters of this report provide more details regarding CEJST. Chapter 5 discusses the selection of indicators and datasets. Chapter 6 reviews indicator integration, and Chapter 7 discusses the validation of tool approaches, measures, and results. CEJST is considered in light of scientific literature and sound practices for indicator construction and EJ tool development to inform recommendations for a future data strategy for EJ tools in Chapter 8.

5

Selecting and Analyzing Indicators and Datasets and CEJST Indicators

Determining how to represent the concept to be measured—either quantitatively or qualitatively—is critical in developing any mapping or screening tool. Three decision points are relevant to this chapter: (1) which burden categories to include, (2) which indicators to represent those burdens, and (3) which individual dataset to represent each of those indicators. Chapter 3 described this process to include defining the concept to be measured, selecting a group of indicators and datasets to represent those indicators, and analyzing the indicators and datasets, all within an iterative process for constant refinement and improvement. This chapter begins with a description of useful practices for quantitatively representing the concept to be measured for individual indicators that contribute to disadvantage (e.g., selecting and analyzing indicators; qualitative representation is described in Chapter 7). It then more specifically considers the burden categories and indicators used in the White House Council on Environmental Quality (CEQ) Climate and Economic Justice Screening Tool (CEJST).[1] The text is intended to provide criteria, parameters, and important considerations for making often difficult choices as to which indicators and datasets are selected.

The committee's statement of task (see Box 1.1) calls on the committee to identify key data gaps that CEQ could address in future iterations of its EJ tool(s). However, without comprehensive documentation of the problem being addressed by their tools, clear definitions of concepts being measured, and descriptions of decision-making rationale, the identification of specific data gaps could be based on erroneous assumptions. Instead, the committee describes several indicators that could be included in a tool such as CEJST based on the committee's understanding of the CEJST objectives, community feedback received at the committee's workshop, and expert judgment. The committee provides examples of datasets and sources in Appendix D that could be considered when

[1] See https://screeningtool.geoplatform.gov/en/#3/33.47/-97.5 (accessed March 3, 2024).

identifying future datasets for tools such as CEJST. New data may become available and incorporated into the tool in the future. CEQ could facilitate the development of new, fit-for-purpose datasets through collaborations with federal agencies, nonprofit research organizations, university-based research groups, and private firms who may have proprietary data sources that fill gaps in publicly available datasets.

An exhaustive list of individual indicators and datasets that could be used in an EJ tool such as CEJST is not provided because these, too, need to be determined by the tool developer through a structured indicator development process as described in this report. Identifying appropriate indicators and datasets through an inclusive, community-informed process could lead to improved and more informed identification of representative data, and scientific and technical advances may drive improvements in data quality and completeness. The structured tool development process recommended in this report could minimize the potential for data gaps in future iterations of the tool.[2]

INDICATOR SELECTION

An indicator is a quantitative proxy for an abstract concept, such as "exposure," "access," or "disadvantage." To select appropriate indicators, it is necessary to understand the relationships between burden, indicator, and dataset (see Figure 5.1). These relationships are also described in Chapter 3 as part of the discussion on the conceptual foundation for constructing composite indicators. Creating a compelling rationale for why different indicators should be included is fundamental to creating composite indicators that are interpretable and useful. For example, various environmental exposures such as particulate matter and heat exposures can interact to affect health. A conceptual model could help frame and explain the complex interrelationships (including potential causal relationships) between the indicators and their impacts on the concept being measured. The model will be dependent on the goal of the tool, the structural approach used to develop the tool (including community input) the data available, and the state of the science that allows understanding of interrelationships, causality, and cumulative impacts related to the problem at hand. A model that hypothesizes the interrelationships of the indicators can help the developer (1) select which domains to be included and why and (2) determine if and how cumulative impacts could be captured.

CEJST includes eight categories of what CEQ labels "burdens"—climate change, energy, health, housing, legacy pollution, transportation, water and wastewater, and workforce development. Each of these categories contains a set of indicators that are intended to represent the burden category. For example, the energy category includes two indicators: energy cost and fine particulate matter ($PM_{2.5}$). The number of indicators differs between burden categories (currently between two and five indicators per category), but each indicator is represented in the tool by one dataset. In general, these datasets are files of numerical quantities that vary in magnitude and in space, intended

[2] Executive Order 14096 authorizes the director of the Office of Science and Technology in conjunction with the Chair of CEQ to "address the need for a coordinated Federal strategy to identify and address gaps in science, data, and research related to environmental justice."

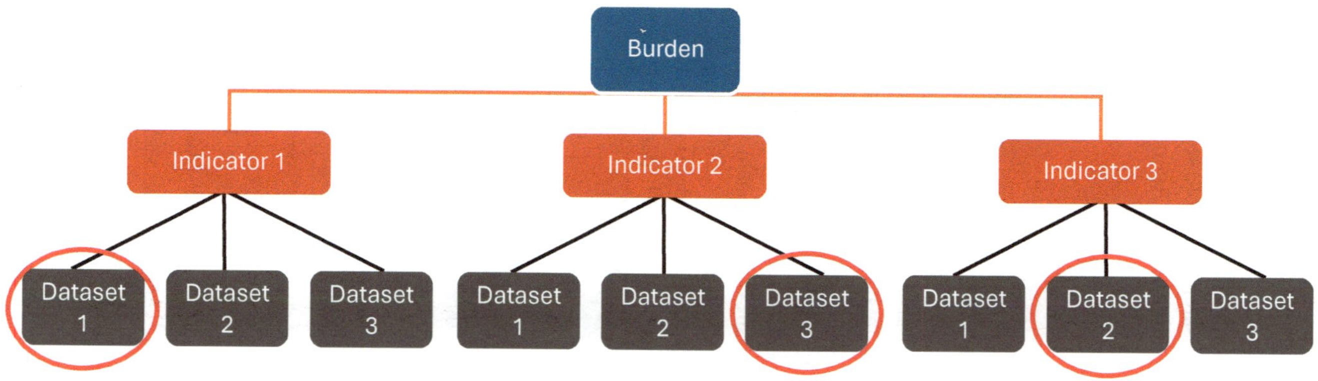

FIGURE 5.1 Illustration of the relationship between burdens, indicators, and datasets. In CEJST, multiple indicators are used for each burden category, and each indicator is represented by a singular dataset (shown by the red circles).

to align with real-world variation in the property being measured. Although multiple datasets are often available for the same indicator, only one dataset is selected and used in CEJST. For example, many annual average ambient particulate matter ($PM_{2.5}$) concentration datasets are used across the federal government and scientific community, and each was developed using different methods and data inputs (e.g., statistical modeling, geophysical modeling, satellite remote sensing); the dataset used in CEJST was developed by Environmental Protection Agency (EPA) using a model-monitor fusion approach.

Criteria

The structured process for indicator construction outlined in Chapter 3 includes a systematic approach to indicator and data selection. Table 5.1 outlines technical and practical indicator characteristics to consider when selecting specific measures, with the aim of optimizing quality and validity. A set of practical questions that tool developers could ask is provided. The criteria are not unique to indicator construction and have been used in other contexts (e.g., comparing federal tools for ranking hazardous waste sites for remedial action; NRC, 1994). Technical characteristics emphasize representational, statistical, and geospatial aspects of indicator data and are typically the focus of analysts and modelers. Practical characteristics, such as data availability and cost, are generally of greater interest to indicator program managers and end users. An important technical criterion is validity—how well the indicator reflects the lived experience. Validity will also be discussed in Chapter 7, particularly as it pertains to community involvement. Whether or not the data are findable, accessible, interoperable, and reusable (FAIR;

TABLE 5.1 Indicator Selection Criteria

Type	Criterion	Description
Technical	Validity	How well does the indicator measure the characteristic it represents?
	Sensitivity	Is the indicator sensitive to changes in the underlying phenomenon?
	Robustness	Is the indicator analytically and statistically sound?
	Reproducibility	Can the data be independently verified and reproduced?
	Scale	Do the geographic and temporal scales of the indicator match those of the characteristic it represents?
Practical	Measurability	How easily can the data be measured quantitatively?
	Availability	How easily can the data be obtained?
	Simplicity	Is the indicator intuitive and easy to interpret?
	Affordability	How reasonable are the data collection costs?
	Credibility	How acceptable is the indicator to the intended audience?
	Relevance	Does the indicator align with the intended goals of the tool?

SOURCE: Adapted from: Tate, 2011.

Wilkinson et al., 2016) may also be part of the data selection criteria, as might whether the data are consistent with CARE principles (collective benefit, authority to control, responsibility, and ethics)[3] and support Indigenous self-determination and innovation (see Box 5.1). Use of FAIR data and CARE principles may serve to enhance the transparency and acceptance of the data as representative of community lived experience. If Indigenous data are used, CARE principles need to be applied so that data are used in a manner in accordance with the rights, knowledge, and values of Indigenous peoples (Carroll et al., 2020).

There are multiple forms of validity, three of which are particularly important for environmental indicators:

- **Construct validity:** how well an indicator measures what it is supposed to. For example, a composite indicator of cumulative disadvantage with high construct validity embodies the principal dimensions and interactions that govern how disadvantage functions.
- **Concurrent validity:** the degree of alignment between two measures that should be related. It is typically evaluated using correlation analysis, for example, testing the statistical association between alternative measures of socioeconomic status.
- **Content validity:** representativeness, essentially the extent to which an indicator includes all principal dimensions of the underlying concept.

Other leading technical criteria include sensitivity, robustness, reproducibility, and spatial or temporal scale. A sensitive indicator will change in direction and magnitude with a change in its real-world proxy. Robustness is a statistical measure of the stability of an indicator to changes in its construction: the indicator should not change substantially with small changes in how it is measured. This is typically assessed using sensitivity analysis. Reproducibility is the ease with which the indicator can be constructed by others independent of the current indicator construction project. The scale criterion is the degree to which the spatial units and time periods of the indicator data align with those of the process or phenomenon being measured. A scalar mismatch can occur when practical considerations of availability, cost, or administrative structures constrain the selection of geographic and temporal scales.

The practical considerations in indicator selection, described in Table 5.1, are often more ambiguous to assess but are no less important. Measurability is the ease of quantitatively manifesting the underlying concept or process. In practice, physical and economic characteristics are easier to measure than more intangible processes, such as marginalization and compounding effects. Failure to incorporate difficult-to-measure concepts can negatively affect the content validity of a single or composite indicator. Availability is the ease of obtaining indicator data for the dimensions, geography, and time frame of interest. Widely available and standardized secondary data are often chosen for indicators used to compare places at the national level, yet they can conflict

[3] See https://www.gida-global.org/care (accessed May 14, 2024).

BOX 5.1
FAIR Data and CARE Principles

To improve infrastructure that supports research data, academics, funding entities, industry, and scholarly publishers have developed a set of principles known as the FAIR (findable, accessible, interoperable, and reusable) Data Principles. Specific guidelines have been established and adopted that are intended to increase data sharing and aid science discovery through guidelines for data and metadata design that enhance the reusability of data by computers and humans (Wilkinson et al., 2016).

The CARE (collective benefit, authority to control, responsibility, and ethics) Principles for Indigenous Data Governance[a] recognize that FAIR data principles ignore the historical contexts of data and power differentials in advancing Indigenous innovation and self-determination. The data of Indigenous Peoples' comprise:

> "*(1) information and knowledge about the environment, lands, skies, resources, and non-humans with which they have relations; (2) information about Indigenous persons such as administrative, census, health, social, commercial, and corporate and, (3) information and knowledge about Indigenous Peoples as collectives, including traditional and cultural information, oral histories, ancestral and clan knowledge, cultural sites, and stories, belongings* (Carroll et al., 2020)"

The CARE Principles consider self-determination by Native Americans and other Indigenous groups through standards to be applied in conjunction with FAIR data guidelines (e.g., Jennings et al., 2023). The principles are established on the rights of Indigenous Peoples to create value from data related to them in ways that are based on their own world views.

[a] See https://www.gida-global.org/care[a]See https://www.gida-global.org/care (accessed May 14, 2024).

with construct and content validity. Simplicity and affordability are perhaps the most straightforward criteria: how understandable is the indicator, and how reasonable are the data acquisition costs in money and time? Credibility is the believability and salience of the indicator for scientific and technical applications, as well as for the public. This is also referred to as community validation and buy-in. Involving community members and other interested and affected parties throughout the indicator selection process can be crucial for building credibility, not only to ensure that it represents people's lived experiences but also to engender trust. This is discussed in more detail in Chapter 7.

Relevance is determined based on alignment between potential indicators and the tool's objectives. As with all other factors of tool development, indicator relevance needs to be determined through a community partnership and transparent process.

With many potential indicators that meet the criteria described above, relevance can also reflect prioritization among the indicators to ensure that each indicator efficiently achieves its objective without accruing unnecessary costs from data identification, storage, and computation. California's CalEnviroScreen[4] is an example of a tool for which indicator relevance is based on community input. Chapter 7 contains information on including community input while also utilizing tools with scientific rigor.

According to the Technical Support Document for CEJST 1.0 (CEQ, 2022a, p. 17), the indicators and data included in CEJST were selected based on the following parameters:

- The indicator is "[relevant] to the goals of Executive Order 14008 and the Justice40 Initiative"; datasets are "related to climate, environmental, energy, and economic justice";
- The indicator data are publicly available (not private or proprietary);
- The indicator data cover all 50 states and the District of Columbia at a minimum, and where possible, the five U.S. territories of Puerto Rico, American Samoa, the Northern Mariana Islands, Guam, and the U.S. Virgin Islands; and
- The indicator data are available at the census-tract scale or finer.

CEJST utilizes 30 indicators across eight categories of burden that meet the above criteria. However, there are many other existing potential indicators that meet their criteria that are not included in the tool. For example, the Environmental Defense Fund's Climate Vulnerability Index includes 184 indicators (Tee Lewis et al., 2023), most of which also meet CEJST criteria for inclusion. In addition, EPA's EJScreen[5] tool uses similar criteria for indicator selection. CEJST documentation (CEQ, 2022a) does not include a rationale for why some EJScreen indicators are included in CEJST, and some are not. That said, more indicators do not indicate a better tool if indicator quality is questionable or if indicators are repetitive or contradictory. At a public workshop organized by the study committee as part of its information gathering for this report (see Appendix B for the workshop agenda), workshop participants commented that using a national scale could be a limitation and that regional, state, municipal, or otherwise non-national data could provide valuable additional information (NASEM, 2023a).

Temporal Changes in Data

Current geospatial screening tools such as CEJST provide only a static or snapshot-in-time approach for exploring the environmental and social characteristics of neighborhoods and evaluating related burdens. There are two limitations of a snapshot approach that do not allow the incorporation of temporal effects in the tool. First, how long vulnerable communities have been exposed to polluted water, air, or other environmental hazards is important and is not considered in estimating risk burdens. Time

[4] See https://oehha.ca.gov/calenviroscreen (accessed March 4, 2024).

[5] See https://www.epa.gov/ejscreen (accessed March 4, 2024).

can be used as a weighting variable for an indicator whose value would be greater, for example, if a pollution source has existed in its present location longer, thus enabling the tool to capture more severe impacts of legacy pollution.

A second limitation of this approach is that it does not allow exploration of how a designation of disadvantage changes over time, for example, whether a tract currently categorized as disadvantaged (or not disadvantaged) was classified similarly in the past. Adding this capability of tracking changes over time could be useful in determining whether investments in disadvantaged communities (DACs) resulted in a reduction of specific climate or socioeconomic burdens or a change in the designation of disadvantage. It may also be useful to examine how spatial patterns and geographic locations of DACs are changing over time in a specific urban area, state, region, or nationwide. Assessing changes over time for specific burdens and overall designation of disadvantage, however, could be problematic and challenging because indicator datasets come from many different time frames and years and are not updated at the same time. If the tool is updated annually, an option could be added for the user to explore whether a particular tract was designated as DAC in previous years or previous versions of the tool. This may require thorough documentation of the updated processes and datasets as well as the resulting changes in the tool output and interpretation. A change in designation can occur for a variety of reasons and may not reflect a change in lived experience in a community. In some cases, a change in designation might be the result of updated data. However, if the data in the previous tool version were old and outdated, then it might not be obvious when changes in the community occurred. In other cases, an update in status might be the result of a change in the data integration approach rather than a change in data.

Temporal Coverage and Frequency of Data

An important consideration when selecting datasets is their temporal coverage and frequency of updating. Datasets are typically available for specific years or sets of years. Datasets are updated at different intervals (e.g., annually), whereas others are updated less frequently (e.g., every 3 years) or have no process for being updated. Additionally, for indicators with high interannual variability (e.g., wildfires and other climate-sensitive indicators), datasets that average over a multiyear period (e.g., 5 years, 10 years) can provide more stable and interpretable estimates. Temporal coverage and updating frequency can, therefore, be driving factors when selecting datasets. Selecting datasets that represent the most recent years, that account for interannual variability (for indicators where interannual variability is high), and are updated frequently can ensure that the screening tool is as current as possible, although some temporal misalignment between indicators is inevitable. Box 5.2 provides a practical example of the potential affects of the frequency of dataset updating on data uncertainty.

IDENTIFYING AND ANALYZING THE DATASETS TO BE USED

Different datasets available for measuring a particular indicator may use different methodological approaches in their development. In some cases, there may be a

BOX 5.2
U.S. Census and American Community Survey Data in CEJST

Socioeconomic data used as burden indicators in CEJST are based on 5-year estimates derived from the American Community Survey (ACS). These estimates include uncertainties that the tool quantifies through margins of error (MOE). Previous sociodemographic research on the ACS illustrates how the 5-year average estimates can have substantial MOEs at subcounty scales (e.g., census tracts) when compared to the decennial U.S. Census data (Bazuin and Fraser, 2013; Folch et al., 2016). To mitigate such measurement errors, researchers have suggested the removal of all census units with small population counts or the use of census units where MOEs of the point estimate are less than half of the point-estimate value (Folch et al., 2016). Since CEJST uses census tracts as the basic analytical unit, it is important to keep in mind that the MOE magnitude generally is considerable for tract-level ACS estimates, particularly in rural areas with lower populations. If the percentage of people living at or below 200 percent of the federal poverty line (i.e., the definition of "low income" in CEJST) in two tracts are compared, the differences between the two may be similar but smaller than the corresponding MOEs. This would seem to indicate no significant difference in the percentage of low-income people in the two tracts, suggesting that the percentiles for the two tracts should be identical and not different. This would be the conclusion if only the ACS point estimates of the percentages of people in the tract at or below 200 percent of the federal poverty line were utilized and the MOE values were not considered. This could result in potentially erroneous estimation of low-income burdens and, consequently, misclassification of community disadvantage. In the case of CEJST, burden calculations rely on data from the ACS 5-year estimates that do not acknowledge or account for uncertainty in the ACS variables. Other national public health and environmental health data products (e.g., National Health and Nutrition Examination Survey [NHANES]) add 95th confidence intervals to the percentiles, an approach that allows incorporation of uncertainty in the raw data and communicates the uncertainty in the percentile values themselves that are linked to the raw data.

"gold standard" methodological approach, but in other cases, multiple approaches may be used and accepted by researchers and practitioners without one approach standing out as most closely representing truth. Many of the same principles for selecting and evaluating indicators also apply to selecting and evaluating datasets used to represent those indicators. Often, datasets that are based on observations are considered of higher quality when compared with estimation approaches. However, when observations are sparse in space or time, combining observations with estimation techniques (e.g., statistical models, process-based models) can provide spatially complete datasets that are informed by observations. Another important consideration is how well the dataset performs upon evaluation against observed quantities (statistical evaluation metrics such as correlation coefficients, bias, and uncertainty). However, multiple datasets can have similar performance against observations while differing in magnitude and spatial distribution of the estimated quantities. The degree to which the dataset is adopted and used by federal agencies or the scientific community, or accepted by communities, can provide additional confidence in the dataset's validity.

Analyzing and comparing different available datasets can reveal insights as to the limitations and implications of the use of the datasets—for example, determining if certain communities are more likely to be missed when using one dataset over another (see Box 5.3). Analyzing indicators and datasets can also ensure that, barring specific stakeholder concerns about representation, the indicators and datasets selected for inclusion are independent of each other, reflect the state of the science, and are trusted by technical experts and validated by lived-experience data of community members. See Chapter 6 for details on integrating multiple indicators and possible correlation between indicators.

Rigorously documenting the process for selecting and analyzing datasets can enhance transparency and highlight areas for further research. Results such as spatial maps, tracts in the top percentile, correlations with other relevant indicators and datasets, and evaluation statistics are important information for agencies, community groups, researchers, and other users with which to understand the rationale and implications of each indicator and dataset selection. The process for obtaining those results, including community engagement, is also relevant for enhancing transparency.

BOX 5.3
Selecting Datasets to Represent Indicators: $PM_{2.5}$ Example

The $PM_{2.5}$ indicator illustrates the challenge of selecting among multiple high-quality datasets that could be used to define the indicator. CEJST currently uses a 12-km gridded annual average $PM_{2.5}$ concentration dataset that was developed by EPA and is also used within EJScreen. A recent comparison of the CEJST dataset with two more highly spatially resolved datasets from the scientific community found that the three datasets differ substantially regarding which tracts are most overburdened within individual urban areas (Carter et al., 2023). As a result, the study identified 335 tracts (representing ~1.5 million people) as disadvantaged (>65th percentile for poverty and >90th percentile $PM_{2.5}$) using both high-resolution datasets but not the 12-km dataset used by CEJST, and 695 tracts (representing ~2.7 million people) as disadvantaged in the 12-km dataset but not the high-resolution datasets. This analysis underscores the challenge of identifying and selecting a single dataset to represent the indicators in the tool. Each dataset also carries uncertainties, which are discussed in Chapter 6 of this report.

CATEGORIES OF BURDEN IN CEJST

CEJST includes eight categories of what CEQ labels burdens—climate change, energy, health, housing, legacy pollution, transportation, water and wastewater, and workforce development. These categories align and expand on the priorities in President Biden's Executive Order (E.O.) 14008 (although E.O. 14008 does not include health in its list of priorities; EOP, 2021). The categories are used in combination with relevant socioeconomic burdens (low income and high school education) to identify DACs. Although the rationale for the E.O. 14008 categories is unstated, the committee did not identify any obvious omissions, given the committee's understanding of the objectives of the tool. The rationale for the inclusion or exclusion of specific indicators within each burden category in CEJST is not provided in the CEJST technical documentation, either in terms of why certain indicators but not others were included or in terms of the categorization of individual indicators in specific burden categories (e.g., the inclusion of $PM_{2.5}$, which has many emission sources, in the energy burden category).

Each of the eight burden categories includes multiple indicators—a different number of indicators for each category—as outlined in Table 5.2. The tool identifies a community as disadvantaged if it is in a census tract that is (1) at or above the threshold for one or more indicators in any burden category and (2) at or above the threshold for an associated socioeconomic burden. For example, the climate change burden category includes five indicators. If a tract meets a threshold for one or more of these indicators, as well as the threshold for the low-income indicator, it is identified as disadvantaged. Census tracts that do not meet any burden thresholds but are at or above the 50th percentile for low income and surrounded by other census tracts that do meet the thresholds for disadvantage are also designated as DACs. Finally, all land within the boundaries of federally recognized Tribes is designated as disadvantaged.

Using this formulation, neither the number of burden categories nor the groupings of indicators within them affect the tool's binary identification of disadvantaged status. However, the tool does have some implicit weighting reflected in the number of indicators included in each burden category. Burden categories with more indicators (e.g., climate, housing, and legacy pollution, each with five indicators) have more chances of triggering the disadvantage identification compared with burden categories with fewer indicators (e.g., energy and water and wastewater, each with only two indicators). Burden categories with more indicators are thus implicitly weighted more heavily than burden categories with fewer indicators. As a result, some categories could be overrepresented and others underrepresented in the tool. In addition, indicator groupings could become important in future iterations of the tool, especially if integration approaches for assessing cumulative impacts are implemented (discussed further in Chapter 6). In addition, community engagement, validation, and transparency in selecting and including burden categories and indicators can help evaluate how well the tool captures burdens that align with the lived experience of communities (discussed further in Chapter 7; Larsen, Gunnarsson-Östling, and Westholm, 2011).

TABLE 5.2 CEJST Burden Categories and Indicators

Burden Categories	Indicators	Socioeconomic Indicator
Climate change	1. Expected agriculture loss rate ≥ 90th percentile OR	Low income*
	2. Expected building loss rate ≥ 90th percentile OR	
	3. Expected population loss rate ≥ 90th percentile OR	
	4. Projected flood risk ≥ 90th percentile (NEW) OR	
	5. Projected wildfire risk ≥ 90th percentile (NEW)	
Energy	1. Energy cost ≥ 90th percentile OR	Low income*
	2. $PM_{2.5}$ in the air ≥ 90th percentile	
Health	1. Asthma ≥ 90th percentile OR	Low income*
	2. Diabetes ≥ 90th percentile OR	
	3. Heart disease ≥ 90th percentile OR	
	4. Low life expectancy ≥ 90th percentile	
Housing	1. Historic underinvestment = Yes (NEW)	Low income*
	2. Housing cost ≥ 90th percentile OR	
	3. Lack of green space ≥ 90th percentile (NEW) OR	
	4. Lack of indoor plumbing ≥ 90th percentile (NEW) OR	
	5. Lead paint ≥ 90th percentile	
Legacy pollution	1. Abandoned mine land present = Yes (NEW) OR	Low income*
	2. Formerly used defense site present = Yes (NEW) OR	
	3. Proximity to hazardous waste facilities ≥ 90th percentile OR	
	4. Proximity to Superfund or National Priorities List (NPL) sites ≥ 90th percentile OR	
	5. Proximity to Risk Management Plan sites ≥ 90th percentile	
Transportation	1. Diesel particulate matter ≥ 90th percentile OR	Low income*
	2. Transportation barriers ≥ 90th percentile (NEW) OR	
	3. Traffic proximity and volume ≥ 90th percentile	
Water and wastewater	1. Underground storage tanks and releases ≥ 90th percentile (NEW) OR	Low income*
	2. Wastewater discharge ≥ 90th percentile	
Workforce development	1. Linguistic isolation ≥ 90th percentile OR	Less than high school education > 10%
	2. Low median income ≥ 90th percentile OR	
	3. Poverty ≥ 90th percentile OR	
	4. Unemployment ≥ 90th percentile	

NOTE: * = percent of a census tract's population in households where household income is at or below 200 percent of the federal poverty level, not including students enrolled in higher education.
SOURCE: Adapted from CEQ, 2022a.

The ensuing subsections expound upon each of the CEJST burden categories with a description of the indicators included in each burden;[6] additional indicators that could be incorporated to meet CEJST's objectives; current data availability, quality, and spatial and temporal resolutions for those indicators; and key data gaps. The text is intended to provide a realistic and practical description of data that could be included in the tool in consideration of those data that meet CEJST's criteria described above. It is not intended to be a comprehensive review of all available indicators and datasets but rather highlights those that the committee considered to be a high priority for relevance based on subject-matter expertise, information gathering (including a public workshop; NASEM, 2023a), and which datasets meet all technical and practical characteristics for inclusion. Potential changes to the burden categories themselves are not addressed; as previously mentioned, categorizations of the indicators only affect the identification of disadvantage to the extent that the number of indicators in each burden category differs, although these categorizations could become important in future iterations of the tool, particularly if an integration approach such as a composite indicator is used.

Climate Change

CEJST includes five indicators in the climate change burden category: expected agriculture loss rate, expected building loss rate, expected population loss rate, projected flood risk, and projected wildfire risk. Agricultural and building loss rate from those natural hazards are economic terms (agricultural value at risk and building value at risk), while population loss rate reports the number of fatalities and injuries caused by the hazard. Flood and wildfire risks are considered within these agricultural, building, and population loss indicators, and are also considered as individual indicators.

Expected agricultural, building, and population loss rates come from FEMA's National Risk Index (NRI)[7] and cover all U.S. states, the District of Columbia, and the five U.S. territories (American Samoa, Commonwealth of the Northern Mariana Islands, Guam, Puerto Rico, and the U.S. Virgin Islands). The NRI includes 18 different natural hazards. CEJST considers 14 of those 18 hazards to be climate-related: avalanche, coastal flooding, cold wave, drought, hail, heat wave, hurricane, ice storm, landslide, riverine flooding, strong wind, tornado, wildfire, and winter weather. A limitation of these composite indicators is that the risk of specific natural hazards varies from location to location. Thus, even when a census tract is identified as being disadvantaged based on a high agricultural, building, or population loss rate due to natural hazards, it is not clear which natural hazard(s) is driving the risk to be high in a specific location.

Projected flood and wildfire risk data originate from the nonprofit First Street Foundation.[8] The flood risk dataset (at the census-tract level) represents how many properties are at risk of floods occurring in the next 30 years from tides, rain, and riverine and

---

[6] Information about which datasets were used for the CEJST indicators can be found in the CEJST Technical Support Documentation (CEQ, 2022a).

[7] See https://hazards.fema.gov/nri/ (accessed February 12, 2024).

[8] See https://firststreet.org/ (accessed February 15, 2024).

storm surges, without considering property values. The wildfire risk dataset (with a 30-meter resolution) represents the chance over 30 years of property in the area burning, considering factors such as fire fuels, weather, human influence, and fire movement. Both indicators are specific to property damage and are insufficient to capture broader impacts on health and livelihoods. For example, floods are associated with population displacement and mental health impacts, disruption in access to medication and health care, water quality issues, and other impacts. Wildfires are similarly associated with population displacement, mental health impacts, and disruption in access to medication and health care, and can cause deterioration of air quality at regional and continental scales, with respiratory, cardiovascular, and other health impacts.

The indicators used within CEJST to address climate change cover multiple relevant community hazards, but they provide a limited view of which communities are most vulnerable to the many impacts of climate change and through which pathways. Community resilience to climate change includes, for example, the impacts of increased heat, disease vectors, mental health, air pollution from sources other than wildfires, water quality, extreme weather events, and drought (USGCRP, 2023). Each of these can have serious downstream consequences for human health and livelihoods. For example, extreme weather events can lead to coastal inundation and inland flooding, leading to contamination of waterways with sewage, animal waste, and chemicals (Cushing et al., 2023b; Erickson et al., 2019), as well as the risk of chemical disasters at industrial facilities (Anenberg and Kalman, 2019). Many climate damage pathways have not yet been well quantified or detailed through qualitative data, and inclusion within CEJST is challenged by the lack of availability of nationwide, high-resolution datasets for each pathway.

Heat, however, is a climate damage pathway for which datasets appropriate for CEJST already exist. Heat is the leading cause of weather-related mortality in the United States over the last several decades (Luber and McGeehin, 2008) and may be underreported (Weinberger et al., 2020). Unless additional measures to protect public health are taken, the frequency and intensity of extreme heat events will increase mortality and morbidity in the future (Shindell et al., 2020). Studies using surface temperature and other heat-related measures have found that heat is inequitably distributed within U.S. urban areas (Mitchell and Chakraborty, 2019; Renteria et al., 2022) and is related to historical redlining and marginalization (Hoffman Shandas, and Pendleton, 2020; Hsu et al., 2021). While surface temperature studies often do not account for humidity, which can modulate the heterogeneity of heat exposure within cities (Chakraborty, T. C., et al., 2022; Keith, Meerow, and Wagner, 2019), communities of color and with lower income levels disproportionately experience moist heat (Chakraborty et al., 2023) as well as heat vulnerability (Manware et al., 2022). Residential air-conditioning prevalence, which can mitigate the health effects of heat exposure, is also inequitable across 115 metropolitan areas (Romitti et al., 2022). Additionally, outdoor workers in many industries across the country, such as agriculture and construction, will still be exposed to extreme heat, even in areas with a high prevalence of residential air-conditioning (Licker, Dahl, and Abatzoglou, 2022).

Future climate risks are another set of potential indicators for which there are datasets appropriate for CEJST. As the geographic area and population affected by climate change are expected to grow, the communities most affected by climate change in the future may be different than those most affected today. Two indicators used within CEJST consider future changes: flood risk and wildfire risk. However, as mentioned above, they capture a limited portion of human health damages. Potentially important indicators to represent community risks from climate change include future changes in extreme temperature, future changes in climate-sensitive natural disasters (only present-day natural disasters are currently included), and wildfire smoke. Further, the future projections for the datasets currently used within CEJST are limited to 30-year time horizons, though projections are available through 2100 from federal agencies. Rather than taking a deterministic approach for projecting future climate changes, which are inherently unknown and therefore uncertain, CEJST could consider a range of possible climate futures. Examples of other federal efforts that use a probabilistic approach considering multiple climate change scenarios include the National Oceanic and Atmospheric Administration's (NOAA) Climate Mapping for Resilience & Adaptation (CMRA) tool[9] and its Fifth National Climate Assessment (USGCRP, 2023), and other efforts across the federal government, such as the U.S. EPA's Climate Change Impacts and Risk Analysis (CIRA) project.[10]

The U.S. government provides a wealth of high-quality national data on climatic conditions, both historical data and future predictions and models. CEJST could take advantage of these extensive resources to provide a more comprehensive set of climate-relevant indicators that are also more transparent and consistent with datasets, tools, and reports from other parts of the federal government. For example, the USGCRP's CMRA tool includes several future climate burden categories that are not included in CEJST: extreme heat, drought, and coastal inundation. The Localized Construct Analogs (LOCA) version 2 dataset,[11] downscaled from the CMIP6 dataset, is available at a 6-km grid resolution and is used in the Fifth National Climate Assessment (USGCRP, 2023). CDC's Environmental Public Health Tracking Network uses two different datasets. The North American Land Data Assimilation System (NLDAS-2)[12] is available at approximately 14-km grid resolution, and the National Oceanic and Atmospheric Administration (NOAA) NCLIMGRID dataset[13] has a 5-km grid resolution. Additional datasets available within the research community have higher spatial resolutions and could be more appropriate for this application (e.g., Funk et al., 2015, 2019, Verdin et al., 2020). The First Street Foundation, which produced the datasets used for the projected flood risk and projected wildfire risk datasets used in CEJST, also provides national-scale data for extreme heat in their Climate Risk dataset. The dataset is at 4-km spatial resolution and covers 2023 to 2053 using the Intergovernmental Panel on Climate Change CMIP5 RCP4.5 greenhouse gas scenario. Available climate datasets are continually advancing

[9] See https://resilience.climate.gov/ (accessed January 29, 2024).

[10] See https://www.epa.gov/cira (accessed January 29, 2024).

[11] See LOCA's website to view details on the dataset: https://loca.ucsd.edu/. (accessed February 15, 2024).

[12] See https://ldas.gsfc.nasa.gov/nldas/v2/forcing (accessed February 15, 2024).

[13] See https://www.ncei.noaa.gov/access/metadata/landing-page/bin/iso?id=gov.noaa.ncdc:C00332 (accessed February 25, 2024).

to higher spatial resolutions and improved accuracy, and reviewing the available datasets regularly would ensure that the datasets used reflect the state of the science.

In the coming years, the USGCRP is expected to develop a new program to provide climate services, including to the rest of the federal government and to the public. Climate services are defined by the Office of Science and Technology Policy as "scientifically-based, usable information and products that enhance knowledge and understanding about the impacts of climate change on potential decisions and actions" (Fast Track Action Committee on Climate Services, 2023). As USGCRP develops its approach for providing climate services, CEQ could be a client for these data, and CEQ and USGCRP could work together to ensure that the climate services data are responsive to the needs of CEJST for future incorporation into the tool.

Potentially important indicators and datasets that could be considered for inclusion in future versions of CEJST are heat and future climate risks. Compared with CEJST's current approach, which uses datasets that assume a single climate projection and 30-year time horizon, using a probabilistic approach and longer time horizons can provide a more holistic view of potential future climate risks. Working with other federal agencies, such as NOAA and USGCRP, to produce and access relevant climate data can ensure that the datasets used in the tool are robust and consistent with other federal efforts.

Energy

The energy burden category includes two indicators: energy cost and $PM_{2.5}$. Energy cost is measured as the average proportion of annual household income spent on energy. Energy cost data come from the Department of Energy's Low-Income Energy Affordability Data (LEAD) Tool[14] and is provided at the census-tract level for all 50 states, the District of Columbia, and Puerto Rico (excluding Pacific Island territories). LEAD Tool estimates of energy cost are modeled from the U.S. Census Bureau's ACS 2020 Public Use Microdata Samples and U.S. Census housing data from the 2016 5-year ACS.[15]

$PM_{2.5}$ is a commonly used metric for air pollution. The $PM_{2.5}$ data are represented as annual average concentrations for 2019, derived from a model-monitor fusion approach implemented by EPA. The data are also used in the EPA's EJScreen tool and are available for all 50 states, the District of Columbia, and Puerto Rico, but not other island territories. $PM_{2.5}$ is a criterion air pollutant that is regulated by the EPA through annual average and 24-hour average National Ambient Air Quality Standards.[16] $PM_{2.5}$ itself is a mixture of chemical components that exist in both solid and liquid form. A small fraction of $PM_{2.5}$ is emitted directly ("primary $PM_{2.5}$"), and a larger fraction is formed in the atmosphere through chemical interactions ("secondary $PM_{2.5}$"), although the specific sources, chemical composition, and fraction that is primary versus secondary

[14] See https://www.energy.gov/scep/slsc/low-income-energy-affordability-data-lead-tool (accessed February 15, 2024).

[15] See https://www.census.gov/programs-surveys/acs (accessed February 15, 2024).

[16] For more on National Ambient Air Quality Standards, see https://www.epa.gov/criteria-air-pollutants/naaqs-table (accessed February 15, 2024).

depend on geographic location, nearby and upwind emissions, and atmospheric conditions. CEJST documentation does not explain why $PM_{2.5}$ is listed under the Energy burden category, given that $PM_{2.5}$ originates from a variety of sources beyond energy generation. Other major emission source sectors for $PM_{2.5}$ and precursor emissions include transportation, agriculture, industry, wildfires, and dust. A portion of $PM_{2.5}$, diesel PM, is included in the Transportation burden category in CEJST. As stated previously, the committee does not focus on indicator placement under specific burden categories because the indicator groupings are irrelevant to the identification of DACs in the CEJST's current formulation.

The rationale for including $PM_{2.5}$ but not other major air pollutants (e.g., ozone, nitrogen dioxide) is not discussed in CEJST technical support documentation (CEQ, 2022a), therefore preventing the committee from understanding this decision and underscoring the need for clear and thorough documentation. While $PM_{2.5}$ is the largest contributor to the burden of disease from ambient air pollution, ambient ozone is another criterion air pollutant linked with premature mortality that is of interest to community members (NASEM, 2023a). National-scale, spatially complete datasets on ozone concentrations are available. For example, EPA's EJScreen tool includes ozone, using the "peak concentration metric" of the annual mean of the 10 highest maximum daily 8-hour concentrations. The ozone concentration data are from the same model-monitor fusion approach used for the $PM_{2.5}$ dataset that is currently included in CEJST. Nitrogen dioxide (NO_2) is another important air pollutant that contributes to ozone formation and is associated with poor health outcomes. As NO_2 is often considered a marker of traffic-related air pollution, this pollutant is discussed in the Transportation burden category section below.

The major sources of $PM_{2.5}$ and the spatial distribution of $PM_{2.5}$ are in a period of flux due to a variety of factors. Stringent emission standards are bringing emissions down within the energy generation (Henneman et al., 2023) and transportation sectors (Anenberg and Kalman, 2019), while climate change is fueling longer, more intense wildfire smoke seasons (Burke et al., 2021; O'Dell et al., 2019) and more airborne soil dust (Achakulwisut, Mickley, and Anenberg, 2018), leading to high interannual variability and, potentially, stagnation of $PM_{2.5}$ declines (Wei et al., 2023). The high interannual variability in $PM_{2.5}$ driven by wildfire severity can change spatial and demographic patterns of exposure. It may be important to use multiyear average $PM_{2.5}$ concentrations to account for this interannual variability. In addition to annual (or multiyear) averages, including an indicator for poor air quality days can capture wildfire smoke episodes.

The dataset used for $PM_{2.5}$ in CEJST has approximately 12-km grid resolution and is too coarse to capture intraurban concentration gradients that might lead to exposure disparities. In addition, newer datasets from the scientific community have advanced beyond the approach used to develop the CEJST dataset by incorporating satellite measurements of aerosol optical depth (e.g., van Donkelaar et al., 2021) and, in some cases, many other data types in a machine learning process (e.g., Amini et al., 2023). These more advanced and higher-resolution exposure assessment approaches (~1-km grid resolution) are increasingly used within the scientific community, including for air pollution epidemiology (Di et al., 2019). As previously mentioned in Box

5.3, a comparison of the CEJST dataset with two higher-resolution datasets from the scientific community found that the three datasets differ substantially regarding which tracts are most overburdened within individual urban areas, though nationwide $PM_{2.5}$ disparities were more consistent between the datasets (Carter et al., 2023). This analysis underscores the importance of analyzing datasets and evaluating them both technically and with community partners to identify the most appropriate dataset to represent the indicator. As satellite-based and community-collected datasets will improve over time with hourly atmospheric composition data from geostationary satellites, such as the TEMPO instrument launched by NASA in April 2023, $PM_{2.5}$ concentrations derived using satellite data as an input are likely to play an increasingly important role in tracking air pollution and associated disparities across federal governmental activities.

Considering other indicators of air pollution beyond annual average $PM_{2.5}$—including ozone and poor air quality days—would capture additional spatial and temporal patterns of exposure to health-harmful air pollution, especially as these pollutants worsen under climate change. The energy burden index used within the DOE's Energy Justice Mapping Tool provides an opportunity to include additional aspects of disadvantage, including the percent of households not connected to gas or electric grids and the number and average duration of power outages.

Other indicators that could be considered in the Energy burden category include those used by the Department of Energy (DOE) in its own Energy Justice Mapping Tool—Disadvantaged Communities Reporter.[17] These include indicators for the percentage of households that use a fuel other than grid-connected gas or electricity, or solar energy as their main heat source (data from DOE LEAD); average duration of power outage events (in minutes) that occurred for all census tracts in each county from 2017 to 2020 (data from DOE Office of Electricity); number of power outage events that occurred for all census tracts in each county from 2017 to 2020 (data from DOE Office of Electricity); and transportation costs as percentage of income for a typical household in the region (data from Center for Neighborhood Technology).

Health

The health burden category currently includes four indicators: asthma, diabetes, heart disease, and life expectancy at birth. Asthma, diabetes (among people ages 18 years and older), and heart disease (among people ages 18 years and older) data come from the CDC's PLACES: Local Data for Better Health project (PLACES) data[18] for 2016–2019. PLACES provides model-based, population-level analysis and community estimates of health measures down to the census tract across all 50 states and the District of Columbia (but not U.S. territories, i.e., American Samoa, the Commonwealth of the Northern Mariana Islands, Guam, Puerto Rico, and the U.S. Virgin Islands). Life expectancy data come from the CDC's U.S. Small-Area Life Expectancy Estimates Project

[17] See https://energyjustice.egs.anl.gov/ (accessed January 10, 2024).

[18] See https://www.cdc.gov/places/index.html (accessed February 15, 2024).

(USALEEP)[19] from 2010 to 2015. The USALEEP project produced estimates for most U.S. census tracts. The rationale for selecting the four measures currently included in the tool is not reflected in the CEJST Technical Documentation (CEQ, 2022a).

E.O. 14008 did not specifically include health burden as a driving factor for focused investments under the Justice40 Initiative. However, poor health makes communities more vulnerable to the health outcomes associated with climate change, pollution, and other CEJST burden categories. Currently, health data and exposure data are separate burden categories in CEJST and are not integrated. Burden disparities can be amplified when considering health risks associated with exposures (as opposed to considering only the exposure itself) since exposure data alone do not adequately indicate who is adversely affected by that exposure and to what degree. For example, many studies have found that air pollution is inequitably distributed, with communities of color experiencing higher exposure levels compared with the national average or the white population. When these higher exposures for communities of color are combined with information on vulnerability to those exposures—driven by higher rates of preexisting disease, lack of access to high-quality health care, and less ability to take action to reduce exposure—disparities are further amplified (e.g., Kerr et al., 2023; Southerland et al., 2021). The converse of adverse effects from simultaneous disproportionate exposure and vulnerability is that the communities with both high exposure and vulnerability are those who benefit most from reducing exposure and vulnerability—the goal of the Justice40 program. While this report does not address methods for determining which communities will benefit from government programs under the Justice40 Initiative and by how much, the report does discuss how consideration of cumulative impacts could be incorporated into CEJST in Chapter 6.

As of December 2023, PLACES has estimates of 36 health measures—13 for health outcomes, 9 for preventive services use, 4 for chronic disease–related health risk behaviors, 7 for disabilities, and 3 for health status.[20] Among those 36, two stand out for their potential to provide unique information compared with the other health outcomes currently included in CEJST. These are cancer among adults 18 years and older and access to health insurance among those 18–64 years. Other health outcomes from PLACES may be less relevant (e.g., all teeth lost, arthritis) or overlap with those currently included in CEJST in terms of biological systems (e.g., chronic obstructive pulmonary disease, stroke, coronary heart disease) and spatial patterns. Cancer is both relevant—it is affected by climate change, legacy pollution, and other indicators in CEJST (Winstead, 2023)—and is unlikely to be spatially aligned with the other CEJST indicators. The lack of health insurance among adults ages 18–64 years is a measure within the PLACES Prevention category. According to the latest PLACES data based on 2020 and 2021 Behavioral Risk Factor Surveillance System data,[21] approximately 11.7 percent of U.S. residents did not have health insurance. However, since lack of health

---

[19] See USALEEP data at https://www.cdc.gov/nchs/nvss/usaleep/usaleep.html (accessed February 15, 2024).

[20] See the full list of PLACES health measures at https://www.cdc.gov/places/measure-definitions/health-outcomes/index.html (accessed February 15, 2024).

[21] See https://www.cdc.gov/brfss/index.html (accessed February 15, 2024).

insurance varies substantially with citizenship status, ethnicity, income, geography, age, and race (Keisler-Starkey, Bunch, and Lindstrom, 2023), analyzing and understanding spatial correlations with other CEJST socioeconomic indicators would be informative.

It is not clear from technical documentation whether the asthma prevalence dataset used in CEJST includes data on individuals of all ages or only adults. The CEJST Technical Support Document (CEQ, 2022a) does not mention age for the asthma indicator, but the PLACES documentation indicates that the asthma prevalence data is for 18 years of age and older. Pediatric asthma prevalence has been shown to be inequitably distributed in major cities across the United States (Kane, 2022; Roberts et al., 2006). In addition, while some of the indicators in CEJST raise the risk of asthma onset, asthma exacerbation may be more greatly affected by differences in the degree of asthma management in different neighborhoods. To date, data on pediatric asthma prevalence and asthma exacerbation are not available at the tract level nationally, but such datasets may be developed in the coming years.

Access to healthcare would be another appropriate indicator since rural communities tend to have far fewer doctors, specialists, and hospitals in their neighborhoods than urban communities. Systemic discrimination within the healthcare system also leads to disparities in healthcare and health outcomes (Williams and Rucker, 2000). The Centers for Medicare and Medicaid Services (CMS) produces nationally available data on the geography of hospital and nonhospital healthcare facilities,[22] as well as clinicians, down to individual street addresses.[23]

In terms of data quality, it is important to note that the PLACES data are modeled using small-area estimation, a multilevel statistical modeling technique, and do not represent observational data. As such, these data are subject to uncertainties but are not restricted for privacy protections as administrative data often are. PLACES provides 95 percent confidence intervals of modeled estimates generated using a Monte Carlo simulation. In addition, the CDC cautions users against using these estimates for program or policy evaluations because the small-area model cannot detect the effects of local interventions.[24] The annual estimates provide a sufficient temporal resolution, and higher-temporal-resolution (e.g., daily, monthly, seasonal) data are not necessary for CEJST's purposes.

Documentation of the rationale for including the four indicators included in CEJST would provide useful information for end users of the tool and other tool developers and enhance tool transparency. Including additional indicators, such as cancer and lack of health insurance, that are distinct from the four already included can capture additional communities who are experiencing disproportionate health burdens. Although no such data currently exist to the committee's knowledge, considering pediatric asthma onset

[22] See CMS's Provider of Services File—Hospital & Non-Hospital Facilities at https://data.cms.gov/provider-characteristics/hospitals-and-other-facilities/provider-of-services-file-hospital-non-hospital-facilities (accessed February 15, 2024).

[23] See CMS's National Downloadable File at https://data.cms.gov/provider-data/dataset/mj5m-pzi6 (accessed February 15, 2024).

[24] Read CDC's full description of its PLACES 2023 data release at https://data.cdc.gov/500-Cities-Places/PLACES-Local-Data-for-Better-Health-County-Data-20/swc5-untb/about_data (accessed February 16, 2024).

and asthma exacerbation, two outcomes that are highly heterogeneous between neighborhoods and broader geographic areas, would complement the adult asthma prevalence indicator that is currently used in CEJST. Integrating health burdens with other burden categories and indicators to identify DACs could more closely align with the goal of the Justice40 program to benefit underserved communities.

Housing

The housing burden category includes five indicators: experienced historic underinvestment, housing cost, lack of green space, lack of indoor plumbing, and lead paint. Historic underinvestment is represented by redlining maps created by the federal government's Home Owners' Loan Corporation (HOLC) between 1935 and 1940.[25] The boundaries in the HOLC maps were converted to census tracts by the National Community Reinvestment Coalition. Within CEJST, census tracts that have National Community Reinvestment Coalition scores of 3.25 or more out of 4 are considered to have experienced historic underinvestment. This indicator is only available for tracts that were included in the original HOLC maps in certain metro areas across the United States. Housing cost is represented by the share of households that are earning less than 80 percent of Housing and Urban Development's (HUD's) Area Median Family Income and are spending more than 30 percent of their income on housing costs. Data are from the Comprehensive Housing Affordability Strategy dataset from 2014 to 2018 and are available for all U.S. states, the District of Columbia, and Puerto Rico. This dataset is also used for the lack of indoor plumbing indicator. Lack of green space is represented by the share of land with developed surfaces covered with artificial materials such as concrete or pavement, excluding cropland used for agricultural purposes. Data are from the Multi-Resolution Land Characteristics Consortium's Percent Developed Imperviousness dataset for 2019[26] and are available for all contiguous U.S. states and the District of Columbia. The lead paint indicator is represented by the share of homes built before 1960, which, according to the CEJST documentation, indicates potential lead paint exposures (CEQ, 2022a). Tracts with median home values above the 90th percentile are excluded as they are considered less likely to face health risks from lead paint exposure. Data on lead paint are from the ACS for 2015 to 2019 and cover all U.S. states, the District of Columbia, and Puerto Rico. The tool does not currently include other important exposures to lead, including lead in drinking water (see Box 5.4) and childhood exposure to contaminated soil, which remains a legacy pollutant due to historic use of leaded gasoline and industrial activity, such as battery incineration (Laidlaw et al., 2017; Laidlaw, Mielke, and Filippelli, 2023; Zartarian et al., 2023).

---

[25] See Chapter 2, Box 2.1 for more on redlining.

[26] See https://www.mrlc.gov/data/nlcd-2019-percent-developed-imperviousness-conus (accessed February 15, 2024).

BOX 5.4
Lead in Drinking Water

An infamous recent drinking water emergency was the result of a poorly maintained drinking water infrastructure—the lead-contaminated public drinking water in Flint, Michigan. Water was contaminated as a result of a change in the source water, increased corrosion of leaded pipes, and chronic inadequacies in maintenance, monitoring, and reporting (Denchak, 2018; MCRC, 2017; Mohai, 2018). Residents had adverse health effects and were advised not to drink the water. Flint is not alone. Other studies show that chronic lead exposure from old, leaded water service lines across the country (Olson and Stubblefield, 2021), as well as breakdowns in urban public water systems resulting in chronic boil water alerts, have disproportionate impacts on lower-income children and communities of color (Greenfield, 2023; Kim, M. et al., 2023). For at least the last decade, researchers have found that poor and, more often, minority communities were more consistently and disproportionately exposed to drinking water contamination (Balazs and Ray, 2014; Balazs et al., 2011; Berberian et al., 2023; Konisky, Reenock, and Conley, 2021; Martinez-Morata et al., 2022; Pullen Fedinick, Taylor, and Roberts, 2019; Ravalli et al., 2022; Schaider et al., 2019; Stillo and MacDonald Gibson, 2017).

Although lead drinking water pipes in community water systems are the primary source of lead in drinking water (EPA, 2016b), there is relatively little data on leaded pipes at a national scale. EPA was mandated by America's Water Infrastructure Act of 2018 to evaluate and report on the cost of replacing lead service lines in its quadrennial Drinking Water Infrastructure Needs Survey and Assessment (DWINSA). In 2021, EPA collected service-line-material information for the seventh DWINSA. That report estimated that there were 9.2 million lead service lines across the country, primarily concentrated in the eastern half of the country, and especially in states in the South, Midwest, and Northeast. However, these data are only available at the state level.[a] EPA is implementing the Lead and Copper Rule Improvements (LCRI), which require water systems to identify and make public the locations of lead service lines,[b] and issued Guidance for Developing and Maintaining a Service Line Inventory in 2022. The number of lead water lines will remain until there are complete inventories of service lines (EPA, 2023c).

---

[a] See EPA's 7th Drinking Water Infrastructure Needs Survey and Assessment dashboard at https://www.epa.gov/dwsrf/epas-7th-drinking-water-infrastructure-needs-survey-and-assessment.

[b] See proposed Lead and Copper Rule Improvements (LCRI) at https://www.epa.gov/ground-water-and-drinking-water/proposed-lead-and-copper-rule-improvements (accessed February 26, 2024).

Legacy Pollution

The Legacy Pollution burden category includes five indicators: abandoned mine land, formerly used defense sites, proximity to hazardous waste facilities, proximity to Superfund sites (National Priorities List),[27] and proximity to Risk Management Plan (RMP) facilities. The abandoned mine land indicator is represented by the presence of an abandoned mine left by legacy coal mining operations, identified using data from the Abandoned Mine Land Inventory System (e-AMLIS) from the Department of the Interior for 2017. The data cover all U.S. states and the District of Columbia. Formerly used defense sites are from the U.S. Army Corps of Engineers for 2019 and cover all U.S. states and the District of Columbia. Proximity to hazardous waste facilities, Superfund sites, and RMP facilities use data from various EPA databases (as compiled by EPA's EJScreen) for 2020 for all U.S. states, the District of Columbia, and Puerto Rico, and all datasets use a 5-km boundary around facility sites as a measure of proximity. These three proximity indicators consider the number of facilities in each indicator category within 5 km divided by the distance in kilometers.

The EPA released a memorandum, "Strengthening Environmental Justice Through Criminal Enforcement" (EPA, 2021), on the need to strengthen tools for detecting environmental crimes[28] in overburdened communities. Information from the EPA's criminal enforcement program might be useful in strengthening EJ tools, including CEQ tools. EPA's ECHO (Enforcement and Compliance History Online)[29] database provides data on enforcement actions for all EPA-regulated facilities, including permit data, inspection/compliance evaluation dates and findings, violations of environmental regulations, enforcement actions, and penalties assessed. However, these data are not easily accessible or interpretable because the violation codes are unclear and often nonspecific, and the specific circumstances underpinning the rationale for violations are often not made public. Other information may be needed to represent the pollution activities.

Transportation

The transportation burden category includes three indicators: diesel particulate matter (PM) exposure, transportation barriers, and traffic proximity and volume. The dataset used for the diesel PM indicator comes from the EPA EJScreen tool and, according to the CEJST documentation (CEQ, 2022a), is originally sourced from the National Air Toxics Assessment from 2014 (EJScreen documentation indicates the source as the 2017 Air Toxics Update).[30] It is available for all 50 states, the District of Columbia,

---

[27] Learn more about the National Priorities List (NPL) on EPA's website at https://www.epa.gov/superfund/superfund-national-priorities-list-npl (accessed February 16, 2024).

[28] Environmental crimes are carried out by "individuals and corporations that have violated laws designed to protect the environment, worker safety, and animal welfare," according to the Environmental Crimes Section of the U.S. Department of Justice: https://www.justice.gov/enrd/environmental-crimes-section/ (accessed February 26, 2024).

[29] See https://echo.epa.gov/ (accessed February 16, 2024).

[30] See EPA Air Toxics Screening Assessment, 2017 Results: https://www.epa.gov/AirToxScreen/2017-airtoxscreen-assessment-results (accessed February 16, 2024).

and Puerto Rico. Transportation barriers represent the average relative cost and time spent on transportation relative to all other tracts and only applies to census tracts with populations greater than 20 people. The source of the transportation barriers dataset is the U.S. Department of Transportation (DOT) Transportation Access Disadvantage category utilized in the DOT Equitable Transportation Community Explorer (ETCE).[31] Traffic proximity and volume are defined as the number of vehicles (average annual daily traffic) at major roads within 500 meters, divided by distance in meters. The data are sourced from EPA's EJScreen tool and are for the year 2017 and all 50 states, the District of Columbia, and Puerto Rico.

While the committee was not tasked with evaluating the DOT ETCE, participants of the committee's public workshop did suggest that transportation metrics be considered for use in CEJST (NASEM, 2023a). The DOT ETCE User Guide suggests that CEJST be used to identify DACs and then the DOT tool be used to better understand the transportation disadvantage component of CEJST and the ETCE's Transportation Insecurity component—which could ensure that DOT's investments address transportation-related causes of disadvantage.[32] Since DOT is using additional indicators beyond those included in CEJST, the committee does not suggest additional transportation access and volume indicators. However, there are two areas that might be considered by CEQ: (1) a closer proxy for transportation-related air pollution than diesel PM, and (2) noise pollution.

There are several limitations with the diesel PM metric when characterizing the impact of transportation on air quality, including that vehicles using other fuels besides diesel are also polluting and diesel PM is difficult to observe on a nationwide basis, requiring reliance on estimation approaches. Compared with diesel PM, NO_2 may be a more appropriate indicator of traffic-related air pollution and is more directly observable using space-based Earth-observing satellites. NO_2 is linked with respiratory effects, asthma development, and premature mortality (HEI, 2022) and reacts with other chemicals in the atmosphere to form both $PM_{2.5}$ and ozone, the two largest contributors to the burden of disease from air pollution in the United States. NO_2 is more spatially heterogeneous and inequitably distributed than $PM_{2.5}$ (Kerr, Goldberg, and Anenberg, 2021; Kerr et al., 2023) since NO_2 has a shorter atmospheric lifetime (i.e., hours compared with days) and more limited influence from regional pollution sources (e.g., agriculture, wildfire smoke, dust). Heavy-duty vehicles are a leading source of NO_2 in urban areas of the United States and contribute to disproportionate NO_2 exposure among communities of color and with lower income and educational attainment levels (Demetillo et al., 2021; Kerr, Goldberg, and Anenberg, 2021).

Since the diesel PM dataset used in CEJST is derived through modeling rather than observation, the time needed to update these datasets introduces a time lag of several years. Furthermore, the spatial pattern of truck traffic and resulting diesel $PM_{2.5}$

[31] See DOT's ETCE and its user guide at https://experience.arcgis.com/experience/0920984aa80a4362b8778d779b090723 (accessed February 12, 2024).

[32] See DOT's webpage on the Justice40 Initiative at https://www.transportation.gov/equity-Justice40 (accessed February 16, 2024).

emissions and concentrations are changing rapidly: warehousing associated with the booming e-commerce industry is increasing truck trips, truck idling, noise, and traffic-related air pollution in some places, which are increasingly nearer to population centers (e.g., Jaller and Pahwa, 2020). Additionally, oil and gas development results in heavy truck traffic (e.g., Adgate, Goldstein, and McKenzie, 2014) in new locations, such as the Permian Basin in Western Texas and the Bakken field in North Dakota. Neither of these rapidly evolving industries would show up in a diesel $PM_{2.5}$ dataset that is several years old. However, these changes are captured by satellite-based NO_2 datasets. A more observational dataset, such as satellite NO_2, could be a valuable metric for CEQ to consider in future CEJST versions, especially as new geostationary satellites, such as TEMPO launched by NASA in 2023,[33] will produce hourly measurements over the United States in the coming years.

It may also be useful to have an indicator for assessing transportation-related noise pollution, which is associated with cardiovascular morbidity and mortality (Münzel, Sørensen, and Daiber, 2021), mental health outcomes (Gong et al., 2022), and other health effects. Transportation-related noise data are available from the DOT Bureau of Transportation Statistics' National Transportation Noise Map and are available for the United States at the tract level (Seto and Huang, 2023). The estimates include noise levels related to aviation, roadway, and rail traffic. These data may meet the criteria for inclusion in CEJST if CEQ and community partners consider them to be relevant for the purposes of CEJST.

Water and Wastewater

The water and wastewater burden category includes two indicators: underground storage tanks (USTs) and releases and wastewater discharge. Both USTs and releases and wastewater discharge indicators use datasets compiled by EPA's EJScreen tool and are available for all U.S. states, the District of Columbia, and Puerto Rico. The UST indicator is drawn from the EPA's UST Finder for 2021,[34] which includes any UST and "any underground piping connected to the tank that has at least 10 percent of its combined volume underground" (EPA, 2024). Federal UST regulations apply only to UST systems above specific thresholds storing either petroleum (e.g., gasoline, diesel, fuel oil) or certain hazardous substances (e.g., ammonia, asbestos, benzene, chromium). More than 99 percent of federally regulated USTs contain petroleum, and most of those are owned or managed by service stations and convenience stores, or by vehicle fleet service operators and local governments. The greatest potential hazard from USTs is the leakage of hazardous substances into the surrounding soil and contamination of groundwater. UST releases are the most common source of groundwater contamination; petroleum is the most common contaminant (EPA, 2023a). This is notable given that nearly one-half of all Americans get their drinking water from groundwater (EPA, 2024). Underground storage tanks and releases are represented by a weighted formula of the

[33] See https://science.nasa.gov/mission/tempo/ (accessed March 8, 2024).

[34] See https://www.epa.gov/ust/ust-finder (accessed February 26, 2024).

density of leaking underground storage tanks and the number of all active underground storage tanks within 1,500 feet of the census tract boundaries.

The wastewater discharge indicator utilizes information from the EPA's Discharge Monitoring Report (DMR) Loading Tool[35] along with the EPA's Risk-Screening Environmental Indicators (RSEI) model[36] to estimate the relative risk to a census-block group from exposure to pollutants in downstream water bodies (EPA, 2023b). The DMR Loading Tool includes data on industrial and municipal point-source wastewater dischargers that are subject to a subset of permits under the National Pollutant Discharge Elimination System (NPDES),[37] as well as wastewater pollutant discharge data from EPA's Toxic Release Inventory (TRI). Wastewater discharge is represented by the RSEI-modeled toxic concentrations at stream segments within 500 meters, divided by the distance in kilometers. The data are for 2020. Both USTs and releases and wastewater discharge indicators use datasets compiled by EPA's EJScreen tool and are available for all U.S. states, the District of Columbia, and Puerto Rico.

Data on USTs and wastewater discharge do not capture the universe of potential pollutant releases to groundwater or surface water. Rather, they include data only on facilities or activities subject to federal regulation, specific reporting requirements that often focus on larger sources and limited ranges of pollutant categories. Federal regulations and data collection around USTs do not apply to smaller, noncommercial farm and residential tanks; heating oil tanks on premises where fuel is used; tanks in underground spaces such as basements or tunnels; septic tanks and storm and wastewater collection systems; flow-through process tanks; tanks small than 110 gallons; and emergency spill and overfill tanks. Several states and local regulatory authorities have more stringent rules around USTs than the federal government and may collect data on a wider range of USTs, but state-specific data may not be included in the UST Finder dataset.

The DMR Loading Tool[38] includes information on discharges for more than 60,000 facilities across the United States. Not all facility, permit, or discharge monitoring data are uploaded to the NPDES database, and data may be reported differently. Pollutants for which discharge permits are not required are not required to be reported, and data related to many regulated sources of wastewater discharge are not available. These include wastewater releases from industrial facilities connected to public treatment works sewerage systems regulated through the Clean Water Act (CWA); CWA Biosolids Program–related biosolid monitoring data; releases related to wet-weather events; construction activity–related discharges; combined and sanitary sewer overflows; and discharges related to concentrated animal feeding operations.

TRI wastewater discharge data mentioned above are limited to industrial facilities with more than 10 employees, and not all industry sectors are included. Reporting emphasizes toxic pollutant discharges. Common wastewater pollutants such as total

---

[35] For more information on the DMR Loading Tool data, see https://echo.epa.gov/trends/loading-tool/resources/about-the-data (accessed February 26, 2024).

[36] See https://www.epa.gov/rsei (accessed February 26, 2024).

[37] See https://www.epa.gov/npdes (accessed February 26, 2024).

[38] See https://echo.epa.gov/trends/loading-tool/resources/about-the-data#:~:text=The%20Loading%20Tool%20contains%20information,under%20the%20Clean%20Water%20Act (accessed March 6, 2024).

suspended solids and biochemical oxygen demand are not included. Data reported are often derived statistically rather than directly measured. Some chemicals are reported as classes rather than individual compounds, the result being that the potential toxicity of releases could be estimated inaccurately given the variation in toxicity of individual compounds in a class (EPA, 2022b).

Drinking water burdens vary geographically and by community and can disproportionately affect poor communities and people of color (see Box 5.5 for examples demonstrating observed relationships). Although USTs and wastewater dischargers are nearly ubiquitous across the country—implying a widespread risk of groundwater and surface water contamination—not all communities are equally dependent on or even exposed to their local ground or surface waters. Many communities, especially urban communities, rely on piped water that may come from sources distant from the community. Their water is often treated but also subject to potential contamination from a distant source or by inadequately maintained infrastructure, treatment, or conveyance (see Box 5.4 on lead in drinking water). Pollution of water sources is not the only way in which communities experience burdens or unequal treatment regarding water. These burdens include water access, water system services, and infrastructure.

Studies of problems and disparities in drinking water systems have relied on data from the EPA Safe Drinking Water Information System (SDWIS) database, which records SDWA violations and provides data for all U.S. states and territories.[39] However, SDWIS does not capture all sources of drinking water, and reporting may be uneven. Although states are required to report drinking water system information to the SDWIS, audits of the system show that states often fail to report many violations. For example, the SDWIS did not include lead violations for Flint, Michigan's lead crisis from 2014 to 2017 (Pullen Fedinick, Taylor, and Roberts, 2019). SDWIS covers a large part of the population, but it is only consistently available at the county level, and it only applies to community drinking water systems that serve at least 25 people or have more than 15 connections. Furthermore, the SDWA does not cover private wells or other noncommunity sources of drinking water (e.g., water taken directly from rivers, streams, or creeks). Private wells supply water for roughly 16 percent of all housing units in the United States (EPA, 2023d). There is no nationwide testing required for those wells, and less testing is done, generally, for private wells than for public wells (Murray et al., 2021). The EPA has developed a mapping system to identify the density of private wells down to the census block group across the country.[40] Dependence on private wells is highest in rural communities, on Tribal lands, in unincorporated places, and for farmworkers living in fields or labor camps. The latter has been shown to be especially susceptible to contamination by chemical and biological hazards from pesticides, fertilizers, and animal and human waste (Balazs et al., 2011; Bischoff et al., 2012; Lohan, 2017). Some communities, such as colonias along the southwestern border and unin-

[39] See https://health.gov/healthypeople/objectives-and-data/data-sources-and-methods/data-sources/safe-drinking-water-information-system-sdwis#:~:text=The%20Safe%20Drinking%20Water%20Information,approximately%20156%2C000%20public%20water%20systems (accessed March 6, 2024).

[40] See EPA Private Domestic Well Map at https://experience.arcgis.com/experience/be9006c30a2148f595693066441fb8eb (accessed March 6, 2024).

BOX 5.5
Relationship Between Race and Clean Water Compliance

In a nationwide study of violations of the Clean Water Act requirement for drinking water quality reports to communities, Bae and Kang (2022) found a statistical relationship between rule violation occurrence and the proportion of Hispanic residents and the poverty rate of host counties. Although these patterns occurred across the country, violations were concentrated in Texas, Oklahoma, and Louisiana. Systematic lack of information on the quality of drinking water sources compromises the ability of residents to understand or respond to risks to their health and reduces the opportunity for holding community water systems accountable, potentially exacerbating other forms of environmental inequities. Bae, Kang, and Lynch (2023) examined the length of time that community water systems across the country were out of compliance with Safe Drinking Water Act (SDWA) regulations from 2015 to 2019 (e.g., mandated treatment techniques or violations of any maximum contaminant levels) and the racial composition of those communities. They found that noncompliant water systems in counties with higher proportions of both Black and Hispanic residents took longer to be returned to compliance than water systems serving a larger percentage of white residents. In general, as the percentage of white residents in an area increased, the time to compliance decreased. Racial differences in noncompliance durations were not explained by differences in the income level or poverty rates of those same communities. The implication is that Black and Hispanic residents systematically experience longer periods of noncompliance for their drinking water systems, and that enforcement of these regulations is unequal. Their findings complement a previous nationwide study conducted by the Natural Resources Defense Council (Pullen Fedinick, Taylor, and Roberts, 2019), which found that SDWA violations were more likely in counties with racial, ethnic, and language vulnerability, lower-quality housing conditions, and less transportation access. Racial, ethnic, and language vulnerability were most strongly related to the length of time out of compliance. More generally, the NRDC analysis observed that community water systems serving fewer than 3,300 people—those more likely to serve low-income, vulnerable populations, face disproportionate hazards and lack resources to address the issues (EPA, 2016a)—account for greater than 80 percent of violations generally and specifically of health-based violations.

corporated communities, lack basic drinking water infrastructure and reliable access to potable water. An estimated 471,000 households—or 1.1 million people—lack a piped water connection, and unplumbed households in cities are more likely to be headed by people of color, earn lower incomes, and live in mobile homes (Meehan et al., 2020).

Workforce Development

A stated goal of E.O. 14008 is to address the economic challenges faced by DACs resulting from disproportionate negative impacts of climate change, pollution, and other burdens (including economic shifts) experienced by these communities. An important focus is on job creation in these communities through federal investment—in workforce development and in building a cleaner and more equitable economy that can offer well-paying jobs and opportunities for equitable economic growth. The workforce development burden category is aimed specifically at identifying communities where these kinds of investments might be beneficial (CEQ, 2022a).

The workforce development burden category in CEJST includes four direct indicators (linguistic isolation, low median income, poverty, and unemployment) plus a socioeconomic indicator specifically associated with indicators in this burden category. The datasets used for all five indicators within this burden category come from the ACS for 2015–2019 (or the 2010 Decennial Census[41] in some island geographies). Two of the direct indicators are income related (median income as a share of area median income and share of people in households where income is at or below 100 percent of the federal poverty level), one measures linguistic isolation (share of households where no one over age 14 speaks English very well) and another directly measures unemployment (number of unemployed people as a part of the labor force).

Several aspects of the workforce development category differ from the other burden categories in CEJST. Firstly, the income-related indicators are treated as direct indicators of the need for workforce-related investment rather than serving as socioeconomic indicators. The socioeconomic indicator used for the workforce development burden category is instead a measure of educational attainment, namely, the percent of people ages 25 years or older whose highest level of education is less than a high school diploma (i.e., low educational attainment). Using low educational attainment as the socioeconomic indicator for workforce development implies that it is a necessary condition for a community to qualify as disadvantaged based on this burden category. To be classified as disadvantaged, a tract needs to be above 10 percent for this socioeconomic indicator (in addition to being at or above the 90th percentile for one of the other four indicators). The CEJST technical documentation does not include a rationale for the different threshold for this socioeconomic indicator (CEQ, 2022a).

Low income is still considered in this burden category, but the low-income indicators differ from the low-income indicator used for the other burden categories. One difference is in the threshold used to define low income—the federal poverty level. For

[41] See https://www.census.gov/programs-surveys/decennial-census/decade.html (accessed February 16, 2024).

the poverty indicator used in the workforce development burden category, the threshold is 100 percent of the federal poverty level, whereas the threshold is 200 percent of the federal poverty level for the socioeconomic indicator in the other burden categories. Thus, the definition of disadvantaged is narrower under the poverty indicator used in the workforce development category than in other burden categories. Again, a rationale for this difference is not clearly stated in the CEJST technical documentation (CEQ, 2022a).

Another difference is the inclusion of a second measure of low income through a measure of median income relative to median income in the area. Unlike the measures of income relative to the federal poverty level, this income measure allows for some differentiation across regions, which could capture, for example, differences in the cost of living. Thus, a community with low educational attainment could qualify as disadvantaged based on the workforce development burden even with an income level that is high relative to the nation as a whole, as long as it is low relative to its area. In this sense, the use of the relative measure can expand the definition of disadvantaged under the workforce development burden category. Again, the CEJST Technical Documentation does not provide a rationale for why a relative (area-specific) measure of low income is used here but not in defining low income through the socioeconomic indicator used for the other burden categories (CEQ, 2022a). Most of the EJ tools described in Chapter 4, including some of the state-level tools, only measure income based on federal poverty levels and therefore do not include area-specific measures that can account for differences in cost of living. However, some state-level tools do provide more localized income measures. For example, the Massachusetts Department of Public Health Environmental Justice Tool (MA-DPH-EJT)[42] includes an indicator of whether the annual median household income in a community is 65 percent or less of the statewide annual median household income.

As noted above, the workforce development indicator is based on information about income, linguistic isolation, unemployment, and educational attainment. While these are possible proxies for measuring where federal and other investments could improve workforce outcomes for DACs, other possible proxies exist as well. For example, in addition to overall unemployment measures, which are based on questions related to employment status in the ACS, the survey also reports information on work status, which includes data on the median earnings for full-time, year-round male and female workers. Given the E.O. 14008 goal of fostering employment in well-paying jobs in DACs, measures of median earnings for full-time workers could provide useful information about the quality of jobs held by residents of a given community.

Another possible indicator of job quality based on the ACS is the percentage of working-age adults (ages 19–64 years) with employer-based health insurance. Employer-based health insurance benefits can constitute a substantial component of overall employee compensation (BLS, 2023), and access can vary by race and other worker characteristics (Lee et al., 2019). Thus, measures of the prevalence of employer-based health insurance for residents of a community can be another indicator of job

[42] See https://matracking.ehs.state.ma.us/Environmental-Data/ej-vulnerable-health/environmental-justice.html (accessed /February 17, 2024).

quality within the community. Notably, several of the EJ tools scanned in Chapter 4 use ACS-based indicators of lack of health insurance coverage (of any form, rather than simply employer-based) as socioeconomic indicators—including Centers for Disease Control and Prevention and Agency for Toxic Substances and Disease Registry Social Vulnerability Index, the FEMA NRI, the Department of Health and Human Services Environmental Justice Index, the DOE's Energy Justice Mapping Tool, the Census Community Resilience Estimates, and the DOT ETCE. Although lack of health insurance coverage from any source can be a significant stressor, especially among disadvantaged populations (Brown et al., 2000), looking specifically at the lack of employer-based health insurance provides different information about the quality of jobs held by residents of a given community.

Some other EJ tools include burden indicators of workforce-related impacts of transitioning away from fossil fuels toward renewable energy. For example, the DOE's Energy Justice Mapping Tool has a burden category for Fossil Dependence, which includes two workforce-related indicators (both from the U.S. Bureau of Labor Statistics): percent of total civilian jobs in the coal sector and percent of total civilian jobs in the fossil energy sector. However, the Workforce Development indicators currently in CEJST do not incorporate any measure of vulnerability or disadvantage related to fossil fuel dependency, except to the extent that employment losses that have already occurred (e.g., from reductions in coal mining or closure of coal-fired power plants) are reflected in the community's unemployment rate.

More direct and inclusive measures of fossil-fuel dependency could be incorporated into CEJST. One example is the designation of "energy communities," a concept used in the Inflation Reduction Act (IRA)[43] to determine eligibility for increased tax credits under the law. This concept is broader than simply the fossil fuel dependency included in DOE's Energy Justice Mapping Tool (coal and fossil energy employment). The IRA includes in the definition of an energy community: (1) any brownfield site, (2) any metropolitan or non-metropolitan statistical areas that have both high fossil fuel employment or tax revenue and an unemployment rate that is above the U.S. national average, and (3) any census tract (or directly adjoining census tract) with a coal mine closed after 1999 or a coal-fired power plant retired after 2009 (Interagency Working Group on Coal & Power Plant Communities & Economic Revitalization, 2023). Satisfying one of these conditions gives a binary classification of a community as either being an energy community or not.[44] With an alignment of geographical scale (for brownfields and statistical areas), this designation could be included in CEJST.

However, as a recent study by Graham and Knittel (2024) points out, the definition of fossil-fuel dependency used in the IRA considers only fossil-fuel extraction and processing sectors and does not consider other sectors where production or consumption is fossil-fuel dependent (like manufacturing). In addition, it does not include fossil-fuel

[43] For more information on the IRA, see https://www.whitehouse.gov/cleanenergy/inflation-reduction-act-guidebook/ (accessed February 27, 2024).

[44] See this classification reflected in DOE's map of Energy Community Tax Credit Bonus at https://arcgis.netl.doe.gov/portal/apps/experiencebuilder/experience/?data_id=dataSource_3-188bf476e26-layer-6%3A1494&id=a2ce47d4721a477a8701bd0e08495e1d (accessed February 27, 2024).

power generation. Moreover, its reliance on the current relative unemployment rate to determine eligibility makes it backward-looking rather than forward-looking in terms of impacts. Graham and Knittel (2024) propose an alternative measure of fossil fuel–related employment vulnerability, termed an employment carbon footprint (ECF), which incorporates both production and consumption channels of vulnerability. The ECF is a continuous index that is calculated at the county level using mostly publicly and nationally available data (Graham and Knittel, 2024). Because the ECF is a continuous index, percentiles can be used to define the counties that are most vulnerable to employment shocks from the transition away from fossil fuels. They compare the results of their analysis to the classification based on the IRA definition of energy communities and show that the IRA definition leads to significant false positives and false negatives, suggesting that the ECF provides a better indicator of fossil fuel–related employment vulnerability. Since the data sources used to calculate the ECF mostly meet the criteria for indicator inclusion in CEJST and the ECF's calculated by Graham and Knittel (2024) are publicly available, it represents an alternative and potentially superior way to incorporate this type of vulnerability as an indicator in the workforce development burden category in CEJST.

Socioeconomic

As described relative to the burden categories above, CEJST combines two socioeconomic indicators with other indicators within these burden categories to determine whether a tract is identified as disadvantaged within CEJST: low income (seven of the eight burden categories) and high school education (workforce development category). Low income is defined as the 65th percentile or above for census tracts that have people in households whose income is less than or equal to twice the federal poverty level, excluding students enrolled in higher education. The indicator for high school education is based on fewer than 10 percent of people ages 25 years or older with a high school education (i.e., graduated with a high school diploma) (CEQ, 2022a). Based on the current formulation of CEJST, no variable is as important as the socioeconomic variable in identifying DACs since socioeconomic status is part of all 30 indicators under the categories of burden (see Box 5.6).

These socioeconomic indicators (as is true for other indicators in CEJST) do not capture heterogeneity within the tract (see Chapter 7 for more discussion on scale). For example, a participant in the committee's information-gathering workshop provided an example of a tract that contained both expensive waterfront housing and low-income housing (NASEM, 2023a). A tract with a wide degree of socioeconomic heterogeneity may not be identified as disadvantaged within the tool because the indicators use tract-level averages. An indicator of socioeconomic inequality within the tract could be used to address this issue. Data on socioeconomic inequality are available at the tract level from the U.S. ACS 5-year estimates: the Gini Index of Income Equality.[45] The Gini

[45] See https://www.census.gov/topics/income-poverty/income-inequality/about/metrics/gini-index.html (accessed February 28, 2024).

BOX 5.6
Low Income as a Socioeconomic Burden Indicator

Since seven of the eight burden categories (i.e., all except workforce development) in CEJST apply the low-income indicator (the percentage of the population in a census tract with income that is less than or equal to twice the federal poverty level) as the key socioeconomic burden indicator, it is important to explore its limitations.[a] First, the CEJST technical documentation does not explain why the 65th percentile for this variable was selected as the threshold value and how this influences the selection of disadvantaged tracts. A sensitivity analysis would be helpful to examine the effects of changing this cutoff threshold to a higher or lower percentile (see Chapter 7). Second and more importantly, there are inherent problems with using the federal poverty level in a national-scale tool. Although the federal poverty level measure is adjusted every year, the value applies to the entire country (except Alaska and Hawaii). Using a single value for all the United States in the tool may result in the characterization of income that is too high or too low because the cost of living across and within regions is not uniform. Changing the socioeconomic burden indicator in CEJST would affect which census tracts are designated as disadvantaged.

[a] While this section focuses mostly on the low-income indictor, the high school education indicator is discussed in more detail in the workforce development section above.

index is a summary measure of income inequality whose values range from 0 (perfect equality with all households in a tract having equal incomes) to 1 (perfect inequality with only one household having an income). This Gini index has been utilized as a measure of neighborhood-level coping capacity and socioeconomic vulnerability in previous environmental justice studies (Chakraborty et al., 2014).

Historically impoverished states may have more DACs than wealthier states. However, the current binary designation does not allow for further characterization of workforce development and education. Another situation not captured well by the indicators is the loss of resources resulting from absentee landowners living outside the tract (resources from the landowner do not circulate back into the tract; NASEM, 2023a).

Using a single, uniform low-income measure in a tool such as CEJST may not accurately reflect lived experiences even after doubling the standard poverty level and accounting for the cost of living. Other indicators have been suggested to inform income measurements. For example, the EPA Science Advisory Board (SAB), in their review of EJScreen, suggested using the criteria of the HUD's Public Housing/Section 8 Income limits for low income,[46] which is 80 percent of the area median income (SAB,

[46] See HUD's FY 2023 methodology for determining Section 8 limits at https://www.huduser.gov/portal/datasets/il//il23/IncomeLimitsMethodology-FY23.pdf (accessed March 8, 2024).

2023). The SAB also suggested an indicator of wealth such as homeownership rate, median home value, or a weighted income metric, acknowledging that income-based measures deserve scrutiny because of the effects of income on all aspects of a person's or household's quality of life (e.g., nutrition, health care, and education).

A related issue is that metrics of income do not necessarily measure wealth. Wealth is a significant reflection of economic security or capacity and is impacted through generational economic movement and the ability to distribute wealth to descendants. As noted by the Horowitz, Igielnik, and Kochhar (2020), income measures the sum of earnings from employment, Social Security, business, or other sources, whereas wealth measures the value of owned assets (e.g., home, savings account) minus outstanding debts (e.g., loans, mortgage). The wealth gap between high-income and low-income households is larger than the income gap and is growing more rapidly (Horowitz, Igielnik, and Kochhar, 2020). One metric of wealth suggested by workshop participants is homeownership or the percentage of homeowners in a community (NASEM, 2023a).

Racism

CEJST does not include indicators of race or ethnicity in its determination of DACs, even though racism is a key driver of climate and economic injustice within the United States. Historical race-based policies in housing, transportation, and other urban development have had lasting impacts on environmental inequality today (see e.g., Ahmed, Scretching, and Lane, 2023; Bonilla-Silva, 1997; Bravo et al., 2022; Bullard, 2001; Bullard et al., 2007; Callahan et al., 2021; Chakraborty, J., et al., 2022; Collins, Nadybal, and Grineski, 2020; Commission on Social Determinants of Health, 2009; Dean and Thorpe, 2022; Dennis et al., 2021; Kodros et al. 2022; Konisky, Reenock, and Conley, 2021; Lane et al., 2022; Martinez-Morata et al., 2022; Mohai and Saha, 2007; O'Shea et al., 2021; Paradies et al., 2015; Trudeau, King, and Guastavino, 2023). Chapter 2 describes how race and ethnicity have been shown to be consistent and statistically independent predictors of a range of social, economic, health, and environmental inequities and are often more significant than economic indicators of socioeconomic status (Bullard et al., 2007; Liu et al., 2021; Mohai and Saha, 2007; Tessum et al., 2021) and provides empirical evidence of racism as a relevant factor in unequal exposures and outcomes. Socioeconomic status is not a substitute for measures of racial or ethnic differences. Although race and ethnicity are often strong predictors of inequity, scholars increasingly recognize that the problem is racism, not race or ethnicity (e.g., Adkins-Jackson et al., 2022; Bailey et al., 2017; Boyd et al., 2020; Braveman et al., 2022; Chadha et al., 2020; Gannon, 2016; Lett et al., 2022; NASEM, 2023b; Payne-Sturges, Gee, and Cory-Slechta, 2021; Smedley and Smedley, 2005).

Advocates and scholars of environmental justice and health inequities argue that measures of racism are necessary to identify and understand inequity or disadvantage. Measures of racism and its relationship to inequity need to be supported by the collection and reporting of disaggregated data on race and ethnicity to monitor the state of racial or ethnic disparities, to properly identify differences in population experiences of racism and inequity, and to avoid perpetuating or exacerbating structural racism through the erasure of real differences between and within population groups (Adkins-Jackson et al., 2022; Braveman et al., 2022; Kauh, Read, and Scheitler, 2021; Polonik et al., 2023; Wang et al., 2022). Disaggregated data on race and ethnicity are readily available through the U.S. Census, and there is a large and growing range of indicators or measures of racism.

Measures of segregation or racism are listed in Appendix D, along with other measures for consideration in EJ tools that have come to the attention of the committee through its scan of tools, experience, and its workshop. Scholars of health inequities and racism increasingly recommend that structural racism is more properly measured through an index approach that better reflects its multidimensional nature (Adkins-Jackson et al., 2022; Dean and Thorpe, 2022; Furtado et al., 2023). Among the various indexes used by health researchers to capture structural racism, Furtado and others (2023) highlight three strategies that use geographic approaches to measure structural racism and which they argue are especially useful at quantifying the magnitude of its impacts: measures of residential segregation, racialized economic segregation, and indexes of disproportionality.

Measures of residential segregation are the most familiar and well-understood metrics of structural racism. A commonly used dataset is redlining data (see Box 2.1), which scholars have shown to be associated with a range of persistent environmental, health, and social inequities (Berberian et al., 2023; Blatt et al., 2024; Bompoti, Coelho, and Pawlowski, 2024; Hoffman, Shandas, and Pendleton, 2020; Kephart, 2022; Lane et al., 2022). CEJST 1.0 incorporates redlining maps as an indicator of "historic underinvestment" in the housing category. However, data on historic federally defined redlining maps are only available for a little over 200 of the largest cities across the country.[47] There are no data for most communities and none for rural areas. There are numerous alternative measures of residential racial segregation and structural racism that are available nationally. Other measures of residential segregation can be calculated using Census data at various geographic scales. They include the Dissimilarity Index and the Isolation Index (housing segregation), the Gini coefficient (income segregation), the Index of Contemporary Mortgage Discrimination (Mendez, Hogan, and Culhane, 2011), the Index of Historical Redlining (Beyer et al., 2016), and dozens of others. The U.S. Census guidance appendix on housing segregation reviews the most common segregation indexes and their calculation (Iceland, Weinberg, and Steinmetz, 2002).

Measures of racialized economic segregation include the Index of Concentration at the Extremes (ICE) (Massey, 2001), which simultaneously evaluates the concentration of deprivation and privilege. Krieger and others (2017) created a modified version of ICE that measures spatial polarizations of race and income by comparing the number of people in the most privileged extreme (i.e., white residents above the 80th income percentile) to the number of people in the most deprived extreme (i.e., Black residents below the 20th income percentile). Unlike unidimensional measures of segregation,

[47] Read more about HOLC redlining maps in the University of Richmond project "Mapping Inequality: Redlining in New Deal America" at https://dsl.richmond.edu/panorama/redlining (accessed February 28, 2024).

measures of racialized economic segregation measure the intersection of both racial concentration and income concentration, which is more aligned theoretically with the multidimensional concept of structural racism and has the benefit of avoiding multicollinearity[48] that occurs when using income and race as two separate indicators.

The Index of Disproportionality refers to a group of related approaches that use racial disparity indicators across an array of domains—political participation, employment and job status, educational attainment, judicial treatment, housing, income, and health care (Furtado et al., 2023). The original approach by Lukachko and others (2014) calculated the Black versus white rate or prevalence ratios for multiple indicators within the domains of political participation, employment and job status, educational attainment, and judicial treatment. The indicators were then input individually into generalized estimating equation models and stratified by race. Since then, researchers have built on this approach to combine multiple indicators of disproportionality into one measure of structural racism using latent variable methods (e.g., factor analysis, cluster analysis, latent class analysis). Dougherty and others (2020) used confirmatory factor analysis to combine Black–white indicators (i.e., prevalence ratios) of differential treatment across the domains of education, housing, employment, criminal justice, and health care into a single metric of structural racism exposure to predict BMI at the county level. Chantarat, Van Riper, and Hardeman (2022) calculated measures of Black–white residential segregation and inequities in education, employment, income, and homeownership at the metropolitan area scales (i.e., Public Use Microdata Areas) and then used latent class modeling to reduce them into one multidimensional measure of structural racism to predict birth outcomes. Similar to measures of racialized economic segregation, latent variable approaches to the Index of Disproportionality have the benefit of operationalizing structural racism as a multidimensional phenomenon and avoiding problems of multicollinearity. These latter approaches have the added advantage of capturing the otherwise invisible intersection of structural racism across multiple domains, although this may come at the cost of easier interpretability.

CHAPTER HIGHLIGHTS

Selecting indicators and datasets to represent them is a critical step in the iterative process for developing screening tools. (See the committee's conceptual framework for indicator construction, Figure 3.2). Indicators in a tool are quantitative proxies for abstract concepts (such as "disadvantage" in the case of CEJST), and the selection of indicators requires consideration of their technical characteristics (validity, sensitivity, robustness, reproducibility, and scale) and practical characteristics (measurability, availability, simplicity, affordability, credibility, and relevance). Engaging community members and other interested and affected parties iteratively throughout the indicator selection and integration process is an essential aspect of building the credibility of

[48] Multicollinearity occurs when two or more independent variables in a model are highly correlated. This can violate the assumptions of statistical models such as linear regression; it can also be an issue when constructing a composite index since it implies the overcounting of a concept.

an EJ tool and engendering trust. The CEQ's indicator selection criteria limit data to those that are publicly available and relevant to E.O. 14088 and Justice40; that cover all 50 states, the District of Columbia, and U.S. territories; and that are available at the census-tract scale or finer. The census-tract scale may not provide the granularity needed to adequately define communities.

The categories of burden selected for CEJST could be used to represent disadvantage. Each burden category includes a different number of indicators, and each indicator is assigned a threshold value. The technical documentation for CEJST (CEQ, 2022a) does not provide the rationale for the choice, number, or thresholds of many of its indicators. These indicators are not explicitly weighted, although the socioeconomic burden category is more heavily weighted because that threshold must be met in addition to the threshold in any other indicator. Indicator interactions and cumulative impacts are not considered in the tool. In the current construction of CEJST, the number of indicators and their categorization under specific burdens do not affect the identification of disadvantage because any single indicator can trigger disadvantage status if the threshold value is met (in addition to the socioeconomic indicator). Using a single, uniform low-income measure in a tool such as CEJST may not accurately reflect lived experiences even after doubling the standard poverty level and accounting for the cost of living.

As currently formulated, CEJST relies on data from existing sources, in some cases using the same data that are also employed within other screening tools available from the federal government and others to map environmental justice and climate vulnerability. Numerous other indicators and datasets are available that could be used instead of or in addition to the indicators now used in CEJST. These meet CEQs criteria and may be able to reflect lived experiences in communities more accurately. Several potential indicators and datasets are described in this chapter and in Appendix D that might be considered, but their inclusion by CEQ would require careful analysis. Community engagement, validation, and transparency in selecting and including burden categories and indicators can help evaluate how well the tool captures burdens that align with the lived experience of communities. Indicator groupings could become important in future iterations of the tool, especially if integration approaches for assessing cumulative impacts are implemented (discussed further in Chapter 6). CEQ could also foster the development of new, fit-for-purpose datasets with various agencies, organization, research groups, and private firms.

The next chapters will broaden this chapter's focus on specific indicators and data gaps in CEJST and return to the process of developing EJ tools. Chapter 6 addresses approaches for integrating indicators and developing composite indicators to align with the concept being measured by the tool. Chapter 7 describes the iterative techniques for validating the robustness and output of the tool.

6

Indicator Integration

The White House Council on Environmental Quality (CEQ) Climate and Economic Justice Screening Tool (CEJST) technical documentation states that "disadvantaged communities face numerous challenges because they have been marginalized by society, overburdened by pollution, and underserved by infrastructure and other key services" (CEQ, 2022a, p. 7). This definition drove the committee's discussion of disadvantaged communities (DACs) and the indicators selected to represent and measure this definition. This definition is also essential when considering CEJST's approach to integrating the selected indicators into a single composite indicator and, ultimately, the designation of census tracts as part of a disadvantaged community. The process of integrating indicators into a composite indicator includes bringing all indicators onto a common scale, applying weights, and aggregating to a single value. Ideally, the process also includes evaluating the resulting composite indicator in terms of internal robustness, which has been shown to improve decision making and transparency in the integration processes (Freudenberg, 2003; Jacobs, Smith, and Goddard, 2004; Mazziotta and Pareto, 2017; Organisation for Economic Co-operation and Development [OECD] and Joint Research Centre [JRC], 2008; Saisana et al., 2019). Constructing a composite indicator is a widely used strategy for measuring cumulative burden and requires intentional decisions during each step of the process. Chapter 3 provided discussion on the construction of composite indicators. This chapter discusses those process components in the context of CEJST and their effect on the designation of census tracts as disadvantaged communities. The discussion also outlines alternative potential configurations that reflect cumulative impacts.

HOW CEJST INTEGRATES INDICATORS

As described in Chapter 5, CEJST comprises 30 indicators, each falling into one of eight categories of burden. CEJST converts most of the indicators into percentiles, with a few exceptions that have natural cutoff values, including the presence or absence of abandoned mines, formerly used defense sites, and historic underinvestment in the form of redlining. The methodology then sets the threshold for each indicator at the 90th percentile. For indicators where a high value would be desirable (e.g., income or life expectancy), the values are inverted so that a high percentile value corresponds to a high disadvantage. In the sections below, the major decisions associated with indicator integration are discussed in detail.

BRINGING INDICATORS ONTO A COMMON SCALE

The goal of calculating a composite indicator is to combine different data types and measurements into a single measure or indicator to obtain as complete a picture of a complex, multidimensional phenomenon as possible (Mazziotta and Pareto, 2017). For CEJST, this means combining data on climate change, energy, health, housing, legacy pollution, transportation, water and wastewater, and workforce development. Within each burden category, multiple indicators are used, many of which use different measurement units. A measurement unit is the quantity used by convention to compare or express the quantified properties of specific types of natural, physical, or social phenomena.[1] To meaningfully combine the indicators within these categories, they must have a common, scaled measurement unit to enable comparison or aggregation. Scaling refers to the process of transforming numbers such that they have a specific range of values (e.g., 0 to 100). The term "normalization" is used here as a general term for transforming numbers into common units with a common measurement scale for comparison or aggregation. Box 6.1 illustrates the process of normalization using the transportation category in CEJST as and example.

There is no one-size-fits-all rule about the best way to normalize indicators, but rather, there are trade-offs that need to be understood when choosing the most appropriate approach based on the concept being measured and the goal of the tool. There are several widely used normalization approaches in indicator construction, including min-max scaling, ranking, z-score standardization, distance from a reference, and categorical scales (often achieved using the calculation of percentiles). Mazziotta and Pareto (2017), the OECD and JRC (2008), Freudenberg (2003), and Jacobs, Smith, and Goddard (2004) explore these options and their trade-offs in a level of detail beyond the scope of this report.

The CEJST developers elected to employ a percentile approach to normalization and explained its rationale in its technical support document (CEQ, 2022a). The reasons provided include that percentiles (a) allow for identifying the relative burden that each census tract experiences compared to the entire United States and territories and (b) can be interpreted and easily understood by a broad range of users. Given the importance of building transparency, trust, and legitimacy as described in Chapter 3 framework for indicator construction, a normalization approach that is easily understood and inter-

[1] See National Institute of Standards and Technology, SI Units at https://www.nist.gov/pml/owm/metric-si/si-units. (accessed January 30, 2024).

BOX 6.1
Illustrating the Process of Normalization

The transportation category of CEJST includes three indicators with different measurement units: diesel particulate matter exposure, transportation barriers, and traffic proximity and volume. Diesel particulate matter exposure is measured in micrograms per cubic meter, while the transportation barriers indicator is calculated from the average relative cost and time spent on transportation relative to all other tracts. Traffic proximity and volume are measured as the number of vehicles on major roads within 500 meters divided by distance in meters. To combine these indicators into a composite, one might try adding or multiplying diesel particular matter exposure, measured in particulate matter per cubic meter (range of 0–1.92), with the traffic proximity and volume indicator, measured as the number of vehicles at major roads divided by distance in meters (range of 0–42,063.59). However, the large range of traffic proximity values would overpower the much lower range of particulate matter values, leading to an unintentional higher weighting of traffic proximity. While each indicator may be a valid measurement of a burden experienced by disadvantaged communities, they cannot be combined in any meaningful way until they have been brought into a common metric with consistent scaling because they are measured in different units and can have vastly different value ranges (Freudenberg, 2003; Mazziotta and Pareto, 2017).

preted can be valuable. The CEJST technical documentation (CEQ, 2022a) features substantial description and justification of how the percentile normalizing approach contributes to those end goals.

However, the CEJST technical documentation (CEQ, 2022a) does acknowledge the disadvantages of using percentiles for normalization. It can be ineffective for indicators that represent measurements that are poorly reflected on a linear scale, such as a bimodal distribution where there is a gap between the "good" and "bad" values of a given indicator. Percentiles can also mask or amplify the magnitude of difference between values. For example, CEJST uses the 90th percentile threshold on the indicator of low median income to determine if a census tract is designated as disadvantaged in the workforce development category. By this logic, a census tract at the 90th percentile is not considered any less disadvantaged than a census tract at the 99th percentile. By contrast, a census tract at the 89th percentile is not considered any more disadvantaged than one at the 0th percentile (i.e., the lowest level of burden). Using percentile normalization and thresholds for designating disadvantaged tracts has two important policy implications. Although the differential degree of burden experienced in tracts in the 89th and 90th percentiles may be slight, the resulting differential access to policy resources may be large. Moreover, the combination of data measurement uncertainties associated with input indicators (e.g., census margin of error) and the aforementioned propensity of percentile normalization to mask differences in magnitude can diminish the capacity of percentiles to distinguish degrees of disadvantage.

Each normalization technique has strengths and weaknesses, and certain indicators and phenomena may benefit from choosing one approach over another. However, the resulting composite indicator is generally less affected by the choice of normalization algorithm than other decisions in the composite indicator construction process (Freudenberg, 2003; Tate, 2012). The lower influence does not mean that the normalization decision does not matter or that robustness testing around normalization techniques is not valuable. It does mean that if the decisions are considered and well documented, they are less likely to significantly influence the resulting composite indicator than other steps in the construction process.

WEIGHTING INDICATORS

As discussed in Chapter 3, once indicators have been chosen and normalized appropriately, the relative importance of each indicator with respect to other indicators in the final composite indicator needs to be determined. As described in Chapter 5, no explicit weighting is employed in CEJST to designate any single indicator as more influential than another on the resulting determination of a census tract as disadvantaged. No indicator is explicitly designated as more important than any other indicator. (Socioeconomic burden is implicitly more important because the socioeconomic threshold must be met in addition to any other indicator for a community to be designated as disadvantaged.) While explicit weighting is not employed in CEJST, it is important to discuss weighting and its impacts when integrating indicators using aggregation, given alternative aggregation approaches that might be applied by CEQ tools in the future. Decisions related to weighting indicators substantially affect the resulting composite indicator, both in terms of its actual values and its acceptance by technical experts and community members with lived expertise (Freudenberg, 2003; Saisana, Saltelli, and Tarantola, 2005).

Explicit Weighting

The calculation of a composite indicator can employ explicit weighting whereby individual indicators (or subindexes) are weighted during the aggregation process such that some indicators have a larger impact on the resulting composite indicator than others. Unequal weights are intended to reflect the differential importance of factors contributing to a composite measure (e.g., disadvantaged community [DAC] designation in the context of CEJST). The degree of relative importance can be derived from numerous sources, including policy objectives, scientific understanding, statistical methods, participatory approaches, and community preferences. Mathematically, weights are typically applied by simply multiplying an indicator before aggregating it with other weighted indicators, thus increasing or decreasing the relative influence of each indicator on the resulting composite measure. There is an art and a science to creating composite indicators and weighting. The methods applied are rarely clearly right or wrong but rather a series of trade-offs and value judgments. The stated goal of the indicator and the definition of what is being measured can act as guiding principles when making these decisions, and the validity of the decisions can be tested through analysis and community engagement.

Many composite indicators do not have explicit weights assigned to each indicator (Freudenberg, 2003), and few of the environmental justice tools described in Chapter 4 use an explicit weighting scheme. Whether explicit (i.e., assigned coefficients) or implicit (hidden statistical structure), weighting is always present. Applying equal thresholds for each environmental indicator in CEJST is, in effect, a decision that each indicator equally contributes to disadvantage. Box 6.2 demonstrates how to explore the impacts of explicit weighting using CEJST downloadable data as an example.

Two common approaches for determining appropriate weights are participatory approaches and data-driven weighting approaches. Participatory approaches can be valuable when the goal of the indicator is to reflect the lived experiences of communities as accurately as possible (Mitra et al., 2013). It is important to the process to receive input from subject-matter and community experts on the impact of the chosen indicators on all their outcomes (discussed further in Chapter 7). There are also many data-driven weighting approaches based on the mathematical characteristics of the chosen indicators.

A common data-driven approach is factor analysis (FA) or principal components analysis (PCA) (Freudenberg, 2003; Mazziotta and Pareto, 2017; OECD and JRC, 2008). The OECD manual points out that while PCA and FA can be valuable for dealing with highly correlated indicators, they are not appropriate when the goal is to measure the theoretical importance of the indicators (OECD and JRC, 2008). The details of each of these methods are beyond the scope of this report but are described in Freudenberg (2003), Mazziotta and Pareto (2017), OECD and JRC (2008).

Each weighting approach has advantages and disadvantages, which require considering the trade-offs, particularly in the context of the purpose of creating the composite indicator and how it will be used. A challenge of explicit weighting is reaching a consensus among interested and affected parties and subject-matter and community experts about the weights used.

Implicit Weighting

Whereas explicit weighting represents a value judgment or policy choice about the relative importance of individual indicators, implicit weighting can occur due to the internal structure of the composite indicator. The magnitudes of intercorrelations and variances among indicators and the arrangement of indicators within the composite can affect the statistical importance of individual indicators. The result is a distortion of the relative importance imposed by explicit weights (Becker et al., 2017). Even when each indicator is assigned an explicit weight of 1, unintended implicit weighting may still occur within the composite indicator. Detecting implicit weighting is often conducted using Pearson correlation matrixes, which show the correlation coefficients between each pair of indicators. The idea is to use intercorrelations to evaluate the degree of

BOX 6.2
Exploring the Impact of Explicit Weighting

The "total categories exceeded" variable included when downloading CEJST "communities list data"[a] can be used to explore the effect of explicit weighting. This variable is the count of the number of burden categories where at least one indicator was exceeded for any given census tract. If a census tract meets the criteria for being designated as disadvantaged based on indicators in a single burden category, it would have a "total categories exceeded" value of 1. If a census tract met the criteria based on indicators in all categories, it would have a value of 8. Although not used to determine if a tract is designated as disadvantaged, this count variable can be used to visualize and understand the spatial distribution of the total categories exceeded variable. Figure 6.2.2 shows each census tract designated as disadvantaged, with darker colors representing tracts that are disadvantaged in more categories. This is currently the default rendering for the Justice40 dataset as shared as part of Esri's Living Atlas,[a] a common place for the GIS community to access data and maps.

With no weighting, each category contributes to the resulting count equally, so the values range from 0 to 8, with each category adding either 0 or 1 to the count. If explicit weighting had been used, and, for example, the climate change category was weighted twice as high as the other categories, then the climate change category would contribute 2 to the resulting count instead of 1. Under these explicit weighting conditions, a census tract exceeding health and housing categories would have a count of 2 (and a tract that only exceeded the climate change category would also have a count of 2). The climate change category would have a larger influence on the resulting composite indicator than the other categories. Using this type of weighting can ensure that a composite indicator reflects the concept being measured, but it requires careful thought. Weights are multiplied by each indicator before aggregation when using an additive approach. When using a multiplicative approach, weights are used as an exponent for each indicator before aggregation, where each indicator is raised to the power of the weight (Mazziotta and Perota, 2017; OECD and JRC, 2008).

alignment between each indicator and the concept to be measured. The results can be used to filter out indicators exhibiting low statistical coherence. Indicators within the same dimension of the concept to be measured should ideally have moderately high positive correlations, while those with very high positive correlations signal potential redundancy (Sherrieb et al., 2010). Indicators with statistically insignificant or low correlations signify indicators that are potentially incoherent with the conceptual framework. An example correlation using a Pearson correlation matrix is provided as Box 6.3.

Correlation matrixes can also be helpful when evaluating the choices of indicators and an appropriate organizational structure. The current model structure of CEJST uses burden categories for thematic organization, but the categories serve no algorithmic purpose in determining DACs. If the aim of the burden categories is to ensure that

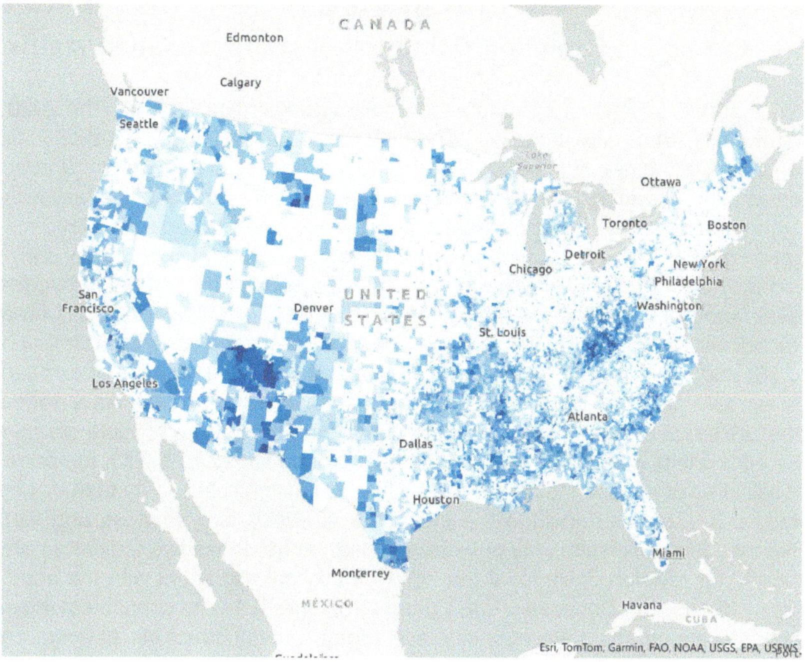

FIGURE 6.2.2 U.S. census tracts designated by CEJST as Disadvantaged Communities, symbolized using the count of the number of categories exceeded. Dark blue tracts exceeded more categories than light blue tracts. SOURCE: Esri "Justice40 by Number of Categories Map" (2022). (accessed January 30, 2024).

[a] See CEJST Downloads at https://screeningtool.geoplatform.gov/en/downloads (accessed January 30, 2024).

[b] Esri's Living Atlas website: https://livingatlas.arcgis.com/en/home/ (accessed February 29, 2024).

the output composite indicator best reflects the concept being measured, alternative approaches to using nominal categories could be considered. This could include using true mathematical subindexes that have undergone statistical evaluation, as discussed above. The evaluation of an appropriate approach to categories includes considering if the structure of the composite indicator is coherent with the concept to be measured, both thematically and mathematically.

When indicators with correlation coefficients outside the desired range are present, there are several potential solutions. The first is to consider different or differently calculated indicators that better align with the conceptual framework (Sherrieb, Norris, and Galea, 2010). One recommended approach is to avoid aggregating negatively correlated indicators (Lindén, 2018). Techniques such as FA or PCA can be applied to statistically

BOX 6.3
Example Correlation Using a Pearson Correlation Matrix

Figure 6.3.1 shows a Pearson correlation matrix of indicators in the 2020 Environmental Performance Index (EPI), which compares countries based on indicators organized into dimensions of climate mitigation, environmental health, and ecosystem vitality (Papadimitriou, Neves, and Saisana, 2020). Because they are within the same dimensions, the indicators assigned to the environmental health dimension (green background in Figure 6.3.1) should be positively and significantly correlated, and they are. The Joint Research Centre (JRC) audit of the 2022 EPI suggests an ideal range of correlation coefficients of 0.3 to 0.92 (Smallenbroek, Caperna, and Papadimitriou, 2023).[a] By this rule of thumb, the correlation coefficients for air quality (0.96) and water quality (0.96) in the environmental health dimension exceed 0.92 (green text in Figure 6.3.1) and are thus overweighted or potentially redundant. The presence of high implicit weights is often referred to as "double counting." In the ecosystem vitality dimension (blue background in Figure 6.3.1), the correlation coefficients of several indicators are statistically insignificant (gray text) or negative (red text). Examples include the indicators of ecosystem services and fisheries. Because the correlations fall outside the ideal range, these two indicators are candidates for revision or removal due to low statistical alignment with the ecosystem vitality dimension. Because CEJST differs from the EPI in its constituent indicators, spatial scale, spatial heterogeneity, and organizational structure, different correlation thresholds may be better suited for evaluating implicit weighting and statistical coherence (Smallenbroek, Caperna, and Papadimitriou, 2023).

collapse multiple highly correlated indicators into a single multidimensional factor that avoids overweighting what they represent (Mazziotta and Pareto, 2017; OECD and JRC, 2008). Another option is to organize highly correlated indicators into subindexes, also referred to as dimensions, for the calculation of a composite indicator.

A subindex is a group of indicators within a composite indicator that have been combined to create an intermediate-level indicator. Subindexes are then combined to create the overall composite indicator (essentially forming composite indexes within the overall composite indicator). For example, if both diabetes and heart disease indicators are included in a composite indicator and are not combined into a subindex, this dimension of health would have a greater influence on the resulting composite indicator due to implicit weighting. Combining the two health indicators into a subindex would limit their outsized impact on the overall composite indicator. However, while using subindexes can address undesired implicit weighting, they can also be sensitive to the same challenges (e.g., if one subindex includes 2 indicators and another has 10, the values

	EPI	Environmental Health	Air Quality	Water Quality	Heavy Metals	Waste Management	Ecosystem Vitality	Biodiversity	Ecosystem services	Fisheries	Climate Change	Air Pollution	Agriculture	Water Resources
Environmental Health	0.94	1.00												
Air Quality	0.89	0.96	1.00											
Water Quality	0.92	0.96	0.86	1.00										
Heavy Metals	0.79	0.83	0.81	0.77	1.00									
Waste Management	0.79	0.81	0.73	0.77	0.66	1.00								
Ecosystem Vitality	0.90	0.69	0.63	0.68	0.59	0.62	1.00							
Biodiversity	0.49	0.29	0.32	0.24	0.28	0.23	0.66	1.00						
Ecosystem Services	-0.06	-0.11	-0.17	-0.02	-0.20	-0.18	0.03	-0.23	1.00					
Fisheries	-0.06	-0.02	-0.03	0.00	0.04	0.00	-0.12	-0.22	0.02	1.00				
Climate Change	0.82	0.70	0.63	0.70	0.59	0.68	0.83	0.26	-0.08	-0.18	1.00			
Air Pollution	0.72	0.63	0.57	0.63	0.55	0.53	0.72	0.29	-0.14	-0.04	0.71	1.00		
Agriculture	0.35	0.26	0.22	0.28	0.22	0.29	0.40	0.26	-0.19	-0.23	0.29	0.29	1.00	
Water Resources	0.82	0.80	0.78	0.76	0.72	0.70	0.69	0.35	-0.14	-0.03	0.59	0.61	0.34	1.00

FIGURE 6.3.1 Example correlation matrix of scaled indicators from the Environmental Performance Index (EPI), organized by composite indicator dimension. Background color indicates the EPI dimension: environmental health (green) and ecosystem vitality (blue), while text color indicates the degree of alignment with the concept to be measured: redundant (purple), good (black), and poor (gray, red). SOURCE: Papadimitriou, Neves, and Saisana, 2020.

[a] The Joint Research Center of the European Commission routinely conducts statistical audits of public policy indices. See https://commission.europa.eu/about-european-commission/departments-and-executive-agencies/joint-research-centre_en (accessed March 10, 2024).

of the individual indicators of the subindex with two indicators influence the resulting composite index more than the individual indicators in the subindex with 10 indicators).

The use of burden categories for thematic organization, but not mathematical subindexes, has implicit weighting implications. If the goal is for each burden category to have equal importance for the resulting designation of tracts, then the number of indicators within each category must be considered. Table 6.1 illustrates the share of indicators within each category of burden. Based on the methodology currently employed in CEJST, the categories of climate change, housing, and legacy pollution have 2.5 times the share of indicators used to designate a census tract as disadvantaged. This is not inherently right or wrong, but understanding the statistical impact of these decisions and justifying them is crucial to ensure that the resulting composite indicator is representative of the concept being measured. Mathematical subindexes would also benefit from a correlation analysis to help assess the alignment of the subindexes with the conceptual design of the composite index. If each subindex is intended to represent

TABLE 6.1 Number and Share of Indicators per Burden Category.

Burden Category	Number of Indicators	Share of indicators
Climate change	5	16.6% (5/30)
Energy	2	6.6% (2/30)
Health	4	13.3% (4/30)
Housing	5	16.6% (5/30)
Legacy pollution	5	16.6% (5/30)
Transportation	3	10% (3/30)
Water and wastewater	2	6.6% (2/30)
Workforce development	4	13.3% (4/30)

a different dimension of disadvantage, then highly correlated subindexes would require further investigation to ensure that they appropriately align with the intended design. For this reason, the conceptual framework outlined in Chapter 3 illustrates the importance of calculating a composite indicator as an iterative process, where decisions on integrating indicators, for instance, might lead to reevaluating the selected indicators and the organizational structure of those indicators.

The current CEJST technical documentation (CEQ, 2022a) does not mention any correlation analysis undertaken. Documenting any such analysis within the technical documentation would be valuable for interested and affected parties when evaluating the resulting DACs. If the analysis has not been done, performing and documenting the assessment would add rigor to the resulting designation of tracts and increase methodological transparency. The CEJST Technical Support Document (CEQ, 2022a, p. 7) states that "CEJST burdens are grouped into categories that were informed by Justice40 investment focus areas," citing the White House Office of Management and Budget (OMB) Memorandum M-21-28 and Interim Implementation Guidance for the Justice40 Initiative (EOP, 2021), but does not offer further elaboration. The documentation also does not explicitly discuss the purpose of the burden categories, whether organizational or methodological.

Overall, indicator weights have significant impacts on the resulting composite indicator. If aggregation (discussed in the next section) will be performed, it is important to consider how each indicator will be weighted during the aggregation process. Explicit weights are essentially a value judgment, while implicit weights are data driven but can be poorly understood. Deriving explicit weights is complicated and sometimes contentious because they represent the relative importance of the different facets of the concept. As such, consensus on explicit weights can be difficult to achieve. Consequently, there is a tendency to apply equal weights to indicators and subindexes. Meanwhile, there is less frequent exploration of implicit weighting. This increases the possibility of producing a composite indicator with obscured statistical redundancy, input indicators with weak statistical alignment with the phenomenon of interest, and differential weights due to a varying number of indicators across categories. Correlation analysis can be applied to evaluate coherence with the concept to be measured, while subindexes can be used to reduce bias introduced by uneven categories (Greco et al., 2019). With any weighting scheme, sensitivity analysis with the weights and visualization of individual indicators and subindexes can improve the understanding of the impacts of weights on model outputs and suggest remedies (Albo, Lanir, and Rafaeli, 2019; Becker et al., 2017; Dobbie and Dail, 2013; Räsänen et al., 2019).

INTEGRATION APPROACHES

Once the indicators have been normalized and weights have been chosen, the next step in constructing a composite indicator is to combine the weighted indicators into the output composite indicator. Two common integration approaches are threshold-based and aggregation approaches. As already described, CEJST employs a threshold-based approach based on a series of baselines that are used to decide if a census tract will be designated as disadvantaged. By contrast, an aggregation approach combines indicators mathematically, for example, by using additive or multiplicative approaches, to produce a continuous (i.e., noninteger) composite indicator. Aggregation approaches will be discussed in detail later in this chapter. Although the CEJST technical documentation (CEQ, 2022a) states that "disadvantaged communities face numerous challenges because they have been marginalized by society, overburdened by pollution, and underserved by infrastructure and other key services," the implementation of CEJST stops short of reflecting those numerous challenges in their designation of census tracts as disadvantaged.

While each census tract has multiple opportunities to be considered disadvantaged based on the 30 individual input indicators (described in Chapter 5), the cumulative nature of burdens (or stressors) is not accounted for in a manner that is consistent with the description of disadvantaged communities above. Although each indicator is assessed based on both meeting the threshold for that indicator in addition to a socioeconomic burden, the cumulative nature of each category of burden (and even within a particular category of burden) plays no role in the determination of DACs. Further, the current designation logic obfuscates differences in the degree of cumulative impacts among the tracts designated as disadvantaged. This limits the capacity of CEJST to identify and prioritize the communities most in need. Redressing cumulative impacts is a policy objective of the White House (EOP, 2021), and the desire for CEJST to incorporate cumulative impact scoring is an objective of the CEQ and a common request of environmental justice advocates and analysts (CEQ, 2022a; WHEJAC, 2021). The closest that the current approach gets to cumulative impact scoring is by counting and recording the indicator criteria exceeded and the categories exceeded. As described in the discussion on weighting, these variables are provided as part of the data download, but they are not used in the final designation of tracts as disadvantaged.

Figure 6.1 shows the frequency distribution of each measure in DACs based on data from CEJST, with the count of indicator criteria exceeded ranging from 1 to 18 (mean = 4.3) and burden category criteria exceeded ranging from 1 to 8 (mean = 3.0). The mode in each cumulative scoring scenario is one, comprising only 20 percent of the DACs based on thresholds exceeded and 24 percent of the DACs based on burden categories exceeded. This means that within the collection of DACs, there are roughly three times as many DACs with multiple burden categories exceeded than with only one exceeded and four times as many tracts with multiple criteria thresholds than with only one. Hence, most DACs are subject to multiple forms of environmental disadvantage. This represents the proportion of census tracts that were designated as disadvantaged based on a single burden indicator or a single category of burdens. As described earlier in the weighting section, the count of categories can also be used to map and explore the number of categories that any given tract exceeded. Tracts that exceed the thresholds within more than one category are likely to experience more cumulative impacts than those that exceed one category. Employing such discrete scoring to model cumulative impacts would require no change to the existing CEJST computational structure but would result in a loss of information from percentile indicator values. It also amplifies the importance of the 90th percentile (burdens) and 65th percentile (income) thresholds.

An integration approach that uses aggregation is a common means to ensure that the presence of multiple stressors or the interaction between stressors is reflected in the resulting index. There are two commonly used aggregation approaches to reflect cumulative burden when calculating composite indicators: additive and multiplicative (OECD and JRC [2008] refer to these as linear and geometric, respectively). These aggregation approaches are more consistent with the research on cumulative burden and the emerging understanding of these issues, discussed in Chapter 2, accounting for the magnitude of stressors, the presence of multiple stressors, and the interactions across stressors. Additive aggregation methods commonly include sum and arithmetic mean, while multiplicative approaches commonly include a product and geometric mean. The technical details of these approaches are documented by Freudenberg (2003), Mazziotta and Pareto (2017), and OECD and JRC (2008) and are beyond the scope of this report. However, some important considerations of potential alternative approaches to be considered for CEJST are provided in the following section.

Existing EJ tools that incorporate cumulative scoring do so via an aggregation approach. For example, CalEnviroScreen[2] employs a weighted sum of normalized indicators within two subindexes of pollution burden and population characteristics. The subindex scores are then multiplied to compute the index (CalEnviroScreen score), with tracts landing within the top quartile (25 percent) designated as disadvantaged. Maryland's EJScreen Mapper[3] adopts a similar mathematical approach. Its two pollution burden subindexes are averaged, as are its two subindexes of population characteristics. The resulting values are then multiplied to compute the index (EJScore). As opposed to the binary classification of indicator criteria or burden categories exceeded in CEJST, these aggregation approaches generate continuous values. If an aggregation methodology was used in CEJST, a threshold for DAC designation would be applied to the aggregation of conditions (like CalEnviroScreen) instead of to individual indicators.

---

[2] See "CalEnviroScreen 4.0." California Office of Environmental Health Hazard Assessment, https://oehha.ca.gov/calenviroscreen/report/calenviroscreen-40 (accessed February 25, 2024).

[3] See "MDE's Environmental Justice Screening Tool Beta 2.0." Maryland Department of the Environment. https://mde.maryland.gov/Environmental_Justice/Pages/EJ-Screening-Tool.aspx (accessed February 25, 2024).

Such an aggregation scheme for scoring aligns more with cumulative impacts than the criteria-based one.

Considerations

When choosing an aggregation approach, a primary consideration is whether a high value in one indicator should offset or compensate for a low value in another indicator. This concept is known as compensability, and it drives many decisions related to aggregation (Mazziotta and Pareto, 2017). Compensability is defined by the OECD manual as "the possibility of offsetting a disadvantage on some indicators by a sufficiently large advantage on other indicators" (OECD and JRC, 2008). Additive approaches for aggregation are generally compensatory, while multiplicative approaches are generally partially noncompensatory.

Table 6.2 offers a hypothetical comparative example of how the aggregation approach affects the output index value. Community A has three indicators, all with a value of 2, and Community B has three indicators with values of 5, 0.5, and 0.5. With an additive approach such as a sum, Community A has a composite index value of 6, and so does Community B. With a multiplicative approach such as a product, Community A has an index value of 8, while Community B has an index value of 1.25. In the additive approach, despite Community B having two indicators with significantly lower values, the higher value for Indicator 3 compensates for the low values and leads to an index value in the middle. With the multiplicative approach, the two low indicators in Community B lower the overall index value substantially. This is because with a multiplicative approach, all values must be high to receive a high index value, and even a small number of indicators with low values will substantially decrease the resulting index.

A feature of noncompensatory approaches is that changes in indicators with low values will have a bigger impact on the resulting index than changes in indicators with a higher value. This means, for example, that a change in an indicator representing park access would have a bigger impact if park access were low. The same change in park access for a community that already has high park access would affect the resulting index less. Both characteristics of noncompensatory (often multiplicative) approaches need to be considered in the context of what is being measured and the goal of the index. In the context of cumulative burden, and specifically the ability of a composite indicator to reflect the presence of multiple stressors and the interaction between stressors, there are clear implications. If a high indicator value can be compensated for by a low

TABLE 6.2 A Hypothetical Aggregation Comparison

Community	Indicator 1	Indicator 2	Indicator 3	Additive Index Value	Multiplicative Index Value
A	2	2	2	6	8
B	5	0.5	0.5	6	1.25

NOTE: Synthetic data are used for illustration purposes.

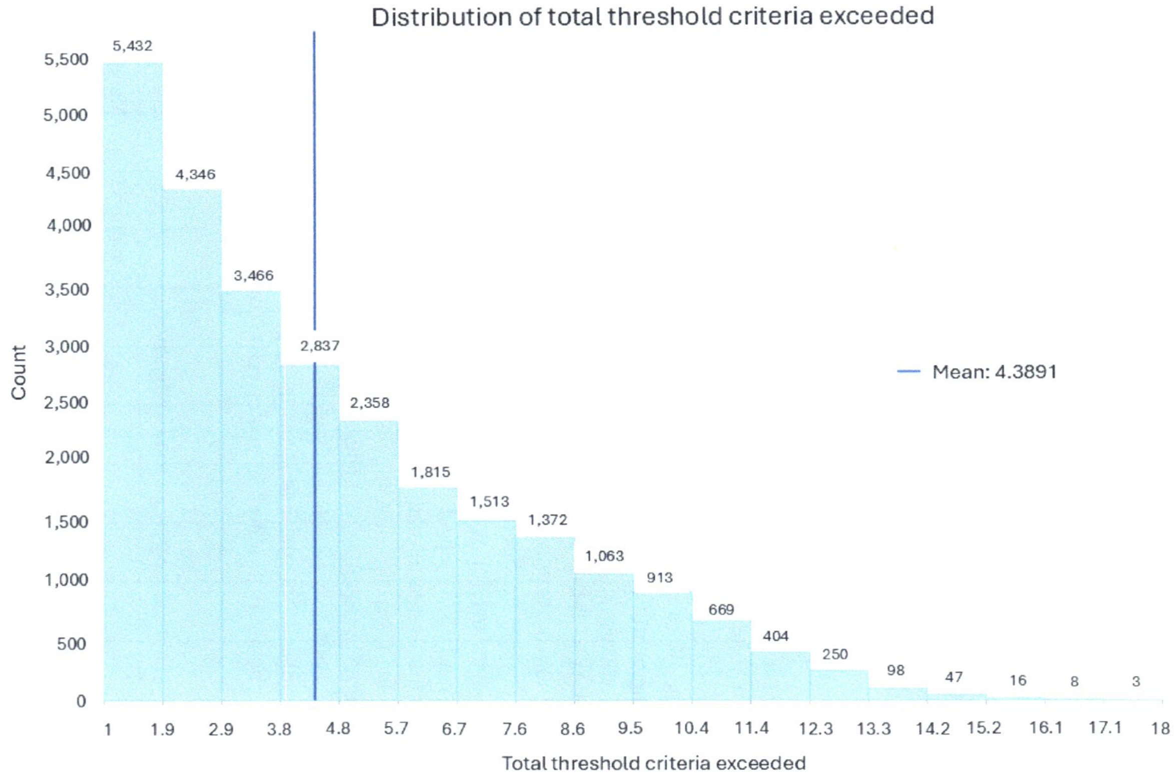

FIGURE 6.1 Frequency distribution of the number of census tracts, with cumulative impacts alternatively represented by the count of indicator threshold criteria exceeded (top), and the count of burden categories exceeded (bottom). SOURCE: Committee's calculation using data from CEJST version 1.0.

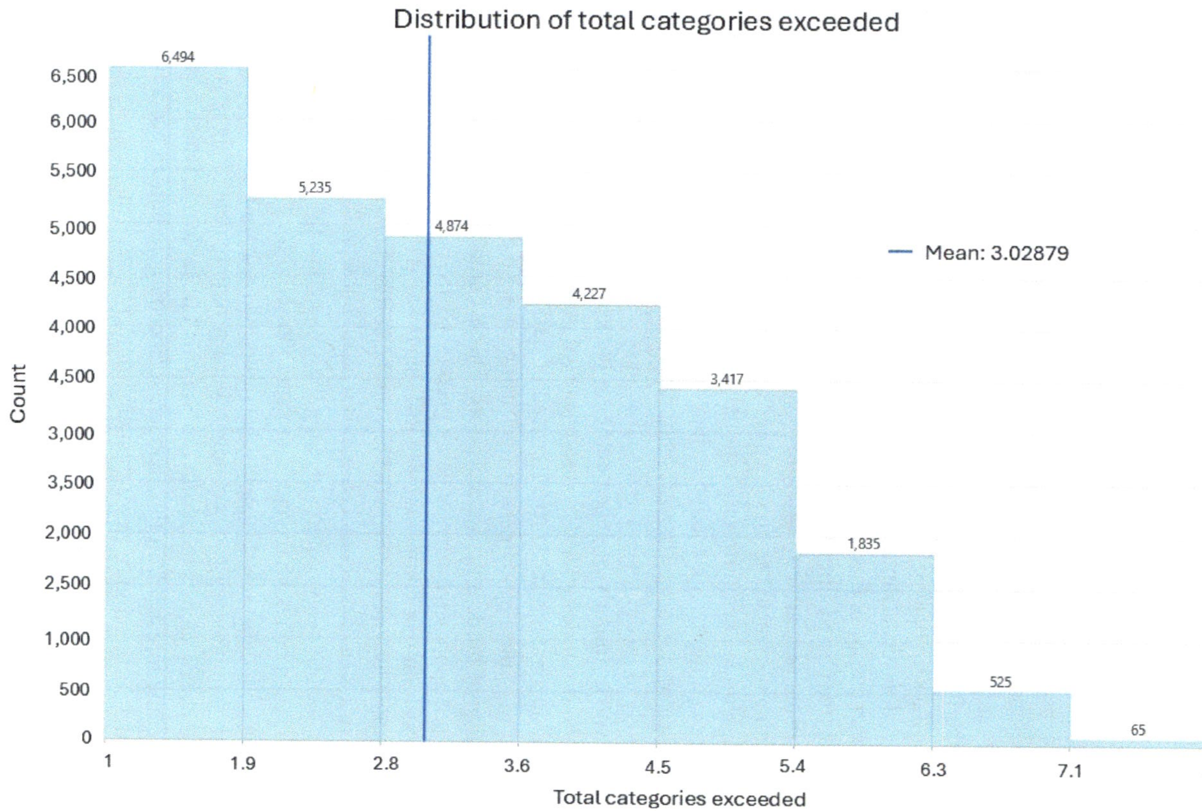

FIGURE 6.1 Continued

indicator value, and changes in low values are more impactful than changes in high values, then the ability to accurately reflect both the presence of multiple stressors and the interaction between stressors may be affected. Alternative approaches to integration that might be applied by CEQ in CEJST or some other tool in the future needs to involve considering the implications of compensability on the resulting composite indicator. For example, if an additive approach were used on the percentiles without an applied threshold, then very low values for a few indicators would dramatically change the resulting composite indicator and could compensate for one or two high indicators.

Table 6.3 explores applying an aggregation-based approach to the calculation for four real Justice40 census tracts where the percentiles for diabetes, asthma, heart disease, low life expectancy, and low income are all aggregated using both a sum (additive, compensatory) and a product (multiplicative, partially noncompensatory). The values for the health category for four distinct census tracts in the CEJST dataset are used to demonstrate the impact of the alternative approaches. The final composite indicator values are based on alternative aggregation approaches (sum and product) to the census tracts designated as disadvantaged based on the health category within the current CEJST implementation. In this example, the presence of a single percentile above the 90th percentile (the CEJST threshold) for any of the environmental indicators combined with a low-income percentile even close to the 65th percentile threshold results in a census tract being designated as disadvantaged, based on the current CEJST logic. The census tract in Florida exemplifies this strongly; the diabetes indicator is high, and the low-income percentile is high enough to meet the threshold, but the other three environmental indicators fall below the threshold. Using an aggregation scheme, the census tract is less likely to be designated as disadvantaged. Alternatively, the North Carolina census tract, with a similarly high percentile for diabetes and higher percentiles for all other indicators, is not designated as disadvantaged under CEJST logic because it just misses the low-income threshold (0.63 instead of 0.65). It would have scored as more disadvantaged than the Florida census tract using an aggregation scheme because of the accumulation of burdens across indicators. This accumulation is not considered in the criteria-based approach currently being used by CEJST.

The demonstration of compensability in Table 6.3 also illustrates the considerable influence that the low-income threshold has on the designation of census tracts as disadvantaged. For seven of the eight categories (i.e., all but workforce development), the current CEJST methodology requires meeting the low-income threshold for designation. A census tract could be above the environmental threshold for every single one of the 26 indicators within those seven categories and still not be designated as disadvantaged because of the simultaneous socioeconomic burden requirement. This is particularly concerning for places such as the North Carolina tract in Table 6.3, which is just below the 65th percentile threshold. In this manner, socioeconomic burden has a far greater influence on the designation of tracts than environmental burden.

TABLE 6.3 Effect of Aggregation Scheme on Burden Thresholds Exceeded

	Category Indicator (CEJST percentile)							
Location (FIPS code)	**Diabetes**	**Asthma**	**Heart Disease**	**Low Life Expectancy**	**Low Income**	**Sum**	**Product**	**Indicator Thresholds Exceeded in Health Category (max = 4)**
Louisiana, Orleans Parish (22071014000)	0.99	0.97	0.99	0.93	0.99	4.87	0.88	4 (designated disadvantaged)
Alabama, Houston County (01069040700)	0.87	0.83	0.87	0.84	0.62	4.03	0.33	0
North Carolina, Forsyth County (37067001601)	0.97	0.86	0.82	0.89	0.63	4.17	0.38	0
Florida, Miami-Dade County (12086000410)	0.96	0.85	0.76	0.34	0.68	3.59	0.14	1 (designated disadvantaged)

NOTE: Darker boxes represent higher scores based on aggregation or thresholds. Federal Information Processing Standard (FIPS) code is represented in parentheses below the location in the first column.
SOURCE: Committee calculation based on CEJST 1.0 data.

Using Composite Indicators to Identify Disadvantaged Communities

A composite indicator calculated using an aggregation approach is almost always a continuous value. The example from Table 6.3 shows several calculations of continuous values for each census tract that could result from using an aggregation approach on the CEJST indicators. However, sometimes the goal of a composite indicator is to produce a binary categorization (e.g., a CEJST designation of disadvantaged or not). While the CEJST approach to integration is threshold based with a binary outcome, choosing to use an aggregation-based approach would not preclude a final post-processing step that allows for census tracts to be ultimately designated as disadvantaged.

For the aggregation sums and products of the four tracts in Table 6.3, a potential post-aggregation step could be a transformation of those sums or products to percentiles. Then, designation as disadvantaged could be based on the percentiles of those aggregated values. Similarly, a cutoff value could be applied to the final range of values based on subject-matter expertise, community input, or other methods. This would allow for an aggregation approach in the calculation of CEJST that results in both a continuous value representing cumulative burden and a designation of tracts as disadvantaged based on that continuous value. CalEnviroScreen is an environmental justice screening tool that uses this approach. After multiplicative aggregation, it designates tracts scoring in the top 25th percentile as disadvantaged.[4]

Any resulting composite indicator will be sensitive to choices made during normalization, weighting, aggregation, and post-processing, including the interactions between them. For example, careful consideration is necessary to understand the interaction between the aggregation approach chosen, the subindexes or categories of indicators being used, and the relationships between individual indicators. For this reason, a well-constructed composite indicator will need to be assessed using uncertainty and sensitivity analyses. Ideally, those analyses evaluate the impact of changes at each decision point to ensure that the goal of the composite indicator is being met. This type of assessment of robustness, along with thorough and transparent documentation of approach decisions and the impacts of those decisions, can also help build trust around the resulting composite indicator among users, decision makers, and community members.

ASSESSING ROBUSTNESS

Assessing the influence of uncertainty sources on policy benefits and costs provides valuable information to decision makers and the public (OMB, 2023). Model outputs are generally associated with two types of uncertainty: aleatory and epistemic (NRC, 1996). Aleatory uncertainty stems from intrinsic and unpredictable randomness in the phenomenon being modeled. Examples relevant to environmental justice (EJ) tools include ambient air pollution and disease incidence. Aleatory uncertainty is irreducible and is therefore represented in models using stochastic variables (Der Kiureghian and Ditlevsen, 2009). Epistemic uncertainty stems from incomplete knowledge of the

[4] See "SB 535 Disadvantaged Communities." California Office of Environmental Health Hazard Assessment. https://oehha.ca.gov/calenviroscreen/sb535 (accessed February 27, 2024).

phenomenon being modeled. An example is imperfect understanding of the nature of interactions among environmental and social processes and the resulting influence on community disadvantage. Epistemic uncertainty can be reduced by integrating better data or improved knowledge into the modeling process. Quantifying and disaggregating uncertainty is the principal focus of robustness analysis for composite indicators.

As has been discussed, the construction of EJ tools, including CEJST, requires numerous modeling decisions for which plausible alternative choices are available. Ideally, such decisions are based on a detailed understanding of the concept to be measured. But with abstract constructs such as disadvantage that stymie direct measurement, composite indicator modelers must often make informed but subjective judgments. These judgments are a source of model-based or epistemic uncertainty; this arises when there is incomplete or imprecise understanding of how well factors such as the model structure, parameter values, spatial resolution, use of expert opinion, and data measurement represent the empirical processes the model is intended to reflect (Helton et al., 2010; Jakeman, Eldred, and Xiu, 2010). High epistemic uncertainty can lead to misalignment of real-world processes with the composite indicators that seek to represent them. Failure to consider and reduce epistemic uncertainty can result in negative policy outcomes and uninformed decision making (Maxim and van der Sluijs, 2011). However, it is often unclear which aspect(s) of a model construction process are the greatest uncertainty contributors.

Robustness analysis is a component of the committee's conceptual framework for composite indicator construction (Figure 3.2) and can be conducted to meet multiple objectives. One objective is to assess the model's statistical soundness or internal validity. Box 6.4 lists different types of analyses that are part of robustness analysis. Placed in the context of EJ tools such as CEJST, robustness assesses how certain the output designation of DAC status is for each census tract. Sensitivity analysis is the principal recommended methodology for assessing model robustness through quantifying overall model uncertainty and identifying the major source(s) of epistemic uncertainty (EPA, 2009). When subjected to sensitivity analysis, a robust model configuration will exhibit only a small change in the output in response to variations among the inputs. Another objective and important benefit of sensitivity analysis is that it can help differentiate the influence that input parameter options have on the composite indicator (e.g., those that greatly influence DAC designation and those that do not). Knowing this can enable the composite indicator designer to focus data collection and methodology development on the input factors that matter most, thereby improving the reliability of the model. Conducting sensitivity analysis on mathematical models is especially important when the model output is used in a policy framework in support of government regulations (Saltelli and Annoni, 2010). Boxes 5.3 and 5.6 in Chapter 5 provide two possible uses of sensitivity analyses in the context of indicator selection in CEJST, examining the $PM_{2.5}$ and low-income variables respectively.

The first step in conducting a sensitivity analysis is determining which parameters will be evaluated (Greco et al., 2019; Munda et al., 2020). Table 6.4 displays examples of eight CEJST model parameters and associated options that could be subjected to sensitivity analysis. The items in boldface italics represent the options selected for the

BOX 6.4
Robustness Analysis Terms Related to Composite Indicator Development

Uncertainty and sensitivity analyses are the leading methods for assessing the robustness of a composite indicator. Robustness analysis is ideally integrated into the construction of a composite index for quality assurance. The following outlines the major aspects of composite indicator robustness analysis and their role in composite indicator construction.

Input parameter (alternatively called input factor in the sensitivity analysis literature) is a stage, step, or component of composite indicator construction. Within each input parameter, composite indicator developers must decide among plausible modeling options or choices.

Uncertainty analysis quantifies the overall variation in composite indicator output(s) based on varying input modeling decisions. Sources of composite indicator modeling uncertainty include parameter choices involving input data, arrangement of indicators into subindexes, imputation of missing values, normalization, weighting schemes, aggregation, and interactions among parameters.

Sensitivity analysis apportions variation in an output indicator to specific modeling decisions. It helps discern which decisions have substantial and minimal influence on the output.

Local sensitivity analysis evaluates the effect on the response of an output indicator to variations of a single input parameter. It is conducted by varying options within the focal parameter one at a time in calculating the composite indicator, while other parameters are held constant. The degree of local sensitivity is typically evaluated by examining statistical correlations or quantifying the percent change between realizations of the composite indicator.

Global sensitivity analysis evaluates the response of an output composite indicator to simultaneous variations among multiple modeling parameters. Uncertainty analysis is first done to determine the probabilistic distribution of the composite indicator, typically using Monte Carlo simulation. Variance-based sensitivity analysis is then applied to apportion the uncertainty to individual modeling parameters and uncover interactions among them.

construction of CEJST. This would be the baseline model configuration. The sensitivity analysis can determine which parameters have the greatest influence on determining whether a census tract is a DAC and which have the least. There are numerous methodological approaches for sensitivity analyses. The most common approaches are local (one at a time) and variance-based global sensitivity analysis (Saltelli et al., 2008). In local sensitivity analysis, the response of the output variable to variation of a single input parameter is evaluated by varying options within the parameter one at a time, while other parameters are held constant (Saltelli and Annoni, 2010; Xu and Gertner, 2008). The evaluation is straightforward to implement and often employs statistical correlation. If the resultant correlation coefficients are high, then one might conclude that the model is robust to variation in the parameter of interest.

Figure 6.2 presents an example schematic of a local sensitivity analysis to assess the robustness of the margin-of-error parameter for CEJST's income variable. Plausible discrete options within the error-margin parameter are evaluated while other parameters (e.g., indicator set, environmental burden threshold) are held constant. In this example, the output measure of interest is the percentage of disadvantaged tracts (*Y*). If the model is robust to changes in the margin of error, *Y* will change little across options. Correlation analysis or the percentage change in *Y* can be used to evaluate the alignment of *Y* across modeling options. The output variable *Y* could also be mapped to identify the tracts for which DAC designation remained constant and for which it varied. The analyst could then focus on the tracts with varying designations to examine if the model is performing as intended. See Jones and Andrey (2007) for a detailed example of local sensitivity analysis.

Local sensitivity tests are well suited for evaluating a single source of model uncertainty. However, they ignore and are unable to detect interactions among uncertainty sources that manifest not independently but jointly (Saltelli and d'Hombres, 2010; Tarantola et al., 2024). Local tests also become inefficient when the number of uncertain factors is large, but only a few are influential. In these instances, the ability to simultaneously vary multiple input parameters would be useful to interrogate the robustness of the current CEJST configuration, as well as potential future changes to CEJST, such as adding new variables or considering alternative percentile thresholds. Global uncertainty and sensitivity analysis are better suited to evaluate the sensitivity of a model to uncertainties in multiple input parameters.

In applying global sensitivity analysis to indicator construction, the objective is to quantify how variation in model output is apportioned to different sources of variation in the input assumptions (Saisana and Saltelli, 2008). This variation, or epistemic uncertainty, arises from subjective decisions in the selection among options for each input parameter. Uncertainty in these decisions propagates through the composite indicator construction to exert a combined impact on the output(s) (Saisana, Saltelli, and Tarantola, 2005). In global sensitivity analysis, options within multiple model construction phases are varied simultaneously via a two-stage process. First, uncertainty analysis is used to quantify the overall variation in model output based on varying modeling decisions. Sensitivity analysis is then applied to apportion the output variance to specific decisions (Razavi et al., 2021). Variance-based global sensitivity analysis is considered

TABLE 6.4 Potential Sources of CEJST Modeling Uncertainty

Parameter	Example Options
Variable selection	***Current 30 variables*** Alternative variable set Alternative variable set #2
Measurement error on census variables	***Ignore*** Sample within range
Imputation for missing data	None ***County average*** State average Regression Spatial interpolation
Normalization method	***Percentile*** Min-max scaling Distance from a reference Z-score standardization Ranking
Environmental burden threshold	Value less than the 90th percentile ***90th percentile*** Value greater than the 90th percentile
Socioeconomic burden threshold	Value less than the 65th percentile ***65th percentile*** Value greater than the 65th percentile
Cumulative burden method	***None*** Count of indicator criteria exceeded Count of burden categories exceeded Additive aggregation Multiplicative aggregation
Disadvantaged community designation	***Count threshold =1*** Count threshold greater than 1 Aggregation percentile threshold = 25 Aggregation percentile threshold = 30 Aggregation percentile threshold = 35

to be the gold standard for assessing the effects of uncertain model inputs on the variability of model outputs (Borgonovo et al., 2016; Saltelli et al., 2019).

In global uncertainty analysis, the model construction algorithm is subjected to a bootstrap analysis, in which the epistemic uncertainty associated with each decision parameter is propagated through the model. Monte Carlo simulation is employed to construct the model repeatedly, with each iteration generating the model output based on a random selection of the input parameter options. Instead of a discrete output value for each analysis unit (e.g., census tracts), the output of the Monte Carlo simulation is a probability distribution with a discrete mean, median, variance, range, and confidence interval. The Office of Management and Budget recommends reporting these statistics when conducting probabilistic uncertainty analysis for regulatory purposes (OMB, 2023). With use of the example in Figure 6.2, the probability distribution for CEJST

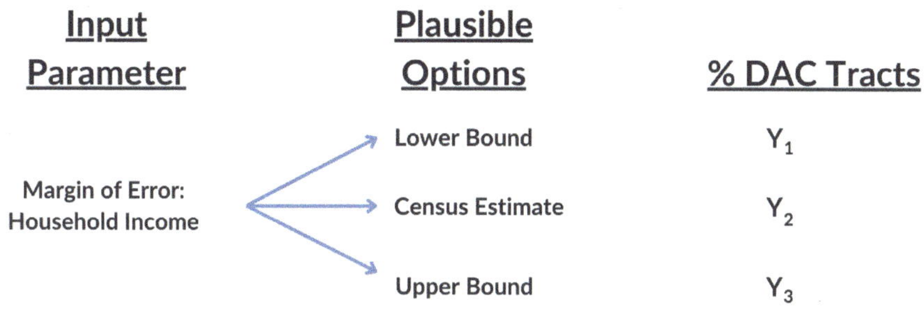

FIGURE 6.2 Example local sensitivity analysis scheme. Each modeling option produces a different output value (*Y*), which is then subjected to correlation analysis to assess robustness of the single input parameter.

could determine the mean, range, and standard deviation of the percentage of census tracts designated as a DAC. Low values of the range and standard deviation would suggest a robust model, while high values would indicate model fragility. The mean value could then be compared with that of the CEJST configuration to assess if it is an outlier compared to the broader universe of potential configurations.

The results of the analysis could also be computed for individual analysis units. Potential descriptive statistics for the percentage of all census tracts designated as disadvantaged (percent DAC) output include the range, median, mean, and standard deviation. Figure 6.3 shows the results of a global uncertainty analysis from the JRC statistical audit of the 2022 EPI (Smallenbroek, Caperna, and Papadimitriou, 2023). The analysis evaluated the effect of varying methodological approaches for normalization, weighting, and aggregation on the output ranking of countries as the unit of analysis. The figure displays the median (blue dots) and 95th confidence interval (vertical lines) of the rankings of each country when all input modeling parameters are simultaneously varied in a Monte Carlo simulation. The countries are ordered from left to right in Figure 6.3 based on their ranking in the EPI. The confidence intervals show that the stability of the index rankings varies substantially across countries, with rankings robust for highly ranked countries but fragile in many other cases. To evaluate the robustness of the index, the audit's authors deemed a country's ranking unstable if the confidence interval exceeds 10 ranking positions, as is the case for 40 percent of the countries. This example demonstrates the utility of uncertainty analysis for both quantifying overall composite indicator fragility to alternative modeling decisions and identifying which analysis units have the most and least reliable ranks.

For CEJST, the large number of census tracts (>80,000) suggests mapping the uncertainty analysis outputs as opposed to charting them. This process of computation and geovisualization in uncertainty analysis would help discern the places with robust and fragile DAC designation. Consider the results of an uncertainty analysis in which overall variance is moderate, but high (low robustness) for disadvantaged communities. The results would help analysts zero in on which places have large uncertainty in the stability of DAC membership and further explore their environmental and demographic

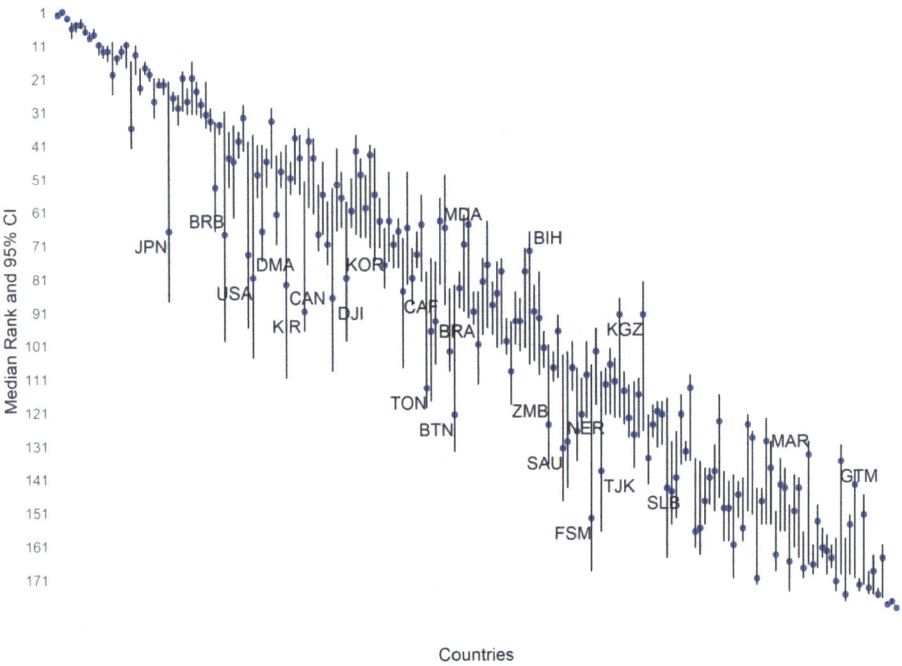

FIGURE 6.3 Example global uncertainty analysis results. The countries, indicated by three letter acronyms (defined in the figure source), are ordered left to right by their index ranking, with the median ranking (blue dots) indicating frequent divergence from the index and the confidence interval (vertical lines) indicating low uncertainty for highly ranked countries and moderate-to-high uncertainty elsewhere. SOURCE: Smallenbroek, Caperna, and Papadimitriou, 2023.

characteristics. Within this smaller subset, other analysis options become available. A common approach is using a scatter plot (Greco et al., 2019), with the output (e.g., mean percent DAC) on the *y* axis and the uncertainty range on the *x* axis. Conversely, for tracts determined to be robust in the uncertainty analysis, there would be greater confidence in their DAC designation.

After epistemic uncertainty is assessed for the entire model or specific geographical areas, the next step is sensitivity analysis. Although uncertainty analysis can be used to estimate the overall degree of variability, it cannot determine which modeling decisions are the greatest contributors. Sensitivity analysis is thus applied to decompose the overall variance and assign proportions to individual input modeling parameters. In this fashion, the analyst could determine, hypothetically, that the choice of environmental burden threshold is responsible for 35 percent of the variation in percent DAC designation, while the normalization parameter is responsible for only 5 percent. In this scenario, the signal for future model development would be to focus on the burden threshold and not worry about the normalization approach. Sensitivity analysis can also be conducted to understand the impact that each individual indicator has on DAC designation using approaches such as a combined Monte Carlo–logistic regression (Merz, Small, and Fischbeck, 1992). Understanding which variables are responsible for DAC designation can provide quantitative insight into whether individual indicators are playing an outsized role in DAC designation. This can support an assessment of the choices of indicators and the formulation of the tool.

Ideally, uncertainty and sensitivity analyses are run in tandem (Saltelli et al., 2008). Unlike local sensitivity analysis, global sensitivity analysis can distinguish between main and interaction effects. An example of main effects would be the relative independent contribution of the socioeconomic threshold parameter to the total uncertainty (variance) in the output percentage of designated DACs. However, a portion of the total uncertainty may arise from interactions of the socioeconomic threshold with other input parameters. Such interactions can occur in composite indicators due to their nonlinear nature. For models with a low degree of interaction effects, understanding the influence of each input parameter is straightforward. However, for models with high interactivity, it is much more difficult. Table 6.5 provides a simplified decision-making framework based on the results of a global sensitivity analysis.

The examples in this subsection use the percent designated DACs as the output measure of interest. However, sensitivity analysis could also be applied to other output measures. These might include the number of indicators or burden categories that exceed environmental and socioeconomic thresholds, the demographic characteristics of DAC-designated tracts, such as total population, income, race, and ethnicity, and the geographic distribution of DAC tracts.

Assessing the impacts of epistemic uncertainties is a core best practice for quality assurance in indicator construction (OECD and JRC, 2008; Saisana et al., 2019) and is part of modeling guidelines adopted by the World Health Organization (WHO/ILO/UNEP, 2008) and the Intergovernmental Panel on Climate Change (Munda et al., 2020). Uncertainty and sensitivity analysis have been applied to prominent environmental and social equality composite indicators, including the EPI (Saisana and Saltelli, 2008; Papadimitriou, Neves, and Saisana, 2020), Human Development Index (Aguña and Kovacevic, 2010), Sustainable Development Goals Index (Papadimitriou, Neves, and Becker, 2019), and Commitment to Reducing Inequality Index (Caperna et al., 2022). Assessing epistemic uncertainty enables a deeper understanding of model structure compared to alternative assessment methods (Saltelli et al., 2008). Uncertainty and sensitivity analysis could help illuminate the implications of decisions made during model construction (WHO/ILO/UNEP, 2008; Tate, 2012). Applied to CEJST, the analyses could help guide future model development by revealing the

- Precision of model outputs,
- Dependency on fragile assumptions (corroboration),
- Modeling decisions that require the most additional consideration (prioritization),
- Modeling decisions that can be simplified,
- Combinations of modeling decisions that generate the most extreme values (exploration),
- Varying combined effects of modeling decisions (interactions),

TABLE 6.5 Sample Decision Framework Based on Global Sensitivity Analysis Findings

Total Uncertainty	Main Effects	Interaction Effects	Model Uncertainties	Finding	Subsequent Modeling Focus
High	High	Low	Additive	Fragile	Reduce epistemic uncertainty for input parameters with high main effects.
High	Low	High	Interactive	Fragile	Reduce epistemic uncertainty for input parameters with high interaction effects.
Low	High	Low	Additive	Robust	Use the model as is or optionally reduce epistemic uncertainty for input parameters with high main effects to further increase robustness.
Low	Low	High	Interactive	Robust	Use the model as is or optionally reduce epistemic uncertainty for input parameters with high interactive effects to further increase robustness.

- Places that would be designated under alternative modeling decisions (spatial distribution),
- Population characteristics of DACs under alternative modeling decisions (impacted populations), and
- Degree and location of geographic clustering of DACs shifts under alternative modeling decisions (spatial heterogeneity).

At present, these characteristics are not known, and uncertainty and sensitivity analyses have yet to be conducted, given the early stages of maturity of many national and state EJ tools. However, failure to consider and remedy uncertainty in modeling processes can result in poorly informed policy decisions. Given the high policy stakes of resource allocation tied to DAC designation, it is important to understand the effect of composite indicator modeling decisions on CEJST outputs. Box 6.5 provides an example workflow for uncertainty and sensitivity analysis that might be useful to CEQ.

Sensitivity analysis techniques are important tools for improving the robustness of models, increasing the transparency of the model construction, and ultimately increasing the validity of the model. However, uncertainty and sensitivity analyses only evaluate the internal statistical fragility of the model. It is possible to develop an analytically and statistically robust model that fails to faithfully represent real-world processes and conditions. As such, sensitivity analysis is ideally complemented by external validation against measurable quantitative or qualitative data. Validation is discussed in Chapter 7.

The approaches described in this chapter for data integration and robustness testing provide a strategy for CEQ based on tenets of composite indicator construction. However, the tool is not "complete" following integration and testing, even though the intended output has been generated. Performing analysis for other measures may help tools like CEJST—those with high degrees of complexity and policy implications—to become accepted by interested and affected parties and represent lived experiences. Performing external analysis demonstrates awareness of important issues associated with CEJST, adds transparency, and helps address concerns raised by affected parties throughout the model development process.

Two examples of external analysis with output measures are evaluating CEJST against race/ethnicity and geographic units. Although many interested and affected parties have advocated for the inclusion of an indicator of race into CEJST, CEQ stated that they do not intend to include race as a factor in determining disadvantaged community status (EOP, 2022). However, there are supplemental analyses that CEQ could produce to show the relationship between burdens and the racial composition of communities. Likewise, analysis of CEJST by geographic units (e.g., states, counties, or legislative districts) can lead to summaries of how disadvantaged communities are distributed throughout the nation, which is valuable for awareness and planning purposes. Supplemental analyses of measures external to CEJST are described in more detail in Chapter 7.

BOX 6.5
Potential Robustness Assessment Workflow for CEJST

Based on guidance for using uncertainty analysis and sensitivity analysis (UA/SA) in evaluating policy-relevant models from Munda and others (2020) and World Health Organization/International Labour Organization/United Nations Environment Programme (2008), the following is a potential workflow to incorporate robustness assessment into the data strategy for CEJST:

1. An advisory group of UA/SA modelers, tool developers, community stakeholders, and scientific experts defines which input modeling choices to vary (treat as uncertain) and assigns a range or probability distribution to each to characterize the uncertainty. Figure 6.5.1 below portrays different probability distributions, while Table 6.4 provides examples of discrete distributions of plausible input modeling choices for CEJST.
2. The advisory group defines the model output(s) of interest and the objectives of the sensitivity analysis.
3. The UA/SA modelers randomly sample the previously defined probability distributions (Monte Carlo) to explore the entire uncertainty space. This corresponds to the Monte Carlo heading in Figure 6.5.1 below.
4. The UA/SA modelers repeatedly run the model in a Monte Carlo simulation, using the values of the input samples. After each run, they record a value of the output(s) of interest.
5. The UA/SA modelers quantify uncertainty measures based on the output distribution from the Monte Carlo simulation. The uncertainty heading in Figure 6.5.1 represents this.

CHAPTER HIGHLIGHTS

The process of integrating indicators in a composite indicator includes normalization, weighting, and aggregation. Normalization entails transforming input indicators into a common measurement scale to facilitate comparison. The primary normalization approaches include min-max scaling, z-score standardization, and percentiles. The CEJST documentation justifies the choice of percentile normalization, and research has shown that the modeling choice of the normalization approach has a negligible impact on a composite indicator.

Weighting is undertaken to inscribe the relative importance of each indicator into the composite indicator. Explicit weights are importance coefficients assigned to each indicator and are essentially value judgments based on scientific understanding, data-driven statistical analysis, participatory community preference, or policy priorities. Most composite indicators employ equal weights due to precedence or limited understanding of the relative importance of dimensions of the concept to be measured. The statistical properties and arrangement of indicators within a composite indicator can also produce

6. The UA/SA modelers identify the most influential modeling choices and examine their potential impact on the CEJST model output(s) of interest. This is reflected by the sensitivity pie chart in Figure 6.5.1.
7. The UA/SA modelers report the conclusions to the advisory group and prioritize areas for model improvement.
8. Document the UA/SA process, including the data, methods, outputs, and interpretation of results, and communicate it using broadly accessible language. Include in the interpretation an assessment of the composite indicator's precision, the credibility of the policy options, and the related policy impact.

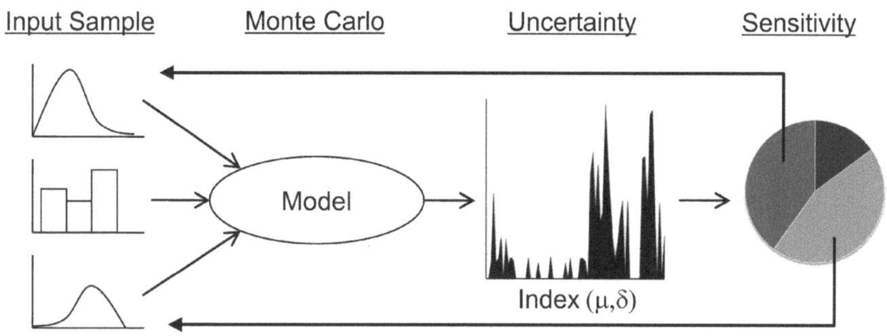

FIGURE 6.5.1 Example global sensitivity analysis scheme. Modelers select input sample distributions for the modeling choices of interest, conduct a Monte Carlo simulation, extract outputs of the resulting uncertainty distribution, and apply sensitivity analysis to quantify the relative influence of each modeling choice. SOURCE: Tate, 2012.

obscured, implicit weights. Implicit weights can be revealed using correlation analysis, with the results used to revise indicators that are statistically redundant due to collinearity or have insignificant or negative correlations that indicate weak statistical coherence with the concept to be measured. Screening for implicit weights is an important element of composite indicator development, although the CEJST documentation provides no evidence that this was undertaken.

In indicator aggregation, scaled and weighted indicators are combined into a single measure. CEJST uses percentile threshold criteria on environmental and socioeconomic indicators to develop a binary measure of community disadvantage. This approach is simple and easy to understand, but the logic of indicator criteria exceeded is limited in its capacity to represent cumulative burden, which is a stated priority of CEQ and EJ advocates. Based on the existing CEJST structure, options for employing cumulative impact scoring include summing the count of indicator criteria exceeded or the count of burden categories exceeded. However, cumulative scoring through additive or multiplicative aggregation to a continuous composite indicator would better represent cumulative impacts, potentially reflect interactions among indicators, and discern disadvantage among communities.

Robustness assessment quantifies the statistical instability of a composite indicator when compared to alternative plausible modeling choices and identify which modeling choices have the greatest influence on that fragility. Uncertainty and sensitivity analyses are employed for these purposes and are a core practice recommended for developing policy-relevant models. The findings of a robustness assessment can iteratively be used to improve a composite indicator through revisions to the indicator set, normalization approach, weighting methodology, and aggregation scheme. CEQ could conduct uncertainty and sensitivity analysis on CEJST to quantify the influence of input composite indicator modeling decisions on the output designation of disadvantaged communities.

7

Importance of and Methods for Tool Validation

Validation is an essential process for elevating the trustworthiness and utility of composite tools across diverse domains of inquiry. The committee's conceptual model for development of a composite tool (see Figure 3.1) is validation through comparison with real-world observations. After the concept to be measured is defined, indicators are selected, scaled, weighted, and integrated, and robustness is evaluated, tool concepts and results need to be validated by tool developers and other interested and affected parties. Validating composite environmental justice (EJ) tools such as the Climate and Economic Justice Screening Tool (CEJST)[1] is a nuanced process, given the breadth of datasets employed. CEJST attempts to characterize disadvantage in the context of Justice40,[2] but disadvantage is a multidimensional construct, and no single outcome (e.g., life expectancy) can accurately validate CEJST results. Concepts such as "disadvantaged" and "overburdened" do not have definitive measures and therefore are challenging to validate.

Validation confirms the accuracy and can increase community acceptance of a tool and its results, but it is also a process for justifying decisions made throughout tool development. Tool validation can also help identify outlier places that do not fit the general pattern, as analyzing them can provide insights into possible data problems and new variables. As such, validation is an important reality-grounding step in EJ tool development (OECD and JRC, 2008). Different validation methods are applied across disciplines, dependent on purpose and function (Stöckl, D'Hondt, and Thienpont, 2009). For example, in epidemiological studies, a health risk assessment is validated by testing the statistical relationship between the variables studied and an independent variable representing the health outcome associated with the contaminant and exposure pathway being studied (Steyerberg et al., 2001). Other types of validation involve data

[1] See https://screeningtool.geoplatform.gov/en/#3/33.47/-97.5 (accessed November 22, 2023).

[2] Read more on the White House's Justice40 Initiative at https://www.whitehouse.gov/environmentaljustice/justice40/ (accessed February 1, 2024).

triangulation, cross-cultural adaptations, and expert consensus to bolster tool credibility and generalizability (Kalkbrenner, 2021; Santos et al., 2020).

Deliberate and intentional community engagement is another form of validation that incorporates perceptions of affected parties and lived experiences (see Box 7.1) into the portrayal of environmental conditions and future policy outcomes (Larsen, Gunnarsson-Östling, and Westholm, 2011). It may help refine definitions of terms such as "overburdened" and "disadvantaged" and provide an opportunity to learn the narratives around data. This, in turn, can reveal the strengths and limitations of an EJ tool. For EJ purposes, it is important to center those who are disproportionately affected by injustices and those with historical legacies of oppression and provide those communities the right to self-representation (Davis and Ramírez-Andreotta, 2021; Liboiron, Zahara, and Schoot, 2018; Wilkins and Schulz, 2023).

Both community-based and noncommunity-based validation methodologies can be applied during EJ tool construction to ensure that the outputs reflect the realities and lived experiences of communities. Methodological decisions are based on a range of criteria, such as precedents set in the development of other tools, statistical uncertainty, data availability, and public input. These criteria are multifaceted and complex, and the decisions related to them are imbued with uncertainty and caveats. After the beta version of CEJST was released in 2021, the White House Council on Environmental Quality (CEQ) solicited and received comments from various interested and affected parties, several of whom questioned the methodologies employed, decisions made, and associated criteria from which those decisions originated.[3] Concerns related to methodologies and assumptions, such as those expressed, can cast doubt on the internal robustness, validity, current relevance, representativeness, and overall responsiveness to community concerns. It may be difficult to systematically identify all burdens within all communities where conditions are over- or underrepresented, and as such a certain amount of error in EJ tools is unavoidable. Tool validation that is transparent to interested and affected parties is valuable in that it is both a mechanism for tool improvement and helps to build trust. To develop a tool that is stable, accepted, scientifically sound, and representative of lived experiences, validation of the overall output is essential. This chapter focuses on approaches for validating tool results (i.e., output).

VALIDATING TO REFLECT REALITY

EJ tools may attempt to characterize millions of people from different geographic regions, representing many cultural and socioeconomic backgrounds. Effective engagement and collaboration to determine appropriate indicators (see Chapter 5) and identify data gaps can ensure that the tool reflects the reality of peoples' diverse experiences. The Yale Equity Research and Innovation Center (ERIC)[4] cites community engagement

[3] As examples, CEQ held numerous listening sessions with the public (see https://screeningtool.geoplatform.gov/en/public-engagement [accessed March 12, 2024] for a listing of events) and received input from the White House Environmental Justice Advisory Council (e.g., WHEJAC, 2021).

[4] See ERIC and its guidebook on community based participatory research at https://medicine.yale.edu/internal-medicine/genmed/eric/ (accessed March 1, 2024).

BOX 7.1
Incorporating Lived Experience into EJ Tools

Lived experience in this report refers to the familiarity and perspectives of community members who experience environmental injustices firsthand and in real time. One may have lived expertise from being a part of a geographic community, or one may have lived experience within a cultural or marginalized community. Lived experience provides information on the who, what, where, and when of burden and underinvestment. Incorporating local knowledge into the EJ construction and validation process complements sampled or modeled data and can result in a better portrayal of the complex environments in which people live (de Onís and Pezzullo, 2017; Dory et al., 2015). Community validation results from active engagement and partnership with disproportionately impacted communities, which are often also disadvantaged communities. A study by Powers, Mowen, and Webster (2023) highlights a potential scale for measuring public perception of environmental justice efforts for the sake of community validation.

Because lived-experience data are often qualitative, some researchers may be reluctant to use them, fearing a lack of rigor, representativeness, and defensibility (e.g., Crooks et al., 2023). Below are example practices provided by Frechette and others (2020) that could assist in incorporating lived experience into tool development, deployment, and evaluation:

- As a researcher, tap into your own humanity when engaging with communities. Pay attention to your body's reactions and rising emotions. This may require engaging the limbic (i.e., emotional) brain, which many researchers may avoid. Being conscious of these reactions may result in better understanding the situational realities and support the validation exercise. Conduct site visits and observe the emotions of community members there. Consider artwork relevant to the research topic and speak with the artists; ask what they wanted the audience to feel or know and compare that with your own human experience.
- When presenting lived-experience data, discuss data points as anecdotal and clearly state the limits of associated findings, especially when the sample size is small and lacks rigorous analytical methods. Propose these points as findings for future research to ensure that the data are included.
- Apply interpretive phenomenology analysis (IPA), a form of qualitative psychological research that favors making sense of situations through individualized and contextualized frameworks. IPA relies on well-defined and pointed research questions, organizations, and guidance for the selection of diverse samples to provide an account of a situation. IPA allows researchers to seek more diverse data collection methods and enhance interpretation of the data beyond description to asking, "so what?"

and community validation as beneficial to both communities and researchers as a means to increase the validity, legitimacy, and effectiveness of interventions. They provide information and a guidebook on a specific type of engagement called community-based participatory research (CBPR), which is used to assist officials, clinicians, and researchers in engaging communities in their research. The process could also be used by tool developers to determine how well the indicators and tool results represent real-world conditions. Tool developers may consider:

- How do the EJ tool results and methodologies compare to other EJ tools addressing similar geographies and topics?
- How well do the EJ tool results represent lived experiences and match the knowledge of those in disadvantaged communities?
- How do the results of the tool change over time, including how the distribution of disadvantaged communities is represented?
- What is the statistical relationship (e.g., correlation) between the EJ tool results and important factors not currently included in the tool (e.g., race), and how does that relationship inform knowledge about disadvantaged communities?

Tool construction decisions may be informed through validity analyses throughout the development and evaluation of the EJ tool. The performance of tools such as CEJST may also be improved through validation conducted between version releases. The validation result could be used to inform the public throughout tool development and may be published in external products (e.g., white papers, journal articles, websites, or the production of downloadable datasets). The following sections detail several of these methods, including ground truthing, convergent validation, community validation, advisory panels, and the incorporation of lived-experience data through mixed methodologies.

Convergent Validation

Convergent validation allows tool developers to evaluate how their tool results compare against other national, state, and local tools and data sets (Collins, Grineski and Nadybal, 2022; Kim and Verweij, 2016; Krabbe, 2017). Data used for validation can have many forms (e.g., satellite data) and numerous methodologies are available can be applied (for example, triangulation [e.g., Fielding, 2012]). An example of convergent validation would be comparing a specific indicator used in CEJST against a similar indicator in a state-level EJ tool, such as CalEnviroScreen. In the case of fine inhalable particulate matter ($PM_{2.5}$), CEJST uses national-scale data from the U.S. Environmental Protection Agency (EPA); this can be compared to California's independently collected data represented in CalEnviroscreen. If CEJST and CalEnviroScreen yield different $PM_{2.5}$ estimates for the same geographic area, questions about the convergent validity of the indicator arise: Are the differences due to data source, temporal range, outcome measured, or some other cause? This type of analysis provides one dimension of tool validation and insight into the tool's strengths, limitations, and applications.

Community Validation

Community validation is a critical component of every step of EJ tool development. This is not a unidirectional sharing of information but rather an iterative process grounded in ongoing collaborative relationships between government agencies and communities. Tool developers, such as CEQ, can validate data, approaches, and tool results through community engagement in a variety of ways: through targeted outreach to interested and affected parties, workshopping in disadvantaged communities via small-group discussions, providing office hours to respond to questions or receive input,[5] or through creative approaches such as data challenges or hack-a-thons. Community validation may also take the form of community peer review (see Box 7.2). Conducting such activities continually during the development and upgrading of a tool will maximize the benefits gained from community validation. When tool developers (e.g., government agencies) directly engage with communities through community-based research and consultation, they can better understand and incorporate community concerns into their research (Shepard, 2002).

Members of disadvantaged communities may live near chemical facilities, landfills, and brownfields; may inhabit areas of food insecurity; or may suffer from living with extreme heat. The committee heard this feedback during its information-gathering workshop (NASEM, 2023a). Lived-experience data—for example, access to air-conditioning or newer air filtration systems—might be collected through community engagement to augment more traditional data sources. Such information provides context for how burdens could be represented in EJ tools. Communities and individuals have long urged federal agencies to incorporate lived experience into their research, programs, and policies.[6] The Assistant Secretary for Planning and Evaluation (ASPE), an advisor to the U.S. Department of Health and Human Services (DHHS), provides a website of information with guidance for federal agencies on how to better incorporate community engagement and lived experience in programming, research, and policy (Skelton-Wilson et al., 2021).

Without ongoing input from community members, EJ tools like CEJST risk selecting irrelevant indicators, mischaracterizing communities, and creating a tool that could negatively affect decisions related to a community. Giving community members regular opportunities to verify or refute data, models, and outputs as consistent with lived observations is a practical approach to tool validation that is fundamental to EJ tool development, acceptance, effectiveness, and sustainability. This could be done, for example, by systematically incorporating mechanisms that allow those most familiar with a region to explore and critique tool results (such as a map showing census tracts designated as disadvantaged) to ensure that designations resulting from a chosen methodology reflect their knowledge and lived experiences. By partnering with community members and subject-matter experts, tool developers can iteratively validate and refine

[5] EPA's EJSCREEN tool holds public bimonthly conversations called "Office Hours," where people can ask EPA EJSCREEN experts questions about many topics, including how to use the tool and technical issues.

[6] ASPE echoes this sentiment on their website on lived experience at https://aspe.hhs.gov/lived-experience (accessed March 1, 2024).

BOX 7.2
Community Peer Review for Informing EJ Tool Development

Community peer review is a growing practice and research tool. It is intended to ensure the opportunity for community consent or refusal to participate in government processes or events if there are concerns these processes or events may cause community harm. Community peer review also provides the right to self-representation and the right to refute false yet established narratives regarding research practices and methodologies (Liboiron, Zahara, and Schoot, 2018). It helps communities determine if the work being done will cause or prevent them harm. Liboiron, Zahara, and Schoot (2018) describe steps in the community peer review process that can be applied to EJ tool development (see list below). They also provided a thorough analysis of their community peer-review sessions, including how to work with different interest groups on a specific issue (e.g., chemical industry and disadvantaged community representatives). Below is an example of a workflow for community peer review.

1. Identify the community. The community may not geographically align with census tracts, census blocks, or other administrative boundaries. The community definition needs to be considered when determining the best way to represent the community (e.g., through indicators and data used in an EJ tool).
2. Examine and analyze the economic, social, and cultural aspects of the community, allowing for increased context of the lived experience surrounding and interacting with observed data. This research need not be limited to the consideration of peer-reviewed studies and datasets. Spend time with community members to learn their histories, points of pride, and concerns.

the tool methodologies to result in a tool that minimizes discrepancies. Community members would be able to help developers determine if the current methodology is missing an important indicator or perhaps is erroneously classifying tracts because a chosen threshold was too high or too low. These partnerships have long-lasting effects, as communities can both advise on future revisions to tools and on targeted mitigation (e.g., types of investment) that could provide the greatest benefit.

Open data forums and events, sometimes referred to as "data challenges," in which anyone can engage with government data and submit feedback, advice, and potential solutions to environmental issues are increasingly common at state[7] and federal[8] levels.

[7] An example of a data challenge at the state level is the California Water Data Challenge: https://waterchallenge.data.ca.gov/background/ (accessed January 31, 2024).

[8] An example of a data challenge at the federal level is DHHS's Environmental Justice Community Innovator Challenge: https://www.hhs.gov/about/news/2023/09/18/hhs-launches-environmental-justice-community-innovator-challenge.html (accessed January 31, 2024).

This moves beyond research and into exploration and can help establish trust. Plan to compensate community members for their participation.

3. Hold a community meeting. A tool developer can identify datasets for which community meetings are part of the data collection process. Assuring that the community meeting can include widespread community representation may mean holding the meeting at a time when most community members can attend, offering childcare and food options, considering in what language to hold the meeting or whether interpreters are needed, and considering transportation access. Meetings need to be viewed as trust-building sessions, where listening to the community outweighs teaching the community. Hold follow-up meetings throughout the tool-development process and be accountable when discussing the results of community input. Community meetings are opportunities to discuss the Spectrum of Community Engagement to Community Ownership (Gonzalez, 2019; also see Chapter 3). The discussion can assist in ensuring interested and affected parties agree with roles and responsibilities.
4. Ensure that communication and information gathering are accomplished in ways that are natural for the community. This may include focus groups, interviews, or other means of information collection (Liboiron, Zahara, and Schoot [2018], for example, used surveys). The goal is to mitigate community stress and begin to build trust-based relationships. It is a good practice to observe and listen. Even observations of the community and meeting environment can help further identify and research the community.
5. Connect community feedback to community research. Ensure that all interpretations and conclusions are rooted in community research and all previous steps. This will be important in tool validation.

Although there are several ways for government to engage with communities to validate EJ tools, community members may be unable to volunteer their time, and local community officials or representatives may not have the resources to engage them. It is becoming more common for government agencies to compensate communities for their time and expertise through stipends, offering food and drinks at meetings, making childcare available, and facilitating grants and technical assistance when available (CSCC, 2022; Daley and Reames, 2015; EPA, 2023e). Governments can work with communities to determine if assistance from technical experts would be more beneficial than assistance from those with lived experience, noting the history of racism as both an input and output from academia (Rahman, Sunder, and Jackson, 2022). Academia is exploring how to institutionalize concepts such as diversity, equity, and inclusion, critical race theory, and community engagement (Asmelash, 2023). However, institutionalizing these concepts and frameworks may have negative results, such as devaluing information about lived experience gained through community validation.

Whereas community engagement is vital, meaningful engagement programs could require significant resources and capacities. Given the scale and scope of national-level EJ tools, tool developers will need to consult experts in community engagement and rely on, for example, advisory panels to help design appropriate programs tailored for an individual tool. Because in-depth engagement cannot realistically occur at the national level with all communities represented by a tool, methods to identify representative communities, to design tool feedback methodologies, and to validate decisions made during indicator construction will necessarily be important in the design of an engagement program.

Ground Truthing

In mapping applications, ground truthing refers to the collection of reference data tied to specific locations on Earth's surface used to gauge the validity of a theoretical model (e.g., Yonto and Schuch, 2020). In the context of EJ tools, the concept includes the traditional testing of theoretical models with empirical data and expands beyond it. It involves a partnership between researchers and community members (Sadd et al., 2014). An effective ground-truthing method is through CBPR, as outlined by ERIC and other research groups (Hacker, 2013; Israel et al., 2005; Leung, Yen, and Minkler, 2004). CBPR involves researcher-community collaborations to collectively select research questions, design studies, collect data, interpret findings, and disseminate results to protect public health and inform public policy (Israel et al., 1994). This approach involves key interested and affected parties in multiple qualitative and quantitative aspects of research to bring about change, such as model refinement. Box 7.3 discusses opportunities for ground truthing in CEJST.

Advisory Panels

Advisory panels (or workgroups) are often established to provide guidance during the construction or management of EJ tools and can be used as another mechanism for validating tools and their results. Advisory groups often include representatives from academic, community, government, nonprofit, and other organizations. U.S. states such as Massachusetts, Minnesota, Connecticut, Pennsylvania, and California have organized their advisory panels to ensure meaningful public participation related to government decisions (Daley and Reames, 2015). For example, the California Environmental Protection Agency formed the Cumulative Impacts and Precautionary Approaches (CIPA) workgroup,[9] an external multidisciplinary stakeholder group that included perspectives from academia, industry, EJ, and community organizations. They provided input on approaches to evaluating cumulative impacts from 2008 to 2013 and contributed to the California Office of Environmental Health Hazard Assessment's report on cumu-

[9] Information about the CIPA work group's meeting on June 5, 2008, can be found on the California Office of Environmental Health Hazard Assessment's website at https://oehha.ca.gov/calenviroscreen/workgroup/cipa-meeting-june-5-2008 (accessed January 31, 2024).

BOX 7.3
Ground Truthing Opportunities for CEJST

Using ground-truthed datasets increases the validity of the tool. The CEJST Technical Support Document (CEQ, 2022a) does not indicate the number of CEJST input datasets that have undergone ground truthing. CEJST draws from several existing data systems and from other EJ tools, including the U.S. Environmental Protection Agency's Environmental Justice Screening and Mapping tool, known as EJScreen.[a] Some of those data have been validated through ground truthing (Rowangould et al., 2019; Sadd et al., 2015). For example, Sadd and others (2015) evaluated three cumulative impacts with EJ tools in California for hazardous waste facility locational accuracy and found location errors of up to 10 km. As CEQ considers validating input datasets and output measures of disadvantage, they can build on existing ground-truthing approaches to achieve greater validity. Detailed documentation regarding ground truthing and other forms of validation will increase transparency and increase the tool's legitimacy.

[a] EJScreen and other EJ tools are discussed in more detail in Chapter 4. See also EJScreen's website at https://www.epa.gov/ejscreen (accessed February 16, 2024).

lative impacts (Alexeef et al., 2012). That report proposed the methodology used in CalEnviroScreen for identifying disproportionately burdened communities. The White House Environmental Justice Advisory Council (WHEJAC)[10] functions as an advisory panel for CEJST, Justice40, and other environmental justice topics. In 2021, WHEJAC submitted a recommendations report to CEQ pertaining to Executive Order 12898, Justice40, and CEJST, which provided recommendations to improve these tools and policies (WHEJAC, 2021). Box 7.4 provides information about WHEJAC and community engagement in CEJST.

Mixed Methods

The methods described above result in a combination of quantitative and qualitative data that need to be integrated to both strengthen data types and offset the limitations of each. Research that combines the use of these data is called mixed-methods research. Mixed-methods approaches are rooted in the social sciences but have expanded into health and other fields. They are used to systematically integrate qualitative and quantitative data within one investigation and can be intertwined throughout the EJ tool

[10] More information on the White House Environmental Justice Advisory Council (WHEJAC) can be found on EPA's website at https://www.epa.gov/environmentaljustice/white-house-environmental-justice-advisory-council (accessed January 31, 2024).

BOX 7.4
CEQ and CEJST Community Engagement

CEQ engaged with communities before and during the development of the first version of CEJST. A major source of engagement has been through WHEJAC. CEQ states that many of these recommendations were adopted between the release of its beta and first versions of the CEJST. Changes included the addition of historical redlining data, the identification of Tribal nations, and the display of demographic information (CEQ, 2022b).

After the beta version of CEJST was publicly released, CEQ hosted a series of training webinars to give members of the public an opportunity to learn how to use the tool.[a] Concurrently, CEQ hosted a series of listening sessions "to seek input and feedback on the beta version of the tool, including on the datasets it includes and the methodology it uses."[a] CEQ also stated on its public engagement page that this feedback would inform future updates of the tool so that it reflects conditions faced by communities. CEQ has promised future engagement opportunities.

[a] See the WHEJAC Charter at https://www.epa.gov/system/files/documents/2023-03/2023%20White%20House%20Environmental%20Justice%20Advisory%20Council%20Charter.pdf (accessed February 16, 2024).

development process. Wisdom and Creswell (2013), as part of the Agency for Healthcare Research and Quality,[11] summarize the use, advantages, and limitations of mixed methods research. Although their research is focused on patient-centered medical home models, their methods can be adapted to EJ tool development and evaluation. They describe five core characteristics of well-designed mixed-methods studies:

1. Collection and analysis of both qualitative and quantitative datasets;
2. Rigorous collection and analysis procedures and criteria for each kind of data (e.g., data meet appropriate sample sizes for each type of data);
3. Integration of data during all stages of research (e.g., during data collection, analysis, and discussion);
4. Procedures that implement analyses of both quantitative and qualitative data types either sequentially or concurrently, and for the same or different datasets; and

[11] See https://www.ahrq.gov/ (accessed March 13, 2024).

5. Framing of research within theoretical or philosophical research models to better understand multiple perspectives of any single issue.

Mixed-methods research can result in comprehensive datasets that are multidisciplinary and it provides methodological flexibility for rigorous community validation of the tool and its results. However, as noted by Wisdom and Creswell (2013), the mixed methods have limitations. Because mixed-methods research is multidisciplinary, research can be more complex to design and implement; it can require more resources, labor, and time to ensure rigor. It may be difficult to locate qualitative experts and collect needed sample sizes, but the available datasets are more diverse. As described in Chapter 6, tool development and indicator construction rely on integrating different kinds of datasets resulting from different kinds of research. But because not all information that would benefit an EJ composite indicator is quantitative, mixed-method approaches provide the creative means to incorporate lived-experience data with statistical techniques. Although mixed methodologies challenge the status quo and may be difficult to plan and execute, their use will result in data interpretation and informed research practices that allow for the incorporation of lived experiences into data analyses while providing a pathway to tool validation.

LIVED EXPERIENCE AND CEJST

Lived experience represents the broader environmental conditions in which people live, work, worship, and play and cannot be understood from a single type of feedback or validation approach. EJ tools like CEJST can represent only aspects of the overall lived experience of communities. Efforts to gather lived-experience data can reveal how multiple indicators of burden may be related. During its public workshop held on June 5, 2023, proceedings of which are summarized in a separate document (NASEM, 2023a; see also Appendix B for the workshop agenda and list of participants), the study committee heard about the lived experiences of invited community members and others interested in and affected by CEJST. The purpose of the workshop was to explore how well data used in CEJST represent the lived experiences of historically marginalized and overburdened communities across the nation.

Input to Define the Concept to Be Measured

Although the Executive Order (E.O.) 14008 mandating the creation of CEJST by CEQ (EOP, 2021) stated that CEQ would develop a tool that would determine community disadvantage, it did not define community disadvantage. CEQ developed a definition that was consistent with the E.O., usable, and scientifically defensible. Multiple workshop participants questioned using census tracts as the spatial definition of community and unit of analysis, indicating that the census-tract scale lacks the granularity necessary to characterize their communities. Participants described that disparities experienced in their local communities were not recognized by CEJST because averaging values across the census tract for indicators (e.g., income) placed the tract above CEJST's low threshold. Community engagement could help refine the definitions of "community" and "community disadvantage," which could then inform choices of scales, indicators, and analysis approaches. Communities would feel a sense of collaboration by having more input into how they are defined, and with better documentation of the engagement, the tool would gain trust, transparency, and legitimacy.

Community Input for Selecting Indicators

Participants of the committee's information-gathering workshop (see Appendix B for the agenda and participant list) discussed issues or presented narratives about the burdens faced by their communities or the communities with which they work. Efforts to seek out such input by CEQ could inform the selection of new or different indicators and datasets that reflect lived experiences. Examples of issues discussed at the workshop are listed below, with more detail provided in the workshop Proceedings in Brief (NASEM, 2023a). Not all these examples can be measured currently with datasets that comply the CEJST data criteria.

- Nayamin Martinez, executive director of the Central California Environmental Justice Network, discussed the cumulative factors of heat, air pollution, pesticide exposure, and the reliance many California Central Valley communities have on evaporative "swamp" cooling systems that do not filter pollutants and exacerbate indoor air pollution.
- Loka Ashwood, associate professor at the University of Kentucky, referenced pollution issues in Burke County, Georgia, and that a census tract in this county hosts four nuclear reactors. That particular tract is not recognized by CEJST as disadvantaged because the CEJST legacy pollution category does not include data on nuclear reactors.
- Vi Waghiyi, member of the WHEJAC, explained that CEJST does not reflect the lived experience of her family and community in the Arctic, citing the impacts of persistent organic pollutants from ocean currents on her people's food and water supplies.

Discussion among multiple workshop participants suggested that economic disparities can be captured with national income and poverty measures but are not currently reflected in CEJST.

Input to Define Indicator Integration

Community input can reveal the burdens of a particular community, measures for those burdens, and how those burdens combine to affect lived experiences. The input can inform tool developers regarding how to weight, aggregate, and analyze indicators. Participants at the committee's information-gathering workshop (NASEM, 2023a) discussed the following issues and needs related to CEJST:

- The tool's ability to understand cumulative impacts;
- Potential issues associated with the exclusive use of a single tool—especially one that employs a binary designation such as CEJST—to inform resource allocation to communities;
- The importance of differentiating among communities with more or fewer burdens;
- The appropriateness and implications of applying a uniform approach for designating a community as disadvantaged to the entire country (without the ability to distinguish between, for example, urban and rural communities);
- The inability to rank, prioritize, or micro-target communities with the greatest needs;
- The incorporation of community-based mapping; and
- Alternative ways of visually representing disadvantaged communities (e.g., the ability to look at results for different categories of burden or having a scale for magnitude of disadvantage).

Future CEQ efforts to engage with communities and gather and respond to their input iteratively could validate integration efforts and yield more legitimate results that reflect the communities being represented.

Input for Validation

Workshop participants discussed how well CEJST results compare to localized mapping efforts (NASEM, 2023a). Mathy Stanislaus and Alexis Schulman of the Environmental Collaboratory at Drexel University demonstrated that CEJST identified more areas as disadvantaged than their own Expanded Environmental Justice Index map for Philadelphia,[12] but workshop participants acknowledged that CEJST may underrepresent disadvantage in other tracts. CEJST does not discern places with the highest disparity. This would be an example of using community engagement and input to "make sense of the data" (OECD and JRC, 2008, Step 9, as described in Chapter 3 of the present report); using data narratives and correlating the indicator with relevant, measurable phenomena to explain similarities and differences. This was described as convergent validation earlier in this chapter. No single tool will be able to represent every community perfectly, but seeking out this kind of input from community members and technical experts can help validate indicators and tool results.

Input Regarding Documentation

The CEJST map interface and documentation are important means of outward communication by CEQ. It is important for CEQ to determine the effectiveness of its communication through community engagement. The committee's workshop included a hands-on CEJST exercise intended to gather input regarding tool results (for validation),

[12] See https://greenlivingphl.com/ (accessed March 13, 2024).

but also regarding the user interface, functionality, and data accessibility (NASEM, 2023a). Some input gathered included that:

- A certain level of education and familiarity with indicators was required to navigate CEJST;
- CEJST could be more interactive when viewing community results, for example, by providing the ability to filter by a specific indicator or category; and
- CEJST could be made more accessible for vision-impaired users.

CEQ could engage with communities to validate its efforts and create greater trust, transparency, and legitimacy. Responding to input regarding user interfaces and the documentation that would be most helpful and informative will help CEQ create a more useful tool.

SUPPLEMENTAL ANALYSIS OF EXTERNAL VARIABLES

Supplemental analysis in the context of EJ tools refers to additional analysis or assessment conducted outside of the tool using independent external datasets. It is distinct from robustness analysis discussed in Chapter 6 for internally validating indicators during tool construction; supplemental analysis is useful for externally validating the outputs of EJ tools. A recent practical example of supplemental analysis in tool development validation was a comparison of COVID-19 testing and positive case rates from a county public health summary dashboard with cumulative impact ratings from the Washington State Environmental Health Disparities Map[13] (Ingram et al., 2022). Analysis combining information from both tools allowed investigators to correlate COVID-19 positivity, low testing rates, census tracts with a greater proportion of people of color, low socioeconomic status outcomes, poor community health outcomes, and higher levels of PM2.5. Whereas this work was a first attempt at understanding the relationship between a particular disease and environmental exposures, it demonstrates how a tool developer might conduct similar techniques to test model validity. A more common example of supplemental analysis as validation are comparisons of metrics of environmental inequality and life expectancy. In these cases, the relationships of measures of burden to premature mortality are validated. The comparisons may be, for example, between pollution and income inequality (Hill et al., 2019) or the built environment (Kim, B. et al., 2023). These types of analysis involve evaluating the output of a measure against life expectancy, an external variable. The findings are consistent with longstanding claims from environmental justice leaders that EJ community members have shorter lifespans (Bullard, 2001).

[13] See https://doh.wa.gov/data-and-statistical-reports/washington-tracking-network-wtn/washington-environmental-health-disparities-map (accessed March 14, 2024).

Purpose of Supplemental Analysis

Supplemental analysis of external variables can lead to multiple benefits, including a greater understanding of sociodemographic composition, determinants of health in communities identified, and in-depth case studies that generate localized narratives (Cushing et al., 2015; Prochaska et al., 2014; Williams et al., 2022). Supplemental analyses can also address fundamental questions about the impacts or implications of tool construction, including questions about the relationship between race/ethnicity and measures of environmental justice. For example, supplemental analysis comparing the distribution of race/ethnicity indicators and CEJST outputs could help CEQ tool developers gain a greater understanding of how well CEJST captures community disadvantage in its current formulation. Researchers at the Bullard Center for Environmental Justice developed an interactive map that does this.[14] The map overlays proportional symbols showing the number of CEJST categories exceeded with choropleth symbols showing the percent people of color in communities. This map is a simple yet powerful example of supplemental analysis of CEJST and race/ethnicity, revealing relationships between indicators of burden and concentrations of people of color.

Because race is an important predictor of some environmental disparities, as discussed in Chapter 2, this analysis could help tool developers check their own indicator data sources for potential gaps or inaccuracies. Chapter 5 of this report describes the range of indicators or measures of racism that could be considered by CEQ tool developers. Future iterations of the tool might be refined in response to supplemental analyses which, in turn, could result in a tool more valid tool. Publication of supplemental analyses results regarding the relationship between race/ethnicity and CEJST would show CEQ responsiveness to public comments, increasing trust in the tool development process and tool results.

The next section provides an example of supplemental analysis applied to CEJST, showing relationships between the tool and racial/ethnic composition across communities in the country. It discusses the use of race and ethnicity data along with CEJST results to show the longstanding relationship between pollution and people of color in this country.

Supplemental Analysis: Disadvantaged Communities and Race/Ethnicity

Race reveals a stronger correlation than commonly used socioeconomic measures such as poverty (Bullard, 1993; Cutter, 1995; Mascarenhas, Grattet, and Mege, 2021; Mohai, Pellow, and Roberts, 2009; Commission for Racial Justice, 1987). Chapter 2 discusses the relationships between racism and unequal exposures and outcomes, as well as measuring racism. Chapter 5 notes that CEJST does not include indicators of race or ethnicity in determination of disadvantaged communities, but including race and ethnicity in CEJST has been raised during CEQ's public comment period (McTarnaghan et al.,

[14] See Bullard Center for Environmental and Climate Justice's Historically Black Colleges and Universities (HBCU) Climate & Environmental Justice Screening Tool at https://cdu-gis.maps.arcgis.com/apps/instant/basic/index.html?appid=de6aa42f3ce24fb7999f2af01540be9f (accessed June 16, 2024).

2022) by environmental justice advocates and organizations (Chemnick, 2022; Shrestha, Rajpurohit, and Saha, 2023; Wang et al., 2023), and during the committee's workshop (NASEM, 2023a). Instead, CEJST uses boundary data from the Home Owner's Loan Corporation (HOLC; Aaronson et al., 2021) as an indicator of racial segregation and inequity. (See Box 2.1 for information on red lining.) However, this dataset is not sufficiently comprehensive to represent contemporary spatial patterns of race or racialized disadvantage (Mallach, 2024; Perry and Harshbarger, 2019). Measuring underlying processes of racism remains challenging due to the paucity of racism indicators, although developing such measures is a recent area of focus (Furtado et al., 2023). Appendix D includes examples of measures of segregation or racism that might be considered.

The California Office of Environmental Health Hazard Assessment (OEHHA) has conducted supplemental analysis of race and ethnicity alongside vintages of CalEnviroScreen since 2013 (CalEPA, 2021). The first version of CalEnviroScreen included race/ethnicity as an indicator, but that indicator was removed in an update to the tool (CalEPA, 2013). The change was made to "facilitate the use of the tool by government entities that may be restricted from considering race/ethnicity when making certain decisions."[15] However, in recognition of the relationship between race and environmental justice, a supplemental chapter was added describing the correlation between race/ethnicity and the pollution burdens of communities and the intention to update and expand that section as new versions of the tool are released. This position was maintained in subsequent versions of the tool.

In the most recent release of CalEnviroScreen, their report states, "...CalEnviroScreen 4.0 does not include indicators of race/ethnicity or age. However, the distribution of the CalEnviroScreen 4.0 cumulative impact scores by race or ethnicity is important. This information can be used to better understand issues related to environmental justice and racial equity in California." They present relationships between the CalEnviroScreen cumulative impact score and Californians by race or ethnicity. Figure 7.1 shows all Californians when assigned a CalEnviroScreen cumulative impact index score (based on the census tract they reside in), grouped by race/ethnicity. All racial/ethnic groups have members living in communities with the lowest and highest overall CalEnviroScreen cumulative impact index scores, but the median CalEnviroScreen cumulative impact index scores are much higher for Latinos, Blacks, and Pacific Islanders than other groups, indicating greater experience of burden for those groups (CalEPA, 2021).

Figure 7.2 portrays supplemental analysis results that indicate the proportions of each racial/ethnic group residing within each decile percentile of CalEnviroScreen scores. Unlike CEJST, CalEnviroScreen uses a cumulative scoring that produces a continuous score for every census tract in the state. The topmost horizontal bar shows the racial/ethnic composition of the first decile, the least impacted census tracts in the state. Meanwhile, the 10th decile at the bottom of the chart shows the racial/ethnic makeup of the most impacted census tracts. The statewide racial/ethnic composition of California is shown at the bottom for reference. If burdens were distributed equally

[15] See https://oehha.ca.gov/calenviroscreen/report-general-info/calenviroscreen-11 (accessed June 16, 2024).

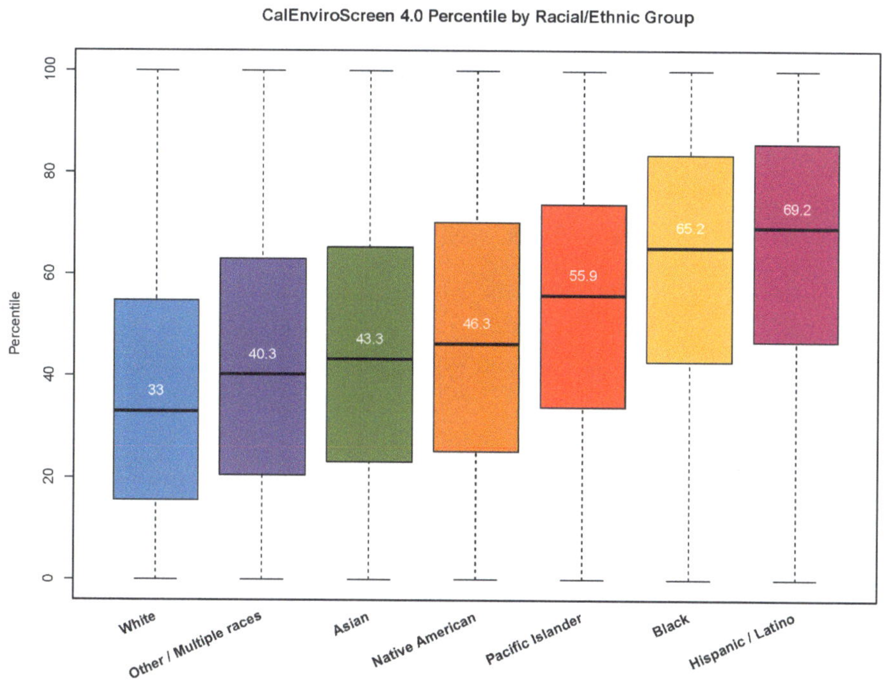

FIGURE 7.1 Distributions of CalEnviroscreen 4.0 score by racial and ethnic population. CalEnviroScreen scores are derived from a cumulative impact methodology that considers exposures, environmental effects, sensitive populations, and socioeconomic conditions. SOURCE: CalEPA, 2021.

among all groups, all bars from 1 to 10 would represent the same population proportions as found in the "CA" bar. However, the figure demonstrates that Latinos and African Americans disproportionately reside in highly impacted communities while other groups reside in less impacted communities. CEJST could conduct similar supplemental analyses. The binary (as opposed to cumulative) approach to designating disadvantage utilized in CEJST would require different analytical approaches than what is shown in CalEnviroScreen, but examining racial and ethnic disparities is still possible by CEQ.

In response to criticisms about the absence of race as an indicator for disadvantaged communities, CEQ argued that their focus on income and environmental burdens would still effectively capture communities of color due to strong correlations between environmental and social inequities and the proportion of non-white residents (Friedman, 2022). Supplementary analyses of CEJST and the relationship of disadvantaged communities and their racial composition by journalists at *Grist* (Sadasivam and Aldern, 2022) and later at *E&E News* (Frank, 2023) lent support to the CEQ argument. These analyses showed a strong correlation between disadvantaged community status and the proportion of non-white residents: the higher the proportion of non-white residents, the higher the likelihood that a tract would be designated as disadvantaged (see Figure 7.3; Sadasivam, 2023). These findings might appear to validate assurances that the indica-

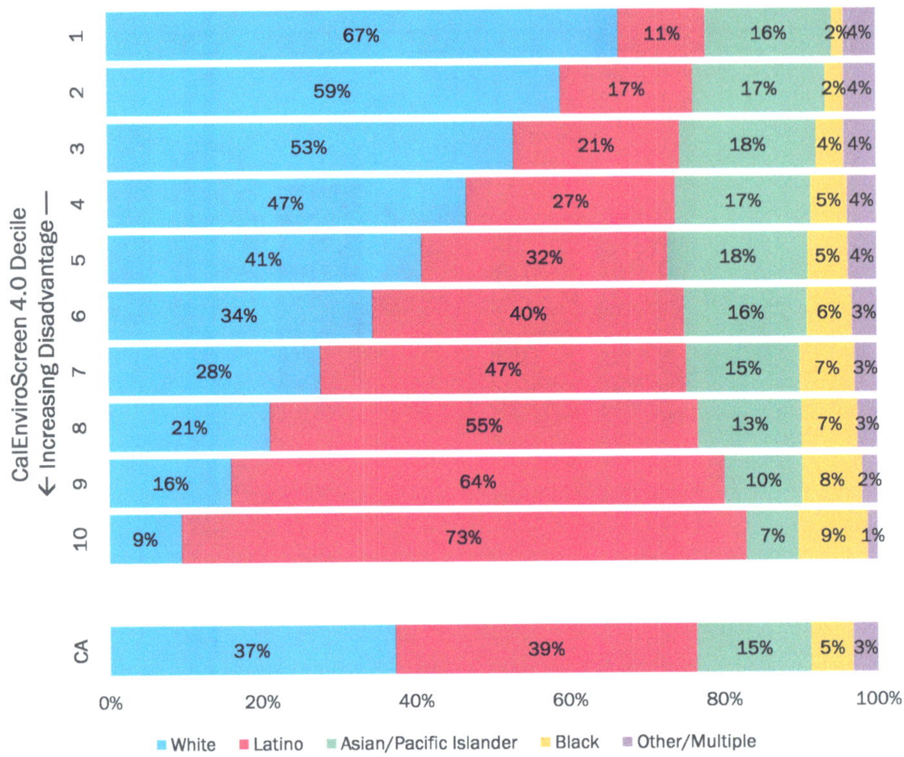

FIGURE 7.2 Proportions of each racial/ethnic group residing within each decile percentile grouping of CalEnviroScreen scores. Each horizontal bar represents a decile of the population, with the uppermost representing the least affected census tracts (labeled "1") and the bottom represents the most impacted (labeled "10"). Each bar is subdivided by race/ethnicity. The statewide racial/ethnic makeup of California is shown at the bottom of the figure for reference. SOURCE: CalEPA, 2021.

tors used by CEJST act as proxies for race without using race, echoing the findings for CalEnviroScreen described above. However, a more recent analysis by the World Resources Institute (WRI; Shrestha, Rajpurohit, and Saha 2023), shows that CEJST's methods still underrepresent the degree of disadvantage and disparity for communities of color.

WRI did its own analysis of CEJST and examined tracts by the number of indicator thresholds exceeded and found a strong correlation between the number of indicator thresholds exceeded and the proportion of non-white residents: the higher the proportion of non-white residents, the greater the number of indicator thresholds exceeded (see Figure 7.4; Shrestha, Rajpurohit, and Saha 2023). The WRI report observed that in disadvantaged communities meeting at least one indicator threshold (i.e., the current methodology in CEJST), 45 percent of the population is white. By contrast, in com-

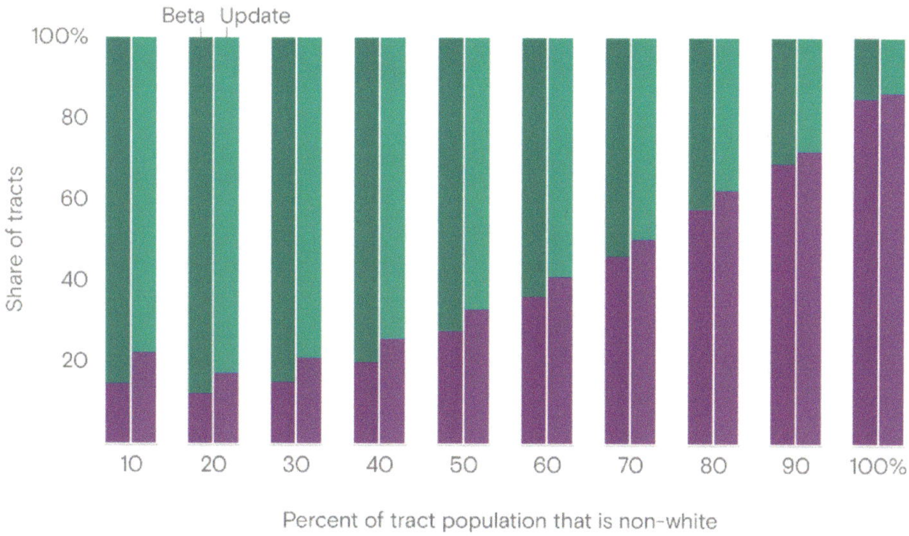

FIGURE 7.3 Correlation between race and CEJST designation as disadvantaged. The green (top portion of the bars) represents not disadvantaged, while the pink (bottom portion) represents disadvantaged. Each 10th percentile of the tract population is represented by two sets of data. The bar on the left represents results from CEQ's beta version of CEJST. The right-hand bar represents results from CEJST's 2022 release. SOURCE: Sadasivam, 2023.

munities that meet the threshold for 10 indicators (which include 7.1 million residents or 7 percent of the population in disadvantaged communities), less than 17 percent of the population is white. The implication is that a cumulative approach to indicator construction would reveal greater disparities between communities that are overwhelmingly non-white and those that are overwhelmingly white. However, this pattern is not apparent or discernible based on how CEJST currently identifies disadvantaged communities. It should be noted that the WRI analysis also shows that this underrepresentation of racial disparity can be exacerbated as more burden indicators are added to the tool in its current formulation (Shrestha, Rajpurohit, and Saha 2023).

CHAPTER HIGHLIGHTS

Systematically identifying all burdens faced by communities throughout the country to determine if those communities are disadvantaged is challenging, and a certain amount of error in EJ tools is unavoidable. Tool validation techniques can be applied to allow tool developers to create a tool that is stable, accepted, and scientifically sound. Different validation approaches are available, for example:

- Convergent validation compares tool components or results with those of similar tools. They can take the form, for example, of correlation analysis of tool results.
- Community validation is an iterative process conducted through collaborative

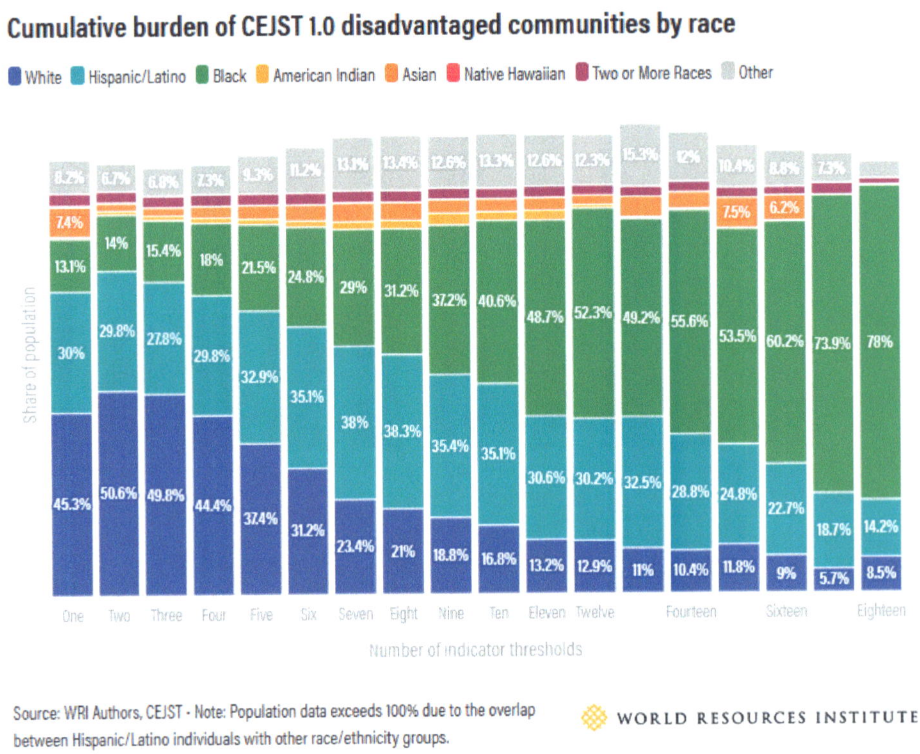

FIGURE 7.4 World Resources Institute analysis showing the relationship between the number of CEJST indicator thresholds exceeded and the percentage of population by racial groups in disadvantaged communities. Each column represents the total population within disadvantaged communities that exceeded a given number of indicator thresholds, shown by the number at the base of the column. Colored sections of each column indicate the proportion of that disadvantaged community population by race or ethnicity. SOURCE: Shrestha, Rajpurohit, and Saha. 2023.

engagement with communities to compare how well the tool reflects lived experiences. Consistent engagement throughout the tool development or upgrading process allows developers to test decisions, approaches, and tool results against community member narratives, while empowering communities to accept or refute definitions being assigned to them and gaining trust in the tool development process.

- Mixed methods that allow collection and analysis of both qualitative and quantitative datasets are framed withing research models to better understand multiple perspectives of any issue and are well suited for tool validation. Although mixed methods challenge the "traditional" scientific mindset focused on quantitative data, their use will result in data interpretation and informed research practices that allow for the incorporation of lived experiences into data analyses.

Supplemental analysis in the context of EJ tools refers to additional analysis or assessment conducted outside of the tool using independent external datasets. Many benefits can derive from supplemental analysis including a greater understanding of sociodemographic composition, determinants of health in communities identified, and generation of localized narratives and in-depth case studies. CEQ might conduct supplemental analysis to, for example, compare the distribution of race/ethnicity indicators and CEJST outputs to test the validity of CEJST's current formulation. Such analysis could help tool developers check indicator data sources for potential gaps or inaccuracies. Future iterations of a tool might be refined in response to analysis findings. Documentation of all validation efforts, including supplemental analysis will increase the transparency, trust, and legitimacy of the tool and show responsiveness to input received during community engagement.

8

Recommendations

A quality environmental justice (EJ) tool has several distinguishing characteristics. It accurately reflects the lived experiences of the communities the tool is intended to represent. The tool outputs are trusted and can be used to inform decision making. Each step of the tool development process has been validated internally for statistical robustness and externally through community engagement to ensure legitimacy. Information about data inputs, indicator construction methods, and community engagement processes are thoroughly documented. Collectively, these characteristics build confidence that the tool is well constructed, its results are accurate, and it can be reliably used to advance policy objectives.

As the committee deliberated its charge, it identified numerous models for the development of geospatial indicators and tools, and some of those models include elements of validation and community engagement. However, the committee came to understand that a data strategy for developing EJ tools needs to embody the concept to be measured and to explicitly integrate community engagement into tool development as an integral part of tool validation. These traits are portrayed in the conceptual framework for the construction of EJ tools introduced in Chapter 3 (as Figure 3.2) and reproduced in this chapter as Figure 8.1. This framework characterizes the desired outcomes of a tool—transparency, trust, and legitimacy—as dependent on substantive and iterative community engagement, validation of indicators and tool results, and detailed documentation of decisions and approaches. A trusted, accurate, and useful EJ tool relies as much on meaningful interchange with communities as it does on technical expertise; it conforms to cultural understanding while being defensible with logic and scientific data. No perfect solution or single approach will satisfy all concerns, but a well-developed tool will be transparent and document uncertainties embedded within it.

The committee was asked to provide recommendations to be incorporated into an overall data strategy for the White House Council on Environmental Quality's (CEQ)

EJ geospatial tool(s) (see Box 1.1). Currently, the Climate and Economic Justice Screening Tool (CEJST) is the only tool that has been developed by CEQ. As described in Chapter 1, the committee concluded early in its deliberations that a good data strategy appropriate for CEQ and CEJST would also be appropriate for other EJ tool developers and tools. Concepts and recommendations in this report and synthesized in this chapter are therefore generalized and applicable to the development or management of any geospatial EJ tool by developers at CEQ and elsewhere. Tool developers may find that some of the recommendations herein could be implemented sooner with few resources. Other recommendations might be incorporated in some number of tool versions in the future. Still others might not fit within existing tool organization and therefore might be part of a new or separate tool to be developed in the future.

The recommendations presented here are organized around the elements of the committee's conceptual diagram for the development of EJ tools (Figure 8.1). As described in Chapter 3, this model evolves beyond current models for indicator construction by emphasizing community engagement as a means of information gathering and model validation in all construction components. Tool developers will be more familiar with the innermost components of the model as important aspects of composite indicator construction, and in fact, the description of the conceptual framework in Chapter 3 discusses those components first. However, because community engagement, documentation, and validation are central to building a transparent, trustworthy, and legitimate tool, this chapter begins with those topics.

TRANSPARENCY, TRUST, AND LEGITIMACY

A transparent tool is one in which the tool's goals, processes, data, and uncertainties are recognized and understood by the tool users and people interested in the results. A legitimate tool is often directly connected with how much acceptance communities will have of the tool and its results. Creating a tool that builds transparency, trust, and legitimacy into aspects of tool construction (i.e., those depicted in Figure 8.1) is a good data strategy for yielding better data selection, integration processes, and outcomes. This is a tool that follows good indicator construction practices (as described in Chapters 3 and 5) and integrates those indicators (Chapter 6) and validates the results (Chapter 7) in an iterative manner through systematic and sustained technical and community peer review.

Incorporating advice from communities and appropriate technical experts in an iterative matter throughout tool development is vital to building transparency, trust, and legitimacy in the process and outputs. Sustained interaction with community experts and advisory boards reflecting community populations is vital so that choices related to the tool can be informed and improved iteratively based on lived experience. This is particularly true for the selection of tool indicators, the data that measure those indicators, and the processes to integrate those indicators. Inadequate indicators or indicator integration techniques may inaccurately represent disproportionate exposure to the social and environmental hazards that lead to community disadvantage.

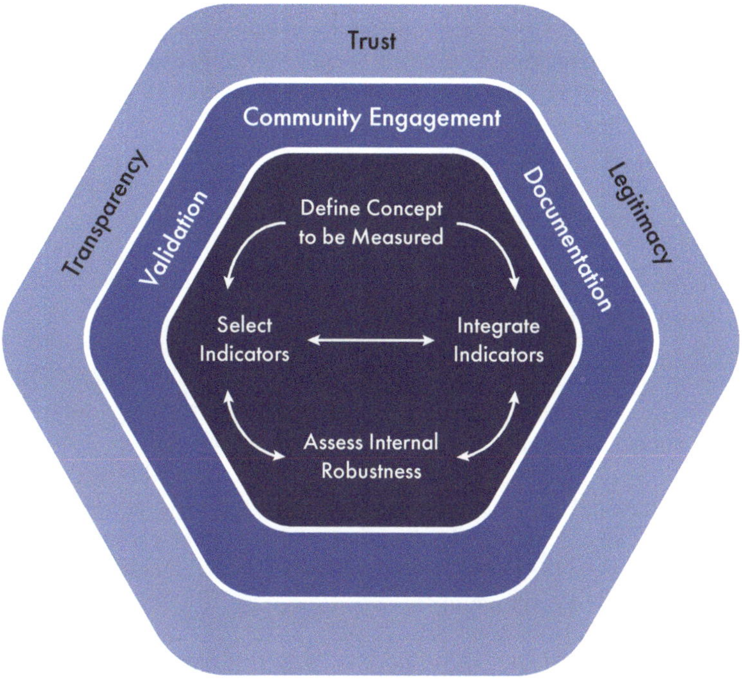

FIGURE 8.1 The committee's conceptual framework for the development of environmental justice tools. The arrows in the innermost ring indicate the direction of influence that each aspect of composite indicator construction has on another (i.e., defining the concept to be measured will influence the selection and integration of indicators, selection of indicators influences integration of indicators and vice versa, assessing internal robustness influences selection and integration of indicators and vice versa).

COMMUNITY ENGAGEMENT

Choosing appropriate indicators, datasets, and integration approaches requires more than statistical robustness to achieve valid results. A good data strategy includes systematic and sustained community engagement in these activities to verify and enhance the legitimacy of and trust in a tool's approaches. Community engagement is also necessary to validate tool results, to understand what kinds of errors are likely; why, where, and how they might be overcome; and how uncertainties in tool results might affect the decisions the tool is meant to inform. Chapter 3 highlights the crucial role of community engagement throughout the entire process of constructing composite indicator-based EJ tools. It also discusses the wide spectrum of community engagement models that could be employed and the need for transparency and honesty in choosing the appropriate model. Chapter 7 stresses the importance of lived experience in tool validation and processes for community engagement in achieving validation.

Recommendation 1: Create and sustain community partnerships that provide forums and opportunities to identify local environmental justice issues, identify the indicators and datasets for measuring them, and determine whether tool results reflect community lived experiences.

Partnering with communities to improve the representation of lived experience with the tool can be achieved through:

- Developing a strategic plan that operationalizes community input regarding indicators, data, and tool approaches;
- Being transparent about how community input will be addressed;
- Seeking continued feedback from communities about the improvement of the tool being developed, particularly focusing on input from those representing marginalized communities; and
- Creating a nonburdensome process for communities that feel they are not well represented in the tool to engage (e.g., reducing economic barriers to meaningful engagement through compensation, childcare, travel assistance, or other appropriate forms of material assistance to members of overburdened or underrepresented communities).

Meaningful community engagement represents a shift in the spectrum of public participation (see Chapter 3). Effective engagement processes allow communities to feel involved in governmental decisions that have local implications and empowered to influence these decisions through their input. Many EJ issues are local in scope, and close community engagement helps bring local issues into context, not only for understanding burdens across communities but also for finding targeted solutions that address unique needs.

Community engagement could require significant resources and capacities to be meaningful. Given the scale and scope of national-level EJ tools intended (a) to define different populations and diverse geographies; (b) to be responsive to different and potentially opposing needs, attitudes, and priorities; and (c) to inform decision making in multiple sectors, tool developers will need to consult experts in community engagement and rely on, for example, advisory panels to help design appropriate programs tailored for an individual tool. Because in-depth engagement cannot realistically occur at the national level with all communities represented by a tool, methods to identify representative communities, to design tool feedback methodologies, and to validate decisions made during indicator construction will necessarily be important in the design of an engagement program. And given that it is common for EJ tools such as CEJST to use indicators from other tools or datasets developed for other purposes, data selection criteria might include guidelines to prioritize those data and indicators that have been validated through community engagement.

DOCUMENTATION

The heterogeneities in the physical and social factors that affect the well-being of communities across the country, as well as in the data available to measure those factors, are large. As a result, there will never be complete consensus regarding the "best" data or methodologies for identifying community disadvantage in an EJ tool, nor will there ever be complete satisfaction in the results derived from such a tool. Furthermore, no matter how good a tool is or how valid its results may be, the transparency, trust, and legitimacy of an EJ tool are more likely to be questioned if the tool is not accompanied by thorough documentation.

The middle ring of the conceptual framework in Figure 8.1 includes community engagement, validation, and documentation; these components are largely driven by communication. The community engagement and validation components provide opportunities for dialog between tool developers and the community, with a large focus of the communication being focused inward to the tool development process. Documentation is an important means of providing information outward to the public. Thorough documentation of all tool components and approaches is vital to ensure proper tool use, to help decision makers understand where and how the tool may be accurate and what kinds of uncertainties should be expected, and to know when tool results need to be supplemented with other types of information. Good documentation makes the strengths and weaknesses of the tool clear to interested and affected tool users or community members and provides guidance regarding how best to use the tool to inform decision making.

Recommendation 2: Provide thorough and accessible documentation of indicator design processes and decisions, including descriptions and rationale for all major indicator construction components.

The current documentation of CEJST methodology and data is laudable. It includes descriptions of the burden categories and datasets, instructions on how to use the tool, and access to the code via a public repository. Less clear are the processes and rationale for decisions regarding indicator selection, data normalization, data treatments, thresholds, indicator aggregation, assessing statistical and conceptual coherence, uncertainty analysis, external validation via community engagement, and the design of the user interface and mapping tool. Documenting this information in formats and via media that are accessible for a variety of technical and nontechnical audiences and users to obtain and interact with will increase trust in and the usefulness of the tool for different decision-making applications. Other EJ tools, such as California's CalEnviroScreen and New Jersey's EJMAP, include information about which indicators are selected, their significance to EJ and human health, and detailed methods that allow users to understand the rationale and thought process that go into tool construction.

As discussed below (in the section on Defining the Concept to be Measured), the committee sees the value of using a structured framework that requires explicit and careful consideration of the interlinked decisions necessary to construct a composite indicator-based EJ tool. A structured composite indicator construction framework can also serve as the template for documenting these decisions.

VALIDATION

Validating composite EJ tools such as CEJST is a nuanced process. Technical decisions related to a tool are often constrained because of external forces (e.g., mandates) or data criteria requirements. For example, the choice to use census tracts to define communities in CEJST is aligned with the Executive Order (E.O.) that mandated the creation of CEJST. The choice was made with the knowledge that census tract boundaries do not always align with community boundaries and that large disparities in community health and well-being within a census tract may exist. However, the choice also takes advantage of national datasets available at that scale. Given that such compromises are inevitable during the development of any tool, and given that no single definitive measure might be available to validate for the purpose of a tool (e.g., "overburden" or "disadvantage" for CEJST), validation methodologies need to be applied throughout the construction of a tool to determine how well the tool relates to real-world conditions.

Recommendation 3: Validate tool development and evaluative processes throughout the construction of an environmental justice tool using approaches such as ground truthing, convergent validation, and community validation to ensure that the tool's indicators and results reflect lived experiences.

Tool validation techniques can be applied to ensure that a tool is stable, accepted, and scientifically sound. Chapter 5 describes effective validation, spanning how well the indicators measure what they are supposed to (construct validity), the degree of alignment among indicators (concurrent validity), and the indicators' representativeness of the underlying concept (content validity). Methodological components and processes are outlined that can be applied during tool construction to ensure that a tool and its findings are rooted in the realities and lived experiences of communities. Validation of indicators and tool results may take the form of a combination of technical, statistical, and community engagement activities. Ground truthing, for example, includes establishing information that can be used to compare with modeled information, and there are multiple ways to do this.

- **Convergent validation** compares tool components or results with those of similar tools. They can take the form, for example, of correlation analysis of tool results.
- **Community validation** is an iterative process conducted through collaborative engagement with communities to compare how well the tool reflects lived experiences. Consistent engagement throughout the tool development or upgrading process allows developers to test decisions, approaches, and tool results against community member narratives, while empowering communities to accept or refute definitions being assigned to them.

- **Mixed methods** that allow collection and analysis of both qualitative and quantitative datasets are framed withing research models to better understand multiple perspectives on any issue and are well suited for tool validation. Although mixed methods challenge the "traditional" scientific mindset focused on quantitative data, their use will result in data interpretation and informed research practices that allow for the incorporation of lived experiences into data analyses.

Tool validation needs to be done in consultation with communities, community experts, and researchers (e.g., environmental health experts) and results in improvements in data quality and increased transparency and trust in the tool development process.

Supplemental analysis conducted outside the tool development process using independent external datasets is an important means of checking indicator data sources for gaps or inaccuracies. They can be used to compare, for example, the spatial correlations between results of different tools for the sake of comparison. CEQ might conduct supplemental analysis to, for example, compare the distribution of race/ethnicity indicators and CEJST outputs to test the validity of CEJST's current formulation. The analysis can result in a greater understanding of sociodemographic composition and determinants of health in communities identified and generate localized narratives to better understand lived experience.

DEFINING THE CONCEPT TO BE MEASURED

Chapter 2 describes the concept of community disadvantage, its complex nature, and the processes and structures (e.g., discrimination and racism) that lead to overburdening of some communities by stressors and underinvestment in them by private and public capital. Chapter 3 discusses the challenge of measuring and identifying concepts such as community disadvantage through EJ tools such as CEJST and the construction of composite indicators—reducing a multidimensional concept into a single value. Composite indicator construction involves a set of carefully considered interlinked decisions, starting with a clear definition of the concept being measured. Use of a structured framework in the construction of a composite indicator helps to improve transparency, trust, and legitimacy by ensuring that all composite indicator construction decisions are considered explicitly and lead to the stated objective of the composite indicator, and then are documented carefully. Multiple frameworks exist to guide composite indicator construction with similar frameworks. The committee found the 10-step Organisation for Economic Cooperation and Development (OECD) pocket guide to be useful for framing composite indicator construction (Saisana et al., 2019). The OECD pocket guide does not specify how to construct a particular tool but rather highlights the decisions that should be made, the interconnectedness of decisions, the need for engagement with community members and other interested and affected parties, and the importance of validation via lived experiences. The first step in composite indicator development in the OECD pocket guide and others is to define the concept that the composite indicator is intended to measure.

Recommendation 4: Initiate environmental justice tool and indicator construction with the development of clear objectives and definitions for the concept(s) to be measured. Follow a structured composite indicator development process that requires explicit consideration and robustness analysis of all major decisions involved with indicator selection and integration; assessment of uncertainties; and validation and visualization of results.

The categories of burden within CEJST reflect the priorities of E.O. 14008, but without transparent roles and relationships within the tool itself. The lack of explicit structure linking the defined concept being measured, its dimensions, and the indicators of those dimensions creates an implicit weighting scheme within CEJST in which the categories of burden with more indicators have greater relative importance since they increase the chance that a census tract will meet one of its indicator conditions. A future data strategy that will incorporate the state of the art and practice in composite indicator construction includes

- Defining the concept to be measured and developing a description of its multiple facets or dimensions.
- Selecting the indicators that measure each dimension. This type of top-down approach promotes conceptual clarity and provides strategies for effectively weighting and aggregating indicators.
- Analyzing, treating, normalizing, and weighting the indicators as appropriate.
- Integrating/aggregating the indicators.
- Assessing statistical and conceptual robustness and coherence and determining the impact of uncertainties.
- Validating the results and presenting them visually.

If future versions of CEJST incorporate more sophisticated indicator integration methods for capturing cumulative burdens (see below), the lack of an explicit conceptual structure linking concept definition, dimensions, and indicators may be problematic.

SELECTING AND ASSESSING INDICATORS

As discussed in Chapter 5, indicators are quantitative proxies for abstract concepts that are developed from existing datasets. Indicators may be selected from those developed by others, or they may be created from existing datasets. The selection of indicators and datasets requires consideration of their technical and practical characteristics and how well they support the tool or indicator objectively. Many indicators may be based on empirical data that can be assessed statistically. However, not all data are empirical, nor are they of equal quality, expressed in the same units or at the same scales, or collected for the same purposes. Given the close interconnection between concept definition, indicator selection, weighting, and methods to ground truth, the decisions related to indicator selection are paramount to a high-quality and accurate tool.

Recommendation 5: Adopt systematic, transparent, and inclusive processes to identify and select indicators and datasets that consider technical criteria (validity, sensitivity, specificity, robustness, reproducibility, and scale) and practicality (measurability, availability, simplicity, affordability, credibility, and relevance). Evaluate measures in consultation with federal agencies, technical experts, and community partners.

Selecting indicators and datasets for any tool requires a careful and structured approach to composite indicator construction, as described in Recommendation 4 and includes a systematic scan of available data. In the case of CEJST, the indicators and corresponding datasets appear reasonable; however, they represent only a small subset of the wide range of possible federal and national datasets that could be used to inform an EJ tool (see Appendix D for examples). A systematic scan of the federal- and national-level landscape, perhaps in partnership with federal agencies and other data providers or a steering committee, could identify other or more appropriate indicators for defining community disadvantage.

After identifying potential indicators, correlation analysis can then inform the selection of indicators and their organization into categories. Analysis that demonstrates highly correlated indicators might also indicate redundancy in the indicator set. The result might be an unintended implicit weighting scheme if the highly correlated datasets are used. Correlations among indicators or datasets that are low, negative, or statistically insignificant signify poor statistical alignment with the concept to be measured. Both results provide an empirical rationale for the targeted revision of the indicator set. Employing statistical analysis to guide indicator selection helps ensure that input indicators in an EJ tool are both thematically and statistically coherent with community disadvantage. Such analysis also increases methodological transparency.

Economic Indicators

The socioeconomic indicator in CEJST is the most influential variable in the tool, given that for a tract to be designated as disadvantaged it must meet the socioeconomic threshold as well as any one of the other 30 indicators across the eight categories of burden. The measure of low income is especially important in identifying communities as disadvantaged as it is coupled with 26 of these indicators (i.e., all but those in the workforce development category). As is the case with any indicator, it is important to consider how well it reflects the lived experience of the community that it is meant to represent. Using the federal poverty level to determine economic burden at the national scale can mischaracterize economic burden because of the varying cost of living around the nation that is not considered.

Recommendation 6: Choose measures of economic burden beyond the federal poverty level that reflect lived experiences, attend to other dimensions of wealth, and consider geographic variations in cost of living.

Using a single, uniform low-income measure in a tool such as CEJST may not accurately reflect lived experiences, even after doubling the standard poverty level and accounting for the cost of living. Other indicators can be used as socioeconomic measures (e.g., U.S. Department of Housing and Urban Development Public Housing/ Section 8 Income limits for low income,[1] homeownership rates, median home values, or a weighted income metric) as long as it is acknowledged that income-based measures deserve scrutiny because of the effects of income on all aspects of a person's or household's quality of life (e.g., nutrition, health care, and education). Metrics of income do not necessarily measure wealth, and the wealth gap between high-income and low-income households is larger than the income gap and is growing more rapidly (Horowitz, Igielnik, and Kochhar, 2020). Tool developers should work alongside communities to identify other dimensions of wealth that would more accurately reflect these differences and to perform sensitivity analyses on these indicators and their thresholds in the process.

Consideration of Racism and Race

There are important distinctions between measures of race or ethnicity and measures of racism. While measures of race simply identify the composition of people living in communities, measures of racism reflect the system of policies and practices that negatively impact specific races or ethnicities and is a key driver of climate and economic injustice within the United States. Chapter 2 of this report describes the disproportionate exposure to hazards in communities largely populated by peoples of color. Incorporating racism as an indicator in an EJ tool can strengthen and add legitimacy to the tool. If a tool developer is unable to incorporate approaches to acknowledge the history of racism and land use policies that have led to the injustices and disparities observed in communities populated by peoples of color, then they should explicitly factor race or ethnicity as an indicator to measure community disadvantage.

Recommendation 7: Use indicators that measure the impacts of racism in policies and practices that have led to the disparities observed today. If indicators of racism are not used, explicitly factor race and ethnicity as indicators when measuring community disadvantage.

Using measures of racism allows tool developers to identify disadvantage being placed on people of color or certain ethnicities. However, if such measures cannot be readily used by tool developers because they do not meet data criteria, disaggregated data on race and ethnicity that are readily available and could be used instead—for example, U.S. Census data on race and ethnicity—until appropriate indicators can be found or developed and incorporated into future iterations of a tool. Because not all people of color have the same lived experiences or histories of discrimination, people

[1] See HUD's FY 2023 methodology for determining Section 8 limits at https://www.huduser.gov/portal/datasets/il//il23/IncomeLimitsMethodology-FY23.pdf (accessed March 8, 2024).

of color should not be treated as a monolithic group. There is a large and growing range of indicators or measures of racism described in Chapter 5 and Appendix D. Tool developers can work with representatives of communities of color and subject-matter experts to revisit existing empirical data and consider the metrics, quantitative data, and qualitative data that reflect community lived experiences.

While CEQ develops measures of racism to be incorporated directly into CEJST, supplemental analysis comparing the distribution of race/ethnicity indicators and CEJST outputs could help CEQ tool developers gain a greater understanding of how well CEJST captures community disadvantage in its current formulation. The results can reveal how the input and output indicators of an EJ tool are distributed by racial and ethnic composition. Such analyses inform understanding of the degree of racial and ethnic disparities in the designation of disadvantaged places, provide insight into possible measures of racism that led to these disparities, and address questions about the ability of CEJST to identify disadvantage without the inclusion of race or ethnicity indicators. Publication of supplemental analyses results regarding the relationship between race/ethnicity and CEJST would show CEQ responsiveness to public comments, increasing trust in the tool development process and tool results.

INTEGRATING INDICATORS AND CUMULATIVE IMPACT SCORING

Measuring and redressing cumulative impacts is a stated objective of E.O. 14008, the CEQ, and EJ advocates. It better reflects the synergism between environmental and socioeconomic burdens and their accumulation over time, which is important if the interplay of multiple concurrent stressors interacting with sociodemographic, environmental, and public health factors leads to the possibility of the total impacts being greater than the sum of the individual stressors.

Recommendation 8: Designate communities as disadvantaged based on cumulative impact scoring approaches that are informed by the state of science; the knowledge, needs, and experiences of agencies, tool developers, and users; and validation efforts conducted in partnership with affected communities. Choose an approach to represent cumulative impacts, such as threshold approaches (e.g., summing thresholds or categories exceeded) or aggregation-based approaches for composite indicator construction (e.g., compensatory, noncompensatory, or a combination of approaches).

CEJST employs a binary approach for identifying disadvantaged communities, one that does not discern communities facing single and multiple burdens. A community is designated as either disadvantaged or not disadvantaged. Cumulative impact scoring is an established practice in state-level EJ tools, including CalEnviroScreen and Maryland's EJ Screening Tool. Such scoring enables clearer comparison of communities and prioritization of investment based on the severity of burden.

Indicator selection, weighting, and aggregation methods for capturing cumulative burdens are intertwined modeling decisions that should be made in an iterative and engaged manner, reflecting scientific knowledge, tenets of indicator construction, perspectives of interested and affected parties, and lived experiences. Although reaching consensus on indicator weights can be difficult, weighting cannot be avoided since it has a major impact on composite indicator results when aggregating. Methods to address this challenge could include interactive methods for varying these decisions, visualizing their effects on the results, and visualizing the composite indicator decomposed into subgroups and individual indicators in both chart and map form. Using the design principles of interactive cartography and spatial decision support system methods will make these analyses and visualization more accessible and understandable to diverse audiences.

As Chapter 3 and Chapter 6 discuss, there are methods for group and collaborative decision making to achieve consensus in indicator selection, weighting, and aggregation decisions. CEQ and other EJ tool developers should pursue collaborative and engaged decision making with interested and affected parties and communities on indicator selection, weighting, and aggregation based on expertise in community engagement practices. Throughout the indicator integration process, developers should partner with communities to understand and reflect cumulative impacts in the integration. Additionally, consulting with other creators of cumulative impact tools (as identified in Chapter 4) will provide insight into lessons learned.

ASSESSING ROBUSTNESS

Uncertainty and sensitivity analyses inform the development of a composite indicator (see Chapter 6). Uncertainty analysis quantifies the variability in model outputs based on changes in model inputs. Sensitivity analysis apportions variability in model outputs to different input parameters or model structures. Both types of analyses can be conducted as a local analysis, in which one parameter is evaluated at a time, and global analysis, in which multiple parameters and their interactions are assessed simultaneously using Monte Carlo simulation.

Recommendation 9: Perform and document uncertainty and sensitivity analyses to evaluate how decisions made during tool development affect tool results. Decisions to be assessed may relate to, for example, the selection of indicators and indicator thresholds; model structure; processes related to the normalization, weighting, and aggregation of indicators; and the criteria used for the final designation or classification of communities.

Constructing a composite indicator requires numerous modeling decisions, each of which includes multiple plausible options based on scientific knowledge, available data, and community preferences. These modeling decisions can independently and conjointly influence which communities the tool identifies as disadvantaged. Particularly for composite indicators such as CEJST that are used for high-consequence resource allocation and project prioritization, it is crucial to understand the degree to which modeling decisions affect the robustness of the outputs. Uncertainty and sensitivity analysis are core best practices for quality assurance in composite indicator construction and should be a part of a data strategy for any tool, including CEQ tools such as CEJST.

Conducting a global uncertainty analysis of CEJST will improve understanding of the precision of disadvantaged community designation when the model is subjected to alternative construction choices. Subsequent global sensitivity analysis can identify which modeling decisions are the major sources of uncertainty. Epistemic uncertainty can then be diminished through subsequent research, targeted data collection, and improved modeling, ultimately reducing statistical fragility and increasing the transparency of the modeling process. Conducting global uncertainty and sensitivity analysis can also provide empirical results to support response to public queries about the certainty of overall and geographically specific designation of community disadvantage.

MOVING FORWARD

Not all these recommendations can be implemented in the next release of CEJST. However, some of the recommendations could be implemented in the short term. For example, even before any changes to the tool construction are made, CEQ could expand the current documentation of CEJST, thereby improving communication with interested and affected parties and tool users by explaining design processes and decisions, including descriptions and rationale for all major indicator construction components, and describing robustness analysis and results. It can begin to expand its community engagement efforts to create the community partnerships that provide forums and opportunities to identify local EJ issues, identify the indicators and datasets for measuring them, and determine whether tool results reflect community lived experiences. CEQ tool developers can work with developers from other agencies and organizations and consider approaches incorporating cumulative impact scoring into its overall approach.

The conceptual framework for indicator and tool construction depicted in Figure 8.1 is focused on a clearly defined objective (i.e., the concept to be measured). A data strategy can only be successful if the concept to be measured is well defined and accepted by those that will be affected by the definition. Measuring something as complex as community disadvantage cannot be a numerical and statistical exercise focused solely on empirical data when the goal is to develop an EJ tool that is transparent, trusted, and legitimate. The tool needs to be grounded, developed, refined, and validated through communication and collaboration with the communities that the tool is intended to define.

References

Aaronson, D., Hartley, D., and Mazumder, B. 2017. "The Effects of the 1930s HOLC 'Redlining' Maps." Federal Reserve Bank of Chicago Working Paper Series (WP-2017-12). https://ideas.repec.org/p/fip/fedhwp/wp-2017-12.html.

Aaronson, D., Hartley, D., and Mazumder, B. 2021. "The Effects of the 1930s HOLC 'Redlining' Maps." *American Economic Journal: Economic Policy* 13 (4): 355-392. https://doi.org/10.1257/pol.20190414.

Aaronson, D., Faber, J., Hartley, D., Mazumder, B., and Sharkey, P. 2021. "The Long-Run Effects of the 1930s HOLC 'Redlining' Maps on Place-Based Measures of Economic Opportunity and Socioeconomic Success." *Regional Science and Urban Economics* 86: 103622. https://doi.org/10.1016/j.regsciurbeco.2020.103622.

Abrams, C. 1955. *Forbidden Neighbors: A Study of Prejudice in Housing*. New York: Harper.

Acevedo-Garcia, D., Lochner, K. A., Osypuk, T. L., and Subramanian, S. V. 2003. "Future Directions in Residential Segregation and Health Research: A Multilevel Approach." *American Journal of Public Health* 93(2): 215-221. https://doi.org/10.2105/AJPH.93.2.215.

Achakulwisut, P., Mickley, L. J., and Anenberg, S. C. 2018. "Drought-Sensitivity of Fine Dust in the US Southwest: Implications for Air Quality and Public Health Under Future Climate Change." *Environmental Research Letters* 13(5): 054025. https://doi.org/10.1088/1748-9326/aabf20.

Adgate, J. L., Goldstein, B. D., and McKenzie, L. M. 2014. "Potential Public Health Hazards, Exposures and Health Effects from Unconventional Natural Gas Development." *Environmental Science & Technology* 48(15): 8307-8320. https://doi.org/10.1021/es404621d.

Adkins-Jackson, P. B., Chantarat, T., Bailey, Z. D., and Ponce, N. A. 2022. "Measuring Structural Racism: A Guide for Epidemiologists and Other Health Researchers." *American Journal of Epidemiology* 191(4): 539-547. https://doi.org/10.1093/aje/kwab239.

Agbai, C. O. 2022. *Wealth Begins at Home: A Historical Analysis of the Role of the 1944 GI Bill in Linking, Race, Place, Wealth, and Health in America*. Ph.D. dissertation, Brown University.

Aguña, C. G., and Kovacevic, M. 2010. "Uncertainty and Sensitivity Analysis of the Human Development Index." United Nations Development Programme. https://hdr.undp.org/content/uncertainty-and-sensitivity-analysis-human-development-index.

Ahmed, M. K., Scretching, D., and Lane, S. D. 2023. "Study Designs, Measures and Indexes Used in Studying the Structural Racism as a Social Determinant of Health in High Income Countries from 2000–2022: Evidence from a Scoping Review." *International Journal for Equity in Health* 22(1): 4. https://doi.org/10.1186/s12939-022-01796-0.

Albo, Y., Lanir, J., and Rafaeli, S. 2019. "A Conceptual Framework for Visualizing Composite Indicators." *Social Indicators Research* 141(1): 1-30. https://doi.org/10.1007/s11205-017-1804-0.

Alexander, M. 2012. *New Jim Crow—Mass Incarceration in the Age of Colorblindness*. New York: The New Press.

Alexeeff, G. V., Faust, J. B., August, L. M., Milanes, C., Randles, K., Zeise, L., and Denton, J. 2012. "A Screening Method for Assessing Cumulative Impacts." *International Journal of Environmental Research and Public Health* 9(2): 648-659. https://doi.org/10.3390/ijerph9020648.

Alkon, A. H. 2018. "Food Justice: An Environmental Justice Approach to Food and Agriculture." In *The Routledge Handbook of Environmental Justice*, edited by R. Holifield, J. Chakraborty, and G. Walker, 412-424. London: Routledge.

Alson, J. G., Robinson, W. R., Pittman, L., and Doll, K. M. 2021. "Incorporating Measures of Structural Racism into Population Studies of Reproductive Health in the United States: A Narrative Review." *Health Equity* 5(1): 49-58. https://doi.org/10.1089/heq.2020.0081.

Amini, H., Danesh-Yazdi, M., Di, Q., Requia, W., Wei, Y., AbuAwad, Y., Shi, L., Franklin, M., Kang, C., Wolfson, M. J., James, P., Habre, R., Zhu, Q., Apte, J. S., Andersen, Z. J., and Xing, X. 2023. Annual Mean PM2.5 Components Trace Elements (TEs) 50m Urban and 1km Non-Urban Area Grids for Contiguous U.S., 2000-2019, v1. Palisades, New York: NASA Socioeconomic Data and Applications Center (SEDAC). https://doi.org/10.7927/1x94-mv38.

Anenberg, S. C., and Kalman, C. 2019. "Extreme Weather, Chemical Facilities, and Vulnerable Communities in the U.S. Gulf Coast: A Disastrous Combination." *GeoHealth* 3(5): 122-126. https://doi.org/10.1029/2019GH000197.

Arcury, T. A., Quandt, S. A., and Russell, G. B. 2002. "Pesticide safety among farmworkers: perceived risk and perceived control as factors reflecting environmental justice." *Environ Health Perspectives* 110 Suppl 2 (Suppl 2): 233-240. https://doi.org/10.1289/ehp.02110s2233.

Arksey, H., and O'Malley, L. 2005. "Scoping Studies: Towards a Methodological Framework." *International Journal of Social Research Methodology* 8 (1): 19-32. https://doi.org/10.1080/1364557032000119616.

Arnstein, S. R. 1969. "A Ladder of Citizen Participation." *Journal of the American Institute of Planners* 35(4): 216-224. https://doi.org/10.1080/01944366908977225.

Arriens, J., Schlesinger, S., and Wilson, S. M. 2022. *Environmental Justice Mapping Tools: Use and Potential in Policy Making to Address Climate Change*. Washington, DC: National Wildlife Federation. https://www.nwf.org/-/media/Documents/PDFs/Environmental-Threats/Environmental-Justice-Mapping-Tools.ashx?la=en&hash=347578719433ACCFCF5C50F1FE56C98AFFD17981.

Asmelash, L. 2023. "DEI Programs in Universities Are Being Cut Across the Country. What Does This Mean for Higher Education?" *CNN*, June 14. https://www.cnn.com/2023/06/14/us/colleges-diversity-equity-inclusion-higher-education-cec/index.html.

Associated Press. 2023. "Exclusion of Race in Federal Climate Justice Screening Tool Could Worsen Disparities, Analysis Says." *U.S. News & World Report*, July 20. https://www.usnews.com/news/news/articles/2023-07-20/exclusion-of-race-in-federal-climate-justice-screening-tool-could-worsen-disparities-analysis-says.

August, L. M., Faust, J. B., Cushing, L., Zeise, L., and Alexeeff, G. V. 2012. "Methodological Considerations in Screening for Cumulative Environmental Health Impacts: Lessons Learned from a Pilot Study in California." *International Journal of Environmental Research and Public Health* 9 (9): 3069-3084. https://doi.org/10.3390/ijerph9093069.

August, L.M., Bangia, K., Plummer, L., Prasad, S., Ranjbar, K., Slocombe, A., and Wieland, W. 2021. *CalEnviroScreen 4.0*. California Environmental Protection Agency, Office of Environmental Health Hazard Assessment. https://oehha.ca.gov/media/downloads/calenviroscreen/report/calenviroscreen40reportf2021.pdf.

Bae, J., and Kang, S. 2022. "Another Injustice? Socio-Spatial Disparity of Drinking Water Information Dissemination Rule Violation in the United States." *Journal of Policy Studies* 37(4): 77-89. https://doi.org/https://doi.org/10.52372/jps37405.

Bae, J., Kang, S., and Lynch, M. J. 2023. "Drinking Water Injustice: Racial Disparity in Regulatory Enforcement of Safe Drinking Water Act Violations." *Race and Justice*. https://doi.org/10.1177/21533687231189854.

Bae, J., and Lynch, M. J. 2023. "Ethnicity, Poverty, Race, and the Unequal Distribution of US Safe Drinking Water Act Violations, 2016-2018." *The Sociological Quarterly* 64(2): 274-295. https://doi.org/10.1080/00380253.2022.2096148.

Baker, E., Carley, S., Castellanos, S., Nock, D., Bozeman, J. F., Konisky, D., Monyei, C. G., Shah, M., and Sovacool, B. 2023. "Metrics for Decision-Making in Energy Justice." *Annual Review of Environment and Resources* 48: 737-760. https://doi.org/10.1146/annurev-environ-112621-063400.

Bakkensen, L. A., Fox-Lent, C., Read, L. K., and Linkov, I. 2017. "Validating Resilience and Vulnerability Indices in the Context of Natural Disasters." *Risk Analysis* 37(5): 982-1004. https://doi.org/10.1111/risa.12677.

Bailey, Z. D., Feldman, J. M., and Bassett, M. T. 2020. "How Structural Racism Works—Racist Policies as a Root Cause of U.S. Racial Health Inequities." *New England Journal of Medicine* 384(8): 768-773. https://doi.org/10.1056/NEJMms2025396.

Bailey, Z. D., Krieger, N., Agénor, M., Graves, J., Linos, N., and Bassett, M. T. 2017. "Structural Racism and Health Inequities in the USA: Evidence and Interventions." *Lancet* 389(10077): 1453-1463. https://doi.org/10.1016/s0140-6736(17)30569-x.

Balakrishnan, C., Su, Y., Axelrod, J., and Fu, S. 2022. *Screening for Environmental Justice: A Framework for Comparing National, State, and Local Data Tools*. Washington, DC: Urban Institute. https://www.urban.org/research/publication/screening-environmental-justice-framework-comparing-national-state-and-local.

Balazs, C. L., and Ray, I. 2014. "The Drinking Water Disparities Framework: On the Origins and Persistence of Inequities in Exposure." *American Journal of Public Health* 104(4): 603-611. https://doi.org/10.2105/ajph.2013.301664.

Balazs, C. L., Morello-Frosch, R., Hubbard, A., and Ray, I. 2011. "Social Disparities in Nitrate-Contaminated Drinking Water in California's San Joaquin Valley." *Environmental Health Perspectives* 119(9): 1272-1278. https://doi.org/10.1289/ehp.1002878.

Bana e Costa, C. A., Oliveira, M. D., Vieira, A. C. L., Freitas, L., Rodrigues, T. C., Bana e Costa, J., Freitas, Â., and Santana, P. 2023. "Collaborative Development of Composite Indices from Qualitative Value Judgements: The EURO-HEALTHY Population Health Index model." *European Journal of Operational Research* 305(1): 475-492. https://doi.org/10.1016/j.ejor.2022.05.037.

Banzhaf, S., Ma, L., and Timmins, C. 2019a. "Environmental Justice: The Economics of Race, Place, and Pollution." *Journal of Economic Perspectives* 33(1): 185-208.

Banzhaf, S., Ma, L., and Timmins, C. 2019b. "Environmental Justice: Establishing Causal Relationships." *Annual Review of Resource Economics* 11(1): 377-398. https://EconPapers.repec.org/RePEc:anr:reseco:v:11:y:2019:p: 377-398.

Baptista, A. I., Perovich, A., Pulido-Velosa, M. F., Valencia, E., Valdez, M., and Ventrella, J. 2022. *Understanding the Evolution of Cumulative Impacts: Definitions and Policies in the U.S.* New York: Tishman Environment and Design Center at the New School. https://www.tishmancenter.org/projects-publications.

Bara, S., Driver, A., Gugssa, W., Hagerty, M., Ravichandran, V., Tellez, V., and Woldu, R. 2018. *A Review of Stakeholder Feedback and Indicator Analysis for the Maryland Environmental Justice Screening Tool*. Partnership for Action Learning in Sustainability (PALS), University of Maryland, College Park. http://hdl.handle.net/1903/21461.

Barnes, A., Luh, A., and Gobin, M. 2021. *Mapping Environmental Justice in the Biden-Harris Administration*. Washington, DC: Center for American Progress. https://www.americanprogress.org/article/mapping-environmental-justice-biden-harris-administration/.

Bassler, A., Brasier, K., Fogle, N., and Taverno, R. 2008. *Developing Effective Citizen Engagement: A How-To Guide for Community Leaders*. Harrisburg: Center for Rural Pennsylvania. https://www.rural.pa.gov/getfile.cfm?file=Resources/PDFs/research-report/archived-report/Effective_Citizen_Engagement.pdf&view=true.

Bazuin, J. T., and Fraser, J. C. 2013. "How the ACS Gets It Wrong: The Story of the American Community Survey and a Small, Inner City Neighborhood." *Applied Geography* 45: 292-302. https://doi.org/10.1016/j.apgeog.2013.08.013.

Becker, W., Saisana, M., Paruolo, P., and Vandecasteele, I. 2017. "Weights and Importance in Composite Indicators: Closing the Gap." *Ecological Indicators* 80: 12-22. https://doi.org/10.1016/j.ecolind.2017.03.056.

Berberian, A. G., Rempel, J., Depsky, N., Bangia, K., Wang, S., and Cushing, L. J. 2023. "Race, Racism, and Drinking Water Contamination Risk from Oil and Gas Wells in Los Angeles County, 2020." *American Journal of Public Health* 113(11): 1191-1200. https://doi.org/10.2105/AJPH.2023.307374.

Berberoglu, B. 1994. "Class, Race and Gender: The Triangle of Oppression." *Race, Sex & Class* 2(1): 69-77. http://www.jstor.org/stable/41680097.

Betts, K. R., and Hinsz, V. B. 2013. "Group Marginalization: Extending Research on Interpersonal Rejection to Small Groups." *Personality and Social Psychology Review* 17(4): 355-370. https://doi.org/10.1177/1088868313497999.

Beyer, K. M., Zhou, Y., Matthews, K., Bemanian, A., Laud, P. W., and Nattinger, A. B. 2016. "New Spatially Continuous Indices of Redlining and Racial Bias in Mortgage Lending: Links to Survival After Breast Cancer Diagnosis and Implications for Health Disparities Research." *Health Place* 40: 34-43. https://doi.org/10.1016/j.healthplace.2016.04.014.

Bhandari, S., Lewis, P. G., Craft, E., Marvel, S. W., Reif, D. M., and Chiu, W. A. 2020. HGBEnviroScreen: Enabling Community Action Through Data Integration in the Houston–Galveston–Brazoria Region. *International Journal of Environmental Research and Public Health* 17(4): 1130. https://doi.org/10.3390/ijerph17041130.

Bischoff, W. E., Weir, M., Summers, P., Chen, H., Quandt, S. A., Liebman, A. K., and Arcury, T. A. 2012. "The Quality of Drinking Water in North Carolina Farmworker Camps." *American Journal of Public Health* 102(10): e49-e54. https://doi.org/10.2105/AJPH.2012.300738.

Blackwood, L., and Cutter, S. L. 2023. "The Application of the Social Vulnerability Index (SoVI) for Geo- targeting of Post-disaster Recovery Resources." *International Journal of Disaster Risk Reduction* 92: 103722. https://doi.org /10.1016/j.ijdrr.2023.103722. https://www.sciencedirect.com/science/article/pii/S2212420923002029.

Blancas, F. J., and Lozano-Oyola, M. 2022. "Sustainable Tourism Evaluation Using a Composite Indicator with Different Compensatory Levels." *Environmental Impact Assessment Review* 93: 106733. https://doi.org/10.1016/j.eiar.2021.106733.

Blatt, L. R., Sadler, R. C., Jones, E. J., Miller, P., Hunter-Rue, D. S., and Votruba-Drzal, E. 2024. "Historical Structural Racism in the Built Environment and Contemporary Children's Opportunities." *Pediatrics* 153(2): e2023063230. https://doi.org/10.1542/peds.2023-063230.

BLS (Bureau of Labor Statistics). 2023. *Employer Costs for Employee Compensation - September 2023*. US Department of Labor (Washington, DC). https://www.bls.gov/news.release/pdf/ecec.pdf.

Bompoti, N. M., Coelho, N., and Pawlowski, L. 2024. "Is Inclusive More Elusive? An Impact Assessment Analysis on Designating Environmental Justice Communities in the US." *Environmental Impact Assessment Review* 104: 107354. https://doi.org/https://doi.org/10.1016/j.eiar.2023.107354.

Bonilla-Silva, E. 1997. "Rethinking Racism: Toward a Structural Interpretation." *American Sociological Review* 62(3): 465-480. https://doi.org/10.2307/2657316.

Borgonovo, E., Hazen, G. B., and Plischke, E. 2016. "A Common Rationale for Global Sensitivity Measures and Their Estimation." *Risk Analysis* 36(10): 1871-1895. https://doi.org/10.1111/risa.12555.

Bose, S., Madrigano, J., and Hansel, N. N. 2022. "When Health Disparities Hit Home: Redlining Practices, Air Pollution, and Asthma." *American Journal of Respiratory and Critical Care Medicine* 206 (7): 803-804. https://doi.org/10.1164/rccm.202206-1063ED.

Boyd, R. W., Lindo, E. G., Weeks, L. D., and McLemore, M. R. 2020. "On Racism: A New Standard for Publishing on Racial Health Inequities." *Health Affairs* (blog). July 2. https://www.healthaffairs.org/content/forefront/racism-new-standard-publishing-racial-health-inequities.

Bradshaw, T. 2008. "The Post-Place Community: Contributions to the Debate About the Definition of Community." *Community Development* 39: 5-16. https://doi.org/10.1080/15575330809489738.

Braveman, P. A., Arkin, E., Proctor, D., Kauh, T., and Holm, N. 2022. "Systemic and Structural Racism: Definitions, Examples, Health Damages, and Approaches to Dismantling." *Health Affairs* 41(2): 171-178. https://doi.org/10.1377/hlthaff.2021.01394. https://doi.org/10.1377/hlthaff.2021.01394.

Bravo, M. A., Anthopolos, R., Bell, M. L., and Miranda, M. L. 2016. "Racial Isolation and Exposure to Airborne Particulate Matter and Ozone in Understudied US Populations: Environmental Justice Applications of Downscaled Numerical Model Output." *Environment International* 92-93: 247-255. https://doi.org/https://doi.org/10.1016/j.envint.2016.04.008.

Bravo, M. A., Warren, J. L., Leong, M. C., Deziel, N. C., Kimbro, R. T., Bell, M. L., and Miranda, M. L. 2022. "Where Is Air Quality Improving, and Who Benefits? A Study of $PM_{2.5}$ and Ozone over 15 Years." *American Journal of Epidemiology* 191(7): 1258-1269. https://doi.org/10.1093/aje/kwac059.

Brown, D. A. 2022. *The Whiteness of Wealth*. New York: Crown.

Brown, E., Ojeda, V. D., Wyn, R., and Levan, R. 2000. *Racial and Ethnic Disparities in Access to Health Insurance and Health Care*. Los Angeles: UCLA Center for Health Policy Research and The Henry J. Kaiser Family Foundation.

Buckley, A., Hardy, P., and Field, K. 2022. "Cartography." In *Springer Handbook of Geographic Information*, edited by W. Kresse and D. Danko, 315-352. Cham: Springer International.

Bullard, R. D. 1993. "Race and Environmental Justice in the United States." *Yale Journal of International Law* 18: 12.

Bullard, R. D. 2001. "Environmental Justice in the 21st Century: Race Still Matters." *Phylon (1960-)* 49(3/4): 151-171. https://doi.org/10.2307/3132626.

Bullard, R. D., and Wright, B. 1987. "Environmentalism and the Politics of Equity: Emergent Trends in the Black Community." *Mid-American Review of Sociology* 12(2): 21-37. http://www.jstor.org/stable/23253043.

Bullard, R. D., and Wright, B. 2009. *Race, Place, and Environmental Justice After Hurricane Katrina: Struggles to Reclaim, Rebuild, and Revitalize New Orleans and the Gulf Coast*. New York: Routledge.

Bullard, R. D., Mohai, P., Saha, R., and Wright, B. 2007. *Toxic Wastes and Race at Twenty 1987-2007: A Report Prepared for the United Church of Christ Justice & Witness Ministries*. Cleveland, OH: United Church of Christ.

Burke, M., Driscoll, A., Heft-Neal, S., Xue, J., Burney, J., and Wara, M. 2021. "The Changing Risk and Burden of Wildfire in the United States." *Proceedings of the National Academy of Sciences of the United States of America* 118(2): e2011048118. https://doi.org/10.1073/pnas.2011048118.

Cain, L., Hernandez-Cortes, D., Timmins, C., and Weber, P. 2024. "Recent Findings and Methodologies in Economics Research in Environmental Justice." *Review of Environmental Economics and Policy* 18(1): 116-142. https://doi.org/10.1086/728100.

CalEPA (California Environmental Protection Agency). 2004. *Environmental Justice Action Plan*. https://calepa.ca.gov/wp-content/uploads/sites/6/2016/10/EnvJustice-ActionPlan-Documents-October2004-ActionPlan.pdf.

CalEPA. 2010. *Cumulative Impacts: Building a Scientific Foundation*. Office of Environmental Health Hazard Assessment. https://oehha.ca.gov/media/downloads/calenviroscreen/report/cireport123110.pdf.

CalEPA. 2013. *California Communities Environmental Health Screening Tool, Version 1 (CalEnviroScreen 1.0): Guidance and Screening Tool*. California Environmental Protection Agency. https://oehha.ca.gov/media/downloads/calenviroscreen/report/042313calenviroscreen1.pdf.

CalEPA. 2021. *Analysis of Race/Ethnicity and CalEnviroScreen 4.0 Scores*. California Environmental Protection Agency. https://oehha.ca.gov/media/downloads/calenviroscreen/document/calenviroscreen40raceanalysisf2021.pdf.

Callahan, C., Coffee, D., DeShazo, J. R., and González, S. R. 2021. *Making Justice40 a Reality for Frontline Communities: Lessons from State Approaches to Climate and Clean Energy Investments*. Los Angeles: UCLA Luskin Center for Innovation.

Caperna, G., Smallenbroek, O., Kovacic, M., and Papadimitriou, E. 2022. *JRC Statistical Audit of the 2022 Commitment to Reducing Inequality Index*. EUR 31259 EN. Luxembourg: Publications Office of the European Union. https://doi.org/10.2760/5642.

Carrillo, I., and Ipsen, A. 2021. "Worksites as Sacrifice Zones: Structural Precarity and COVID-19 in U.S. Meatpacking." *Sociological Perspectives* 64: 073112142110120. https://doi.org/10.1177/07311214211012025.

Carrino, L. 2017. "The Role of Normalisation in Building Composite Indicators. Rationale and Consequences of Different Strategies, Applied to Social Inclusion." In *Complexity in Society: from Indicators Construction to Their Synthesis*, edited by F. Maggino, 251-289. Cham: Springer International.

Carroll, S. R., Garba, I., Figueroa-Rodríguez, O. L., Holbrook, J., Lovett, R., Materechera, S., Parsons, M., Raseroka, K., Rodriguez-Lonebear, D., Rowe, R., Sara, R., Walker, J. D., Anderson, J., and Hudson, M. 2020. "The CARE Principles for Indigenous Data Governance." *Data Science Journal*. https://doi.org/10.5334/dsj-2020-043.

Casey, J. A., Daouda, M., Babadi, R. S., Do, V., Flores, N. M., Berzansky, I., González, D. J. X., Van Horne, Y. O., and James-Todd, T. 2023. "Methods in Public Health Environmental Justice Research: A Scoping Review from 2018 to 2021." *Current Environmental Health Reports* 10(3): 312-336. https://doi.org/10.1007/s40572-023-00406-7.

Carter, T. S., Kerr, G. H., Amini, H., Martin, R. V., Ovienmhada, U., Schwartz, J., van Donkelaar, A., and Anenberg, S. 2023. "$PM_{2.5}$ Data Inputs Alter Identification of Disadvantaged Communities." *Environmental Research Letters* 18 (11): 114008. https://doi.org/10.1088/1748-9326/ad0066.

CDC/ATSDR (Centers for Disease Control and Prevention/Agency for Toxic Substances and Disease Registry). 1997. *Principles of Community Engagement*, 1st ed. Atlanta, GA: CDC/ATSDR Committee on Community Engagement.

CDC/ATSDR. 2022. *SVI 2020 Documentation*. https://www.atsdr.cdc.gov/placeandhealth/svi/documentation/pdf/SVI2020Documentation_08.05.22.pdf (accessed February 28, 2024).

CDC/ATSDR. 2023. 2022 Environmental Justice Index. https://www.atsdr.cdc.gov/placeandhealth/eji/index.html (accessed November 5, 2023).

CEQ (Council on Environmental Quality). 1978. *National Environmental Policy Act Implementing Regulations. CFR* 40, § Chapter V Subchapter A.

CEQ. 1997. *Guidance Under the National Environmental Policy Act*. Washington, DC: Executive Office of the President.

CEQ. 2022a. *Climate and Economic Justice Screening Tool Technical Support Document—*Version 1.0. Washington, DC. https://static-data-screeningtool.geoplatform.gov/data-versions/1.0/data/score/downloadable/1.0-cejst-technical-support-document.pdf.

CEQ. 2022b, "Biden-Harris Administration Launches Version 1.0 of Climate and Economic Justice Screening Tool, Key Step in Implementing President Biden's Justice40 Initiative." November 22. https://www.whitehouse.gov/ceq/news-updates/2022/11/22/biden-harris-administration-launches-version-1-0-of-climate-and-economic-justice-screening-tool-key-step-in-implementing-president-bidens-justice40-initiative/.

Chadha, N., Lim, B., Kane, M., and Rowland, B. 2020. *Toward the Abolition of Biological Race in Medicine*. UC Berkeley Othering & Belonging Institute. https://belonging.berkeley.edu/toward-abolition-biological-race-medicine-8.

Chakraborty, J. 2019. "Proximity to Extremely Hazardous Substances for People with Disabilities: A Case Study in Houston, Texas." *Disability and Health Journal* 12 (1): 121-125. https://doi.org/10.1016/j.dhjo.2018.08.004.

Chakraborty, J. 2020. "Unequal Proximity to Environmental Pollution: An Intersectional Analysis of People with Disabilities in Harris County, Texas." *The Professional Geographer* 72: 1-14. https://doi.org/10.1080/00330124.2020.1787181.

Chakraborty, J. 2022. "Disparities in Exposure to Fine Particulate Air Pollution for People with Disabilities in the US." *Science of the Total Environment* 842: 156791. https://doi.org/10.1016/j.scitotenv.2022.156791.

Chakraborty, J., Collins, T. W., Grineski, S. E., Montgomery, M. C., and Hernandez, M. 2014. "Comparing Disproportionate Exposure to Acute and Chronic Pollution Risks: A Case Study in Houston, Texas." *Risk Analysis* 34 (11): 2005-2020. https://doi.org/10.1111/risa.12224.

Chakraborty, J., Collins, T. W., and Grineski, S. E. 2019. "Exploring the Environmental Justice Implications of Hurricane Harvey Flooding in Greater Houston, Texas." *American Journal of Public Health* 109(2): 244-250. https://doi.org/10.2105/ajph.2018.304846.

Chakraborty, J., Maantay, J. A., and Brender, J. D. 2011. "Disproportionate Proximity to Environmental Health Hazards: Methods, Models, and Measurement." *American Journal of Public Health* 101(Suppl 1): S27-S36. https://doi.org/10.2105/AJPH.2010.300109.

Chakraborty, J., McAfee, A., Collins, T., and Grineski, S. 2021. "Exposure to Hurricane Harvey flooding for subsidized housing residents of Harris County, Texas." *Natural Hazards* 106. https://doi.org/10.1007/s11069-021-04536-9.

Chakraborty, J., Collins, T., Grineski, S., and Aun, J. 2022. "Air Pollution Exposure Disparities in US Public Housing Developments." *Scientific Reports* 12: 9887. https://doi.org/10.1038/s41598-022-13942-3.

Chakraborty, T. C., Venter, Z. S., Qian, Y., and Lee, X. 2022. "Lower Urban Humidity Moderates Outdoor Heat Stress." *AGU Advances* 3 (5): e2022AV000729. https://doi.org/10.1029/2022AV000729.

Chakraborty, T. C., Newman, A. J., Qian, Y., Hsu, A., and Sheriff, G. 2023. "Residential Segregation and Outdoor Urban Moist Heat Stress Disparities in the United States." *One Earth* 6(6): 738-750. https://doi.org/10.1016/j.oneear.2023.05.016.

Chambers, B. D., Arabia, S. E., Arega, H. A., Altman, M. R., Berkowitz, R., Feuer, S. K., Franck, L. S., Gomez, A. M., Kober, K., Pacheco-Werner, T., Paynter, R. A., Prather, A. A., Spellen, S. A., Stanley, D., Jelliffe-Pawlowski, L. L., and McLemore, M. R. 2020. "Exposures to Structural Racism and Racial Discrimination Among Pregnant and Early Post-partum Black Women Living in Oakland, California." *Stress and Health* 36(2): 213-219. https://doi.org/10.1002/smi.2922.

Chantarat, T., Van Riper, D. C., and Hardeman, R. R. 2022. "Multidimensional Structural Racism Predicts Birth Outcomes for Black and White Minnesotans." *Health Services Research* 57 (3): 448-457. https://doi.org/https://doi.org/10.1111/1475-6773.13976.

Chelwa, G., Hamilton, D., and Stewart, J. 2022. "Stratification Economics: Core Constructs and Policy Implications." *Journal of Economic Literature* 60(2): 377-399. https://doi.org/10.1257/jel.20211687.

Chemnick, J. 2022. "Experts to White House: EJ Screening Tool Should Consider Race." *Climatewire*, June 1. https://www.eenews.net/articles/experts-to-white-house-ej-screening-tool-should-consider-race/ (accessed October 4, 2023).

City of Columbus Historical Data. 1936. Redlining – Engaging Columbus. Ohio Wesleyan University. Available at: https://engagingcolumbus.owu.edu/redlining/. Accessed July 24, 2024.

Clark, G. E., Moser, S. C., Ratick, S. J., Dow, K., Meyer, W. B., Emani, S., Jin, W., Kasperson, J. X., Kasperson, R. E., and Schwarz, H. E. 1998. "Assessing the Vulnerability of Coastal Communities to Extreme Storms: The Case of Revere, MA., USA." *Mitigation and Adaptation Strategies for Global Change* 3(1): 59-82. https://doi.org/10.1023/A:1009609710795.

Collins, C. A., and Williams, D. R. 1999. "Segregation and Mortality: The Deadly Effects of Racism?" *Sociological Forum* 14(3): 495-523. https://doi.org/10.1023/A:1021403820451.

Collins, H. N., Johnson, P. I., Calderon, N. M., Clark, P. Y., Gillis, A. D., Le, A. M., Nguyen, D., Nguyen, C., Fu, L., O'Dwyer, T., and Harley, K. G. 2023. "Differences in Personal Care Product Use by Race/Ethnicity Among Women in California: Implications for Chemical Exposures." *Journal of Exposure Science & Environmental Epidemiology* 33(2): 292-300. https://doi.org/10.1038/s41370-021-00404-7.

Collins, T. W., Grineski, S. E., Chakraborty, J., Montgomery, M. C., and Hernandez, M. 2015. "Downscaling Environmental Justice Analysis: Determinants of Household-Level Hazardous Air Pollutant Exposure in Greater Houston." *Annals of the Association of American Geographers* 105(4): 684-703. https://doi.org/10.1080/00045608.2015.1050754.

Collins, T. W., Grineski, S. E., and Nadybal, S. M. 2022. "A Comparative Approach for Environmental Justice Analysis: Explaining Divergent Societal Distributions of Particulate Matter and Ozone Pollution Across U.S. Neighborhoods." *Annals of the American Association of Geographers* 112(2): 522-541. https://doi.org/10.1080/24694452.2021.1935690.

Collins, T. W., Nadybal, S., and Grineski, S. E. 2020. "Sonic Injustice: Disparate Residential Exposures to Transport Noise from Road and Aviation Sources in the Continental United States." *Journal of Transport Geography* 82: 102604. https://doi.org/10.1016/j.jtrangeo.2019.102604.

Coming Clean, Inc. n.d. "Cumulative Impacts and Mandatory Emissions Reductions Team" https://comingcleaninc.org/projects/ci-mer (accessed December 18, 2023).

Commission for Racial Justice. 1987. *Toxic Wastes and Race in the United States: A National Report on the Racial and Socio-Economic Characteristics of Communities with Hazardous Waste Sites*. (New York: United Church of Christ).

Commission on Social Determinants of Health. 2008. *Closing the Gap in a Generation: Health Equity Through Action on the Social Determinants of Health*. World Health Organization. https://www.who.int/publications/i/item/WHO-IER-CSDH-08.1.

Costley, D. 2022. "Race Excluded as WH Rolls Out Climate Justice Screening Tool." *AP News*, February 18. https://apnews.com/article/climate-science-environment-pollution-fe337100f1bb1f5fa86f081af8a3b69f (accessed October 5, 2023).

Crenshaw, K. 1991. "Mapping the Margins: Intersectionality, Identity Politics, and Violence Against Women of Color." *Stanford Law Review* 43(6): 1241-1299. https://doi.org/10.2307/1229039.

Crooks, J., Flemming, K., Shulman, C., and Hudson, B. 2023. "Opportunities and Challenges in Involving People with Lived Experience of Inclusion Health as Co-researchers in Palliative and End of Life Research: A Rapid Review and Thematic Synthesis." *Research Involvement and Engagement* 9 (1): 25. https://doi.org/10.1186/s40900-023-00436-3.

CSCC (California State Coastal Conservancy). 2022. *Tips for Meaningful Community Engagement*. State of California. https://scc.ca.gov/justice-equity-diversity-and-inclusion-jedi/tips-for-meaningful-community-engagement-2/.

Cushing, L. J., Faust, J., August, L. M., Cendak, R., Wieland, W., and Alexeeff, G. 2015. "Racial/Ethnic Disparities in Cumulative Environmental Health Impacts in California: Evidence from a Statewide Environmental Justice Screening Tool (CalEnviroScreen 1.1)." *American Journal of Public Health* 105(11): 2341-2348. https://doi.org/10.2105/ajph.2015.302643.

Cushing, L. J., Li, S., Steiger, B. B., and Casey, J. A. 2023a. "Historical Red-Lining is Associated with Fossil Fuel Power Plant Siting and Present-Day Inequalities in Air Pollutant Emissions." *Nature Energy* 8(1): 52-61. https://doi.org/10.1038/s41560-022-01162-y.

Cushing, L. J., Ju, Y., Kulp, S., Depsky, N., Karasaki, S., Jaeger, J., Raval, A., Strauss, B., and Morello-Frosch, R. 2023b. "Toxic Tides and Environmental Injustice: Social Vulnerability to Sea Level Rise and Flooding of Hazardous Sites in Coastal California." *Environmental Science & Technology* 57(19): 7370-7381. https://doi.org/10.1021/acs.est.2c07481.

Cutter, S. L. 1995. "Race, Class and Environmental Justice." *Progress in Human Geography* 19(1): 111-122. https://doi.org/10.1177/030913259501900111.

Cutter, S. L. 2024. "The Origin and Diffusion of the Social Vulnerability Index (SoVI)." *International Journal of Disaster Risk Reduction* 109: 104576. https://doi.org/10.1016/j.ijdrr.2024.104576.

Cutter, S. L., and Finch, C. 2008. "Temporal and Spatial Changes in Social Vulnerability to Natural Hazards." *Proceedings of the National Academy of Sciences of the United States of America* 105(7): 2301-2306. https://doi.org/10.1073/pnas.0710375105.

Cutter, S. L., and Morath, D. P. 2013. "The Evolution of the Social Vulnerability Index (SoVI)." In *Measuring Vulnerability to Natural Hazards: Towards Disaster Resilient Societies*, edited by Joern Birkmann, 304-321. New York: United Nations University Press.

Cutter, S. L., Boruff, B. J., and Shirley, W. L. 2003. "Social Vulnerability to Environmental Hazards." *Social Science Quarterly* 84(2): 242-261. https://doi.org/10.1111/1540-6237.8402002.

Cutter, S. L., Boruff, B. J., and Shirley, W. L. 2012. "Social Vulnerability to Environmental Hazards." In *Hazards Vulnerability and Environmental Justice*, 143-160. Routledge.

Cyril, S., Smith, B. J., Possamai-Inesedy, A., and Renzaho, A. M. 2015. "Exploring the Role of Community Engagement in Improving the Health of Disadvantaged Populations: A Systematic Review." *Global Health Action* 8: 29842. https://doi.org/10.3402/gha.v8.29842.

Daley, D., and Reames, T. 2015. "Public Participation and Environmental Justice: Access to Federal Decision Making." In *Failed Promises: Evaluating the Federal Government's Response to Environmental Justice*, edited by D. M. Konisky, 143-172. Cambridge, MA: MIT Press.

Daniels, G. R. 2020. *Uncounted: The Crisis of Vote Suppression in America*. New York: NYU Press.

Davis, L. F., and Ramírez-Andreotta, M. D. 2021. "Participatory Research for Environmental Justice: A Critical Interpretive Synthesis." *Environ Health Perspectives* 129(2): 26001. https://doi.org/10.1289/ehp6274.

Dawes, D. E. 2020. *The Political Determinants of Health*. Baltimore, MD: Johns Hopkins University Press.

de Onís, C. M., and Pezzullo, P. C. 2017. "The Ethics of Embodied Engagement: Ethnographies of Environmental Justice." In *The Routledge Handbook of Environmental Justice Justice, edited by R. Holifield, J. Chakraborty, and G. Walker*, 231-240. London: Routledge.

Dean, L. T., and Thorpe, R. J., Jr. 2022. "What Structural Racism Is (or Is Not) and How to Measure It: Clarity for Public Health and Medical Researchers." *American Journal of Epidemiology* 191(9): 1521-1526. https://doi.org/10.1093/aje/kwac112.

Deas, I., Robson, B., Wong, C., and Bradford, M. 2003. "Measuring Neighbourhood Deprivation: A Critique of the Index of Multiple Deprivation." *Environment and Planning C: Government and Policy* 21(6): 883-903. https://doi.org/10.1068/c0240.

Demetillo, M. A. G., Harkins, C., McDonald, B. C., Chodrow, P. S., Sun, K., and Pusede, S. E. 2021. "Space-Based Observational Constraints on NO_2 Air Pollution Inequality from Diesel Traffic in Major US Cities." *Geophysical Research Letters* 48(17): e2021GL094333. https://doi.org/10.1029/2021GL094333.

Denchak, M. 2018. "Flint Water Crisis: Everything You Need to Know." National Resources Defense Council. https://www.nrdc.org/stories/flint-water-crisis-everything-you-need-know (accessed February 17, 2024).

Dennis, A. C., Chung, E. O., Lodge, E. K., Martinez, R. A., and Wilbur, R. E. 2021. "Looking Back to Leap Forward: A Framework for Operationalizing the Structural Racism Construct in Minority Health Research." *Ethnicity & Disease* 31(Suppl 1): 301-310. https://doi.org/10.18865/ed.31.S1.301.

Denton, F. 2002. "Climate Change Vulnerability, Impacts, and Adaptation: Why Does Gender Matter?" *Gender & Development* 10(2): 10-20. https://doi.org/10.1080/13552070215903.

Der Kiureghian, A., and Ditlevsen, O. 2009. "Aleatory or Epistemic? Does It Matter?" *Structural Safety* 31(2): 105-112. https://doi.org/10.1016/j.strusafe.2008.06.020.

Di, Q., Amini, H., Shi, L., Kloog, I., Silvern, R., Kelly, J., Sabath, M. B., Choirat, C., Koutrakis, P., Lyapustin, A., Wang, Y., Mickley, L. J., and Schwartz, J. 2019. "An Ensemble-Based Model of $PM_{2.5}$ Concentration Across the Contiguous United States with High Spatiotemporal Resolution." *Environment International* 130: 104909. https://doi.org/10.1016/j.envint.2019.104909.

Di Fonzo, D., Fabri, A., and Pasetto, R. 2022. "Distributive Justice in Environmental Health Hazards from Industrial Contamination: A Systematic Review of National and Near-National Assessments of Social Inequalities." *Social Science & Medicine* 297: 114834. https://doi.org/10.1016/j.socscimed.2022.114834.

Dietz, R. D. 2002. "The Estimation of Neighborhood Effects in the Social Sciences: An Interdisciplinary Approach." *Social Science Research* 31(4): 539-575. https://doi.org/ 10.1016/S0049-089X(02)00005-4.

Diez Roux, A. V., and Mair, C. 2010. "Neighborhoods and Health." *Annals of the New York Academy of Sciences* 1186: 125-145. https://doi.org/10.1111/j.1749-6632.2009.05333.x.

Dobbie, M., and Dail, D. 2013. "Robustness and Sensitivity of Weighting and Aggregation in Constructing Composite Indices." *Ecological Indicators* 29: 270–277. https://doi.org/10.1016/j.ecolind.2012.12.025

Donaghy, T., Healy, N., Jiang, C., and Battle, C. 2023. "Fossil Fuel Racism in the United States: How Phasing Out Coal, Oil, and Gas Can Protect Communities." *Energy Research & Social Science* 100: 103104. https://doi.org/10.1016/j.erss.2023.103104.

Doremus, J. M., Jacqz, I., and Johnston, S. 2022. "Sweating the Energy Bill: Extreme Weather, Poor Households, and the Energy Spending Gap." *Journal of Environmental Economics and Management* 112: 102609. https://doi.org/10.1016/j.jeem.2022.102609.

Dory, G., Qiu, Z., Qiu, C., Fu, M. R., and Ryan, C. E. 2015. "Lived Experiences of Reducing Environmental Risks in an Environmental Justice Community." *Proceedings of the International Academy of Ecology and Environmental Sciences* 5(4): 128-141.

Dougherty, G. B., Golden, S. H., Gross, A. L., Colantuoni, E., and Dean, L. T. 2020. "Measuring Structural Racism and Its Association with BMI." *American Journal of Preventive Medicine* 59(4): 530-537. https://doi.org/10.1016/j.amepre.2020.05.019.

Driver, A., Mehdizadeh, C., Bara-Garcia, S., Bodenreider, C., Lewis, J., and Wilson, S. 2019. Utilization of the Maryland Environmental Justice Screening Tool: A Bladensburg, Maryland Case Study. *International Journal of Environmental Research and Public Health* 16(3): 348. https://doi.org/10.3390/ijerph16030348.

Du Bois, W. E. B. 1889. *The Philadelphia Negro: A Social Study*. Philadelphia: University of Pennsylvania Press.

Du Bois, W. E. B. 1935. *Black Reconstruction: An Essay Toward a History of the Part Which Black Folk Played in the Attempt to Reconstruct Democracy in America, 1860–1880*. New York: Harcourt, Brace & Co.

Elliott, J., and Pais, J. 2006. "Race, Class, and Hurricane Katrina: Social Differences in Human Responses to Disaster." *Social Science Research* 35: 295-321. https://doi.org/10.1016/j.ssresearch.2006.02.003.

El Gibari, S., Gómez, T., and Ruiz, F. 2019. "Building Composite Indicators Using Multicriteria Methods: A Review." *Journal of Business Economics* 89(1): 1-24. https://doi.org/10.1007/s11573-018-0902-z.

Emrich, C. T., and Cutter, S. L. 2011. "Social Vulnerability to Climate-Sensitive Hazards in the Southern United States." *Weather, Climate, and Society* 3 (3): 193-208. https://doi.org/10.1175/2011WCAS1092.1.

EOP (Executive Office of the President). 1994. "Executive Order 12898 of February 11, 1994, Federal Actions to Address Environmental Justice in Minority Populations and Low-Income Populations." *Federal Register* 59(32): 7629-7633. https://www.archives.gov/files/federal-register/executive-orders/pdf/12898.pdf (accessed December 17, 2023).

EOP. 2021. "Executive Order 14008 of January 27, 2021, Tackling the Climate Crisis at Home and Abroad." *Federal Register* 86 (19): 7619-7633

EOP. 2022. "Climate and Economic Justice Screening Tool: Frequently Asked Questions." Washington, DC. https://www.whitehouse.gov/wp-content/uploads/2022/02/CEQ-CEJST-QandA.pdf.

EOP. 2023. "Addendum to the Interim Implementation Guidance for the Justice40 Initiative, M-21-28, on using the Climate and Economic Justice Screening Tool (CEJST)." Memorandum M-23-09, January 27. https://www.whitehouse.gov/wp-content/uploads/2023/01/M-23-09_Signed_CEQ_CPO.pdf (accessed February 29, 2024).

EPA (U.S. Environmental Protection Agency). 2009. *Guidance on the Development, Evaluation, and Application of Environmental Models* EPA/100/K-09/003, Office of the Science Advisor, Council for Regulatory Environmental Modeling, U.S. Environmental Protection Agency (Washington, DC). https://www.epa.gov/measurements-modeling/guidance-development-evaluation-and-application-environmental-models.

EPA. 2016a. *Environmental Justice and Water Infrastructure Finance and Capacity (Charge to the National Environmental Justice Advisory Council)*. https://www.epa.gov/sites/default/files/2016-12/documents/nejac_environmental_justice_and_water_infrastructure_finance_and_capacity_final_charge.pdf.

EPA. 2016b. *Lead and Copper Rule Revisions White Paper.* Washington, DC: EPA Office of Water. https://www.epa.gov/sdwa/lead-and-copper-rule-revisions-white-paper.

EPA. 2021. *Memorandum: Strengthening Environmental Justice Through Criminal Enforcement.* https://www.epa.gov/system/files/documents/2021-07/strengtheningejthroughcriminal062121.pdf.

EPA. 2022a. *Cumulative Impacts Research: Recommendations for EPA's Office of Research and Development.* EPA/600/R-22/014a. Washington, DC.

EPA. 2022b. *Factors to Consider When Using Toxics Release Inventory Data.* Washington, DC. https://www.epa.gov/toxics-release-inventory-tri-program/factors-consider-when-using-toxics-release-inventory-data.

EPA. 2023a. *Underground Storage Tank Program Facts.* Washington, DC: Office of Underground Storage Tanks. https://www.epa.gov/system/files/documents/2023-11/ust-programfacts-nov2023.pdf.

EPA. 2023b. *EJScreen Technical Documentation for Version 2.2.* Washington, DC. https://www.epa.gov/system/files/documents/2023-06/ejscreen-tech-doc-version-2-2.pdf.

EPA. 2023c. *Drinking Water Infrastructure Needs Survey and Assessment: 7th Report to Congress.* Washington, DC: Office of Water. https://www.epa.gov/dwsrf/epas-7th-drinking-water-infrastructure-needs-survey-and-assessment.

EPA. 2023d. *U.S. Private Well Estimates (2020).* SHC.404.2.1.1. Washington, DC: Office of Research and Development. https://epa.maps.arcgis.com/home/item.html?id=034d058a2bd94c389c608719ccde1182.

EPA. 2023e. *Capacity Building Through Effective Meaningful Engagement: A Tool for Local and State Governments.* EPA 440B23001, Washington, DC: EPA. https://www.epa.gov/system/files/documents/2023-09/epa-capacity-building-through-effective-meaningful-engagement-booklet_0.pdf.

EPA. 2024. "Learn About Underground Storage Tanks." https://www.epa.gov/ust/learn-about-underground-storage-tanks (accessed February 16, 2024).

Erickson, T. B., Brooks, J., Nilles, E. J., Pham, P. N., and Vinck, P. 2019. "Environmental Health Effects Attributed to Toxic and Infectious Agents Following Hurricanes, Cyclones, Flash Floods and Major Hydrometeorological Events." *Journal of Toxicology and Environmental Health, Part B: Critical Reviews* 22(5-6): 157-171. https://doi.org/10.1080/10937404.2019.1654422.

Esri. 2022. "Justice40 by Number of Categories Map". (accessed January 30, 2024). Certain Esri Basemaps in this work are owned by Esri and its data contributors and are used herein with permission. Copyright © 2024 Esri and its data contributors. All rights reserved.

Evergreen Collaborative. 2020. *Designing a New National Equity Mapping Program.* Seattle, WA: Demos, Evergreen Collaborative. https://collaborative.evergreenaction.com/policy-hub/EquityMapping.pdf.

Faber, J. W. 2020. "We Built This: Consequences of New Deal Era Intervention in America's Racial Geography." *American Sociological Review* 85(5): 739-775. https://doi.org/10.1177/0003122420948464.

Failing, L., and Gregory, R. 2003. "Ten Common Mistakes in Designing Biodiversity Indicators for Forest Policy." *Journal of Environmental Management* 68(2): 121-132. https://doi.org/10.1016/s0301-4797(03)00014-8.

Farrell, J., Burow, P. B., McConnell, K., Bayham, J., Whyte, K., and Koss, G. 2021. "Effects of Land Dispossession and Forced Migration on Indigenous Peoples in North America." *Science* 374(6567): eabe4943. https://doi.org/10.1126/science.abe4943.

Fast Track Action Committee on Climate Services. 2023. *A Federal Framework and Action Plan for Climate Services*. Washington, DC: National Science and Technology Council, Executive Office of the President. https://www.whitehouse.gov/wp-content/uploads/2023/03/FTAC_Report_03222023_508.pdf.

Faust, J., August, L. M., Slocombe, A., Prasad, S., Wieland, W., Cogliano, V., and Monahan Cummings, C. 2021. "California's Environmental Justice Mapping Tool: Lessons and Insights from CalEnviroScreen." *Environmental Law Reporter* 51(8): 10684-10687. https://www.eli.org/sites/default/files/files-pdf/ELR%20_0821_copyright_0.pdf.

Fears, D. 2023. "Without Focus on Race, Biden Effort on Air Pollution Disparities Will Fail, Report Says." *The Washington Post*, July 20. https://www.washingtonpost.com/climate-environment/2023/07/20/without-focus-race-biden-effort-air-pollution-disparities-will-fail-report-says/.

Fekete, A. 2012. "Spatial Disaster Vulnerability and Risk Assessments: Challenges in Their Quality and Acceptance." *Natural Hazards* 61: 1161–1178. https://doi.org/10.1007/s11069-011-9973-7.

Feldman, J. M., and Bassett, M. T. 2021. "Variation in COVID-19 Mortality in the US by Race and Ethnicity and Educational Attainment." *JAMA Network Open* 4(11): e2135967. https://doi.org/10.1001/jamanetworkopen.2021.35967.

Feldmeyer, D., Wilden, D., Jamshed, A., and Birkmann, J. 2020. "Regional Climate Resilience Index: A Novel Multimethod Comparative Approach for Indicator Development, Empirical Validation and Implementation." *Ecological Indicators* 119: 106861. https://doi.org/10.1016/j.ecolind.2020.106861.

FEMA (Federal Emergency Management Agency). 2022. Notice of Funding Opportunity for the Fiscal Year 2022 Flood Mitigation Assistance Program. https://www.fema.gov/sites/default/files/documents/fema_fy22-fma-nofo-fact-sheet_092022.pdf.

FEMA. 2024. National Risk Index. FEMA.gov. [online] hazards.fema.gov. Available at: https://hazards.fema.gov/nri (accessed 16 Jul. 2024).

Fernandez, M., Harris, B., and Rose, J. 2021. "Greensplaining Environmental Justice: A Narrative of Race, Ethnicity, and Justice in Urban Greenspace Development." *Journal of Race, Ethnicity and the City* 2(2): 210-231. https://doi.org/10.1080/26884674.2021.1921634.

Fielding, N. G. 2012. "Triangulation and Mixed Methods Designs: Data Integration with New Research Technologies." *Journal of Mixed Methods Research* 6 (2): 124-136. https://doi.org/10.1177/1558689812437101.

Finch, C., Emrich, C. T., and Cutter, S. L. 2010. "Disaster Disparities and Differential Recovery in New Orleans." *Population and Environment* 31(4): 179-202. https://doi.org/10.1007/s11111-009-0099-8.

Finio, N. 2022. "Measurement and Definition of Gentrification in Urban Studies and Planning." *Journal of Planning Literature* 37(2): 249-264. https://doi.org/10.1177/08854122211051603.

Fishback, P., Lavoice, J., Shertzer, A., and Walsh, R. 2020. "The HOLC Maps: How Race and Poverty Influenced Real Estate Professionals' Evaluation of Lending Risk in the 1930s." *SSRN Electronic Journal*. https://doi.org/10.2139/ssrn.3739643.

Fishback, P. V., Rose, J., Snowden, K. A., and Storrs, T. 2021. "New Evidence on Redlining by Federal Housing Programs in the 1930s." National Bureau of Economic Research Working Paper Series 29244. http://www.nber.org/papers/w29244.

Flanagan, B. E., Hallisey, E. J., Adams, E., and Lavery, A. 2018. "Measuring Community Vulnerability to Natural and Anthropogenic Hazards: The Centers for Disease Control and Prevention's Social Vulnerability Index." *Journal of Environmental Health* 80(10): 34-36.

Folch, D. C., Arribas-Bel, D., Koschinsky, J., and Spielman, S. E. 2016. "Spatial Variation in the Quality of American Community Survey Estimates." *Demography* 53(5): 1535-1554. https://doi.org/10.1007/s13524-016-0499-1.

Fothergill, A., Maestas, E. G., and Darlington, J. D. 1999. "Race, Ethnicity and Disasters in the United States: A Review of the Literature." *Disasters* 23(2): 156-173. https://doi.org/10.1111/1467-7717.00111.

Fothergill, A., and Peek, L. A. 2004. "Poverty and Disasters in the United States: A Review of Recent Sociological Findings." *Natural Hazards* 32(1): 89-110. https://doi.org/10.1023/B:NHAZ.0000026792.76181.d9.

Fotheringham, A.S., and Sachdeva, M. 2022. "Scale and Local Modeling: New Perspectives on the Modifiable Areal Unit Problem and Simpson's Paradox." *Journal of Geographical Systems* 24: 475-499. https://doi.org/10.1007/s10109-021-00371-5.

Fotheringham, A. S., and Wong, D. W. S. 1991. "The Modifiable Areal Unit Problem in Multivariate Statistical Analysis." *Environment and Planning A: Economy and Space* 23(7): 1025-1044. https://doi.org/10.1068/a231025.

Frank, T. 2023. "How the White House Found EJ Areas Without Using Race." *ClimateWire*, January 24, 2023. https://www.eenews.net/articles/how-the-white-house-found-ej-areas-without-using-race/.

Frechette, J., Bitzas, V., Aubry, M., Kilpatrick, K., and Lavoie-Tremblay, M. 2020. "Capturing Lived Experience: Methodological Considerations for Interpretive Phenomenological Inquiry." *International Journal of Qualitative Methods* 19: https://doi.org/10.1177/1609406920907254.

Freudenberg, M. 2003. *Composite Indicators of Country Performance: A Critical Assessment*. No. 2003/16, OECD Science, Technology and Industry Working Papers. Paris: OECD Publishing. https://www.oecd-ilibrary.org/content/paper/405566708255.

Friedman, L. 2022. "White House Takes Aim at Environmental Racism, but Won't Mention Race." *The New York Times*, February 15. https://www.nytimes.com/2022/02/15/climate/biden-environment-race-pollution.html.

Funk, C., Peterson, P., Landsfeld, M., Pedreros, D., Verdin, J., Shukla, S., Husak, G., Rowland, J., Harrison, L., Hoell, A., and Michaelsen, J. 2015. "The Climate Hazards Infrared Precipitation with Stations—A New Environmental Record for Monitoring Extremes." *Scientific Data* 2(1): 150066. https://doi.org/10.1038/sdata.2015.66.

Funk, C., Peterson, P., Peterson, S., Shukla, S., Davenport, F., Michaelsen, J., Knapp, K. R., Landsfeld, M., Husak, G., Harrison, L., Rowland, J., Budde, M., Meiburg, A., Dinku, T., Pedreros, D., and Mata, N. 2019. "A High-Resolution 1983–2016 T_{max} Climate Data Record Based on Infrared Temperatures and Stations by the Climate Hazard Center." *Journal of Climate* 32(17): 5639-5658. https://doi.org/10.1175/JCLI-D-18-0698.1. Furtado, K., Rao, N., Payton, M., Brown, K., Balu, R., and Dubay, L. 2023. *Measuring Structural Racism: Approaches from the Health Literature*. Washington, DC: Urban Institute. https://www.urban.org/research/publication/measuring-structural-racism.

Furtado, K., Rao, N., Payton, M., Brown, K., Balu, R., and Dubay, L. 2023. *Measuring Structural Racism: Approaches from the Health Literature*. Washington, DC: Urban Institute. https://www.urban.org/research/publication/measuring-structural-racism.

Fusco, E., Vidoli, F., and Sahoo, B. K. 2018. "Spatial Heterogeneity in Composite Indicator: A Methodological Proposal." *Omega* 77: 1-14. https://doi.org/10.1016/j.omega.2017.04.007.

Fusco, E., Libório, M. P., Rabiei-Dastjerdi, H., Vidoli, F., Brunsdon, C., and Ekel, P. I. 2023. "Harnessing Spatial Heterogeneity in Composite Indicators Through the Ordered Geographically Weighted Averaging (OGWA) Operator." *Geographical Analysis*. https://doi.org/https://doi.org/10.1111/gean.12384.

Galster, G., and Sharkey, P. 2017. "Spatial Foundations of Inequality: A Conceptual Model and Empirical Overview." *RSF: The Russell Sage Foundation Journal of the Social Sciences* 3(2): 1-33. https://doi.org/10.7758/rsf.2017.3.2.01.

Gan, X., Fernandez, I. C., Guo, J., Wilson, M., Zhao, Y., Zhou, B., and Wu, J. 2017. "When to Use What: Methods for Weighting and Aggregating Sustainability Indicators." *Ecological Indicators* 81: 491-502. https://doi.org/10.1016/j.ecolind.2017.05.068.

Gannon, M. 2016. "Race Is a Social Construct, Scientists Argue." Livescience. https://www.scientificamerican.com/article/race-is-a-social-construct-scientists-argue/.

GAO (General Accounting Office). 1983. *Siting of Hazardous Waste Landfills and Their Correlation with Racial and Economic Status of Surrounding Communities*. RCED-83-168. https://www.gao.gov/products/rced-83-168.

Gaynor, N. 2013. "The Tyranny of Participation Revisited: International Support to Local Governance in Burundi." *Community Development Journal* 49: 295-310. https://doi.org/10.1093/cdj/bst031.

Giarratano, G., Harville, E. W., Barcelona de Mendoza, V., Savage, J., and Parent, C. M. 2015. "Healthy Start: Description of a Safety Net for Perinatal Support During Disaster Recovery." *Maternal and Child Health Journal* 19(4): 819-827. https://doi.org/10.1007/s10995-014-1579-8.

Goldsmith, L., Raditz, V., and Méndez, M. 2022. "Queer and Present Danger: Understanding the Disparate Impacts of Disasters on LGBTQ+ Communities." *Disasters* 46(4): 946-973. https://doi.org/https://doi.org/10.1111/disa.12509.

Gómez-Limón, J. A., Arriaza, M., and Guerrero-Baena, M. D. 2020. Building a Composite Indicator to Measure Environmental Sustainability Using Alternative Weighting Methods. *Sustainability* 12(11): 4398. https://doi.org/10.3390/su12114398.

Gong, X., Fenech, B., Blackmore, C., Chen, Y., Rodgers, G., Gulliver, J., and Hansell, A. L. 2022. Association between Noise Annoyance and Mental Health Outcomes: A Systematic Review and Meta-Analysis. *International Journal of Environmental Research and Public Health* 19(5). 2696. https://doi.org/10.3390/ijerph19052696.

Gonzalez, R. 2019. *The Spectrum of Community Engagement to Ownership*. Movement Strategy Center. https://movementstrategy.org/resources/the-spectrum-of-community-engagement-to-ownership/.

Graham, K., and Knittel, C. R. 2024. "Assessing the Distribution of Employment Vulnerability to the Energy Transition Using Employment Carbon Footprints." *Proceedings of the National Academy of Sciences* 121 (7): e2314773121. https://doi.org/10.1073/pnas.2314773121.

Greco, S., Ishizaka, A., Tasiou, M., and Torrisi, G. 2019. "On the Methodological Framework of Composite Indices: A Review of the Issues of Weighting, Aggregation, and Robustness." *Social Indicators Research* 141(1): 61-94. https://doi.org/10.1007/s11205-017-1832-9.

Greenfield, N. 2023. "America's Failing Drinking Water System." National Resources Defense Council. https://www.nrdc.org/stories/americas-failing-drinking-water-system (accessed February 17, 2024).

Grier, L., Mayor, D., Zeuner, B., and Mohai, P. 2022. "Community Input on State Environmental Justice Screening Tools." *Environmental Law Reporter* 52: 10441-10452. https://www.eli.org/sites/default/files/files-pdf/52.10441.pdf.

Groos, M., Wallace, M., Hardeman, R., and Theall, K. P. 2018. "Measuring Inequity: A Systematic Review of Methods Used to Quantify Structural Racism." *Journal of Health Disparities Research and Practice* 11(2): 13. https://digitalscholarship.unlv.edu/jhdrp/vol11/iss2/13.

Gudi-Mindermann, H., White, M., Roczen, J., Riedel, N., Dreger, S., and Bolte, G. 2023. "Integrating the Social Environment with an Equity Perspective into the Exposome Paradigm: A New Conceptual Framework of the Social Exposome." *Environmental Research* 233: 116485. https://doi.org/10.1016/j.envres.2023.116485.

Gustafsson, P. E., San Sebastian, M., Janlert, U., Theorell, T., Westerlund, H., and Hammarström, A. 2014. "Life-Course Accumulation of Neighborhood Disadvantage and Allostatic Load: Empirical Integration of Three Social Determinants of Health Frameworks." *American Journal of Public Health* 104(5): 904-910. https://doi.org/10.2105/ajph.2013.301707.

Hacker, K. 2013. *Community-Based Participatory Research*. London: Sage. https://doi.org/10.4135/9781452244181.

Haldane, V., Chuah, F. L. H., Srivastava, A., Singh, S. R., Koh, G. C. H., Seng, C. K., and Legido-Quigley, H. 2019. "Community Participation in Health Services Development, Implementation, and Evaluation: A Systematic Review of Empowerment, Health, Community, and Process Outcomes." *PLoS ONE* 14(5): e0216112. https://doi.org/10.1371/journal.pone.0216112.

Hammonds, E. M., and Herzig, R. M. 2009. *The Nature of Difference: Sciences of Race in the United States from Jefferson to Genomics*. Cambridge, MA: MIT Press.

Hankerson, S. H., Suite, D., and Bailey, R. K. 2015. "Treatment Disparities Among African American Men with Depression: Implications for Clinical Practice." *Journal of Health Care for the Poor and Underserved* 26(1): 21-34. https://doi.org/10.1353/hpu.2015.0012.

Hansen, K. L. 2015. "Ethnic Discrimination and Health: The Relationship Between Experienced Ethnic Discrimination and Multiple Health Domains in Norway's Rural Sami Population." *International Journal of Circumpolar Health* 74: 25125. https://doi.org/10.3402/ijch.v74.25125.

Hardy-Fanta, C., Lien, P., Pinderhughes, D. M., and Sierra, C. M. 2006. "Gender, Race, and Descriptive Representation in the United States: Findings from the Gender and Multicultural Leadership Project." *Journal of Women, Politics & Policy* 28(3-4): 7-41. https://doi.org/10.1300/J501v28n03_02.

Harnois, C. E., Bastos, J. L., Campbell, M. E., and Keith, V. M. 2019. "Measuring Perceived Mistreatment across Diverse Social Groups: An Evaluation of the Everyday Discrimination Scale." *Social Science & Medicine* 232: 298-306. https://doi.org/10.1016/j.socscimed.2019.05.011.

HEI (Health Effects Institute Panel on the Health Effects of Long-Term Exposure to Traffic-Related Air Pollution). 2022. *Systematic Review and Meta-analysis of Selected Health Effects of Long-Term Exposure to Traffic-Related Air Pollution*. Special Report 23. Boston, MA: Health Effects Institute.

Helton, J., Johnson, J., Oberkampf, W., and Sallaberry, C. 2010. "Representation of Analysis Results Involving Aleatory and Epistemic Uncertainty." *International Journal of General Systems* 39: 605-646. https://doi.org/10.1080/03081079.2010.486664.

Henneman, L. R. F., Rasel, M. M., Choirat, C., Anenberg S. C., and Zigler, C. 2023. "Inequitable Exposures to U.S. Coal Power Plant–Related $PM_{2.5}$: 22 Years and Counting." *Environmental Health Perspectives* 131(3): 037005. https://doi.org/10.1289/EHP11605.

Hezri, A., and Dovers, S. 2006. "Sustainability Indicators, Policy and Governance: Issues for Ecological Economics." *Ecological Economics* 60: 86-99. https://doi.org/10.1016/j.ecolecon.2005.11.019.

Hill, T. D., Jorgenson, A. K., Ore, P., Balistreri, K. S., and Clark, B. 2019. "Air Quality and Life Expectancy in the United States: An Analysis of the Moderating Effect of Income Inequality." *SSM - Population Health* 7: 100346. https://doi.org/10.1016/j.ssmph.2018.100346.

Hinkel, J. 2011. "Indicators of Vulnerability and Adaptive Capacity: Towards a Clarification of the Science-Policy Interface." *Global Environmental Change* 21: 198-208. https://doi.org/10.1016/j.gloenvcha.2010.08.002.

Hinton, E. 2016. *From the War on Poverty to the War on Crime: The Making of Mass Incarceration in America*. Cambridge, MA: Harvard University Press.

Hoffman, J. S., Shandas, V., and Pendleton, N. 2020. "The Effects of Historical Housing Policies on Resident Exposure to Intra-Urban Heat: A Study of 108 US Urban Areas." *Climate* 8(1): 12. https://doi.org/10.3390/cli8010012.

Horowitz, J. M., Igielnik, R., and Kochhar, R. 2020. "Most Americans Say There Is Too Much Economic Inequality in the U.S., but Fewer Than Half Call It a Top Priority." Washington, DC: Pew Research Center.

Howell, J., and Korver-Glenn, E. 2021. "The Increasing Effect of Neighborhood Racial Composition on Housing Values, 1980–2015." *Social Problems* 68(4): 1051-1071. https://doi.org/10.1093/socpro/spaa033.

Hsu, A., Sheriff, G., Chakraborty, T., and Manya, D. 2021. "Disproportionate Exposure to Urban Heat Island Intensity Across Major US Cities." *Nature Communications* 12(1): 2721. https://doi.org/10.1038/s41467-021-22799-5.

Huang, G., and London, J. 2016. "Mapping In and Out of "Messes": An Adaptive, Participatory, and Transdisciplinary Approach to Assessing Cumulative Environmental Justice Impacts." *Landscape and Urban Planning* 154: 57-67. https://doi.org/10.1016/j.landurbplan.2016.02.014.

Huyer, S., Acosta, M., Gumucio, T., and Ilham, J. 2020. "Can We Turn the Tide? Confronting Gender Inequality in Climate Policy." *Gender & Development* 28 (3): 571-591. https://doi.org/10.1080/13552074.2020.1836817.

Iceland, J., Weinberg, D., and Steinmetz, E. 2002. *Racial and Ethnics Residential Segregation in the United States: 1980-2000, Appendix B: Measures of Residential Segregation*. Washington, DC: U.S. Census Bureau. https://www.census.gov/library/publications/2002/dec/censr-3.html.

Indiana University. 2019. "Hoosier Resilience Index." https://hri.eri.iu.edu/index.html (accessed May 31, 2024).

Ingram, C., Min, E., Seto, E., Cummings, B. J., and Farquhar, S. 2022. "Cumulative Impacts and COVID-19: Implications for Low-Income, Minoritized, and Health-Compromised Communities in King County, WA." *Journal of Racial and Ethnic Health Disparities* 9(4): 1210-1224. https://doi.org/10.1007/s40615-021-01063-y.

Interagency Working Group on Coal & Power Plant Communities & Economic Revitalization. 2023. "Energy Community Tax Credit Bonus." US DOE National Energy Technology Laboratory. https://energycommunities.gov/energy-community-tax-credit-bonus/#:~:text=The%20IRA%20defines%20energy%20communities,at%20any%20time%20after%202009 (accessed February 27, 2024).

Israel, B. A., Checkoway, B., Schulz, A., and Zimmerman, M. 1994. "Health Education and Community Empowerment: Conceptualizing and Measuring Perceptions of Individual, Organizational, and Community Control." *Health Education Quarterly* 21(2): 149-170. https://doi.org/10.1177/109019819402100203.

Israel, B. A., Eng, E., Schulz, A. J., and Parker, E. A., Eds. 2005. *Methods in Community-Based Participatory Research for Health*. Jossey-Bass.

Jackson, K. T. 1987. *Crabgrass Frontier: The Suburbanization of the United States*. Oxford University Press.

Jacobs, R., Smith, P., and Goddard, M. 2004. *Measuring Performance: An Examination of Composite Performance Indicators*. CHE Technical Paper Series 29, York, UK: University of York Centre for Health Economics . https://www.york.ac.uk/che/pdf/tp29.pdf.

Jakeman, J., Eldred, M., and Xiu, D. 2010. "Numerical Approach for Quantification of Epistemic Uncertainty." *Journal of Computational Physics* 229(12): 4648-4663. https://doi.org/10.1016/j.jcp.2010.03.003.

Jaller, M., and Pahwa, A. 2020. "Evaluating the Environmental Impacts of Online Shopping: A Behavioral and Transportation Approach." *Transportation Research Part D: Transport and Environment* 80: 102223. https://doi.org/10.1016/j.trd.2020.102223.

Jampel, C. 2018. "Intersections of Disability Justice, Racial Justice and Environmental Justice." *Environmental Sociology* 4: 1-14. https://doi.org/10.1080/23251042.2018.1424497.

Jankowski, P., Nyerges, T. L., Smith, A., Moore, T. J., and Horvath, E. 1997. "Spatial Group Choice: A SDSS Tool for Collaborative Spatial Decision Making." *International Journal of Geographical Information Science* 11: 566-602. https://doi.org/10.1080/136588197242202.

Jankowski, P. 2008. *Encyclopedia of Geographic Information Science*. Thousand Oaks, California: SAGE. https://doi.org/10.4135/9781412953962.

Jennings, L., Anderson, T., Martinez, A., Sterling, R., Chavez, D. D., Garba, I., Hudson, M., Garrison, N. A., and Carroll, S. R. 2023. "Applying the 'CARE Principles for Indigenous Data Governance' to Ecology and Biodiversity Research." *Nature Ecology & Evolution* 7(10): 1547-1551. https://doi.org/10.1038/s41559-023-02161-2.

Johnston, J., and Cushing, L. 2020. "Chemical Exposures, Health, and Environmental Justice in Communities Living on the Fenceline of Industry." *Current Environmental Health Reports* 7(1): 48-57. https://doi.org/10.1007/s40572-020-00263-8.

Johnston, J. E., Chau, K., Franklin, M., and Cushing, L. 2020. "Environmental Justice Dimensions of Oil and Gas Flaring in South Texas: Disproportionate Exposure among Hispanic Communities." *Environmental Science & Technology* 54(10): 6289-6298. https://doi.org/10.1021/acs.est.0c00410.

Jones, B., and Andrey, J., 2007. "Vulnerability Index Construction: Methodological Choices and Their Influence on Identifying Vulnerable Neighbourhoods." *International Journal of Emergency Management* 4(2): 269-295. https://doi.org/10.1504/IJEM.2007.013994.

Jones, M. R., Diez-Roux, A. V., Hajat, A., Kershaw, K. N., O'Neill, M. S., Guallar, E., Post, W. S., Kaufman, J. D., and Navas-Acien, A. 2014. "Race/Ethnicity, Residential Segregation, and Exposure to Ambient Air Pollution: The Multi-Ethnic Study of Atherosclerosis (MESA)." *American Journal of Public Health* 104(11): 2130-2137. https://doi.org/10.2105/ajph.2014.302135.

Jurjevich, J., Griffin, A., Spielman, S., Folch, D., Merrick, M., and Nagle, N. 2018. "Navigating Statistical Uncertainty: How Urban and Regional Planners Understand and Work with American Community Survey (ACS) Data for Guiding Policy." *Journal of the American Planning Association* 84: 112-126. https://doi.org/10.1080/01944363.2018.1440182.

Kalkbrenner, M. 2021. "A Practical Guide to Instrument Development and Score Validation in the Social Sciences: The MEASURE Approach." *Practical Assessment, Research, and Evaluation* 26: 1. https://doi.org/10.7275/SVG4-E671.

Kane, N. 2022. "Revealing the Racial and Spatial Disparity in Pediatric Asthma: A Kansas City Case Study." *Social Science & Medicine* 292: 114543. https://doi.org/10.1016/j.socscimed.2021.114543.

Karas, D. P. 2015. "Highway to Inequity: The Disparate Impact of the Interstate Highway System on Poor and Minority Communities in American Cities." *New Visions for Public Affairs* 7: 9-21.

Kauh, T. J., Read, J. G., and Scheitler, A. J. 2021. "The Critical Role of Racial/Ethnic Data Disaggregation for Health Equity." *Population Research and Policy Review* 40(1): 1-7. https://doi.org/10.1007/s11113-020-09631-6.

Keenan, P. B., and Jankowski, P. 2019. "Spatial Decision Support Systems: Three Decades On." *Decision Support Systems* 116: 64-76. https://doi.org/10.1016/j.dss.2018.10.010.

Keisler-Starkey, K., Bunch, L. N., and Lindstrom, R. A. 2023. *Health Insurance Coverage in the United States: 2022*. September 12. Washington, DC: U.S. Government Printing Office. https://www.census.gov/library/publications/2023/demo/p60-281.html.

Keith, L., Meerow, S., and Wagner, T. 2019. "Planning for Extreme Heat: A Review." *Journal of Extreme Events* 6(3-4): 2050003. https://doi.org/10.1142/S2345737620500037.

Kephart, L. 2022. "How Racial Residential Segregation Structures Access and Exposure to Greenness and Green Space: A Review." *Environmental Justice* 15(4): 204-213. https://doi.org/10.1089/env.2021.0039.

Kerr, G. H., Goldberg, D. L., and Anenberg, S. C. 2021. "COVID-19 Pandemic Reveals Persistent Disparities in Nitrogen Dioxide Pollution." *Proceedings of the National Academy of Sciences of the United States of America* 118(30): e2022409118. https://doi.org/10.1073/pnas.2022409118.

Kerr, G. H., Goldberg, D. L., Harris, M. H., Henderson, B. H., Hystad, P., Roy, A., and Anenberg, S. C. 2023. "Ethnoracial Disparities in Nitrogen Dioxide Pollution in the United States: Comparing Data Sets from Satellites, Models, and Monitors." *Environmental Science & Technology* 57(48): 19532-19544. https://doi.org/10.1021/acs.est.3c03999.

Kershaw, K. N., Diez Roux, A. V., Burgard, S. A., Lisabeth, L. D., Mujahid, M. S., and Schulz, A. J. 2011. "Metropolitan-Level Racial Residential Segregation and Black-White Disparities in Hypertension." *American Journal of Epidemiology* 174(5): 537-545. https://doi.org/10.1093/aje/kwr116.

Kim, Y., and Verweij, S. 2016. "Two Effective Causal Paths That Explain the Adoption of US State Environmental Justice Policy." *Policy Sciences* 49(4): 505-523. https://doi.org/10.1007/s11077-016-9249-x.

Kim, B., Spoer, B. R., Titus, A. R., Chen, A., Thurston, G. D., Gourevitch, M. N., and Thorpe, L. E. 2023. "Life Expectancy and Built Environments in the U.S.: A Multilevel Analysis." *American Journal of Preventive Medicine* 64 (4): 468-476. https://doi.org/10.1016/j.amepre.2022.10.008.

Kim, M., De Vito, R., Duarte, F., Tieskens, K., Luna, M., Salazar-Miranda, A., Mazzarello, M., Showalter Otts, S., Etzel, C., Burks, S., Crossley, K., Franzen Lee, N., and Walker, E. D. 2023. "Boil Water Alerts and Their Impact on the Unexcused Absence Rate in Public Schools in Jackson, Mississippi." *Nature Water* 1(4): 359-369. https://doi.org/10.1038/s44221-023-00062-z.

Kodros, J. K., Bell, M. L., Dominici, F., L'Orange, C., Godri Pollitt, K. J., Weichenthal, S., Wu, X., and Volckens, J. 2022. "Unequal Airborne Exposure to Toxic Metals Associated with Race, Ethnicity, and Segregation in the USA." *Nature Communications* 13(1): 6329. https://doi.org/10.1038/s41467-022-33372-z.

Konisky, D., Gonzalez, D., and Leatherman, K. 2021. *Mapping for Environmental Justice: An Analysis of State Level Tools*. Bloomington, IN: Environmental Resilience Institute, Indiana University. https://eri.iu.edu/research/projects/environmental-justice-mapping-tools.html.

Konisky, D. M., Reenock, C., and Conley, S. 2021. "Environmental Injustice in Clean Water Act Enforcement: Racial and Income Disparities in Inspection Time." *Environmental Research Letters* 16(8): 084020. https://doi.org/10.1088/1748-9326/ac1225.

Kosanic, A., Petzold, J., Martín-López, B., and Razanajatovo, M. 2022. "An Inclusive Future: Disabled Populations in the Context of Climate and Environmental Change." *Current Opinion in Environmental Sustainability* 55: 101159. https://doi.org/10.1016/j.cosust.2022.101159.

Krabbe, P. F. M. 2017. "Validity." In *The Measurement of Health and Health Status*, edited by P. F. M. Krabbe, 113-134. San Diego: Academic Press.

Kramer, M. R., and Hogue, C. R. 2009. "Is Segregation Bad for Your Health?" *Epidemiologic Reviews* 31: 178-194. https://doi.org/10.1093/epirev/mxp001.

Krieger, N., Rowley, D. L., Herman, A. A., Avery, B., and Phillips, M. T. 1993. "Racism, Sexism, and Social Class: Implications for Studies of Health, Disease, and Well-Being." *American Journal of Preventive Medicine* 9(6 Suppl): 82-122.

Krieger, N., Feldman, J. M., Waterman, P. D., Chen, J. T., Coull, B. A., and Hemenway, D. 2017. "Local Residential Segregation Matters: Stronger Association of Census Tract Compared to Conventional City-Level Measures with Fatal and Non-Fatal Assaults (Total and Firearm Related), Using the Index of Concentration at the Extremes (ICE) for Racial, Economic, and Racialized Economic Segregation, Massachusetts (US), 1995–2010." *Journal of Urban Health* 94(2): 244-258. https://doi.org/10.1007/s11524-016-0116-z.

Krivo, L. J., Byron, R. A., Calder, C. A., Peterson, R. D., Browning, C. R., Kwan, M.-P., and Lee, J. Y. 2015. "Patterns of Local Segregation: Do They Matter for Neighborhood Crime?" *Social Science Research* 54: 303-318. https://doi.org/10.1016/j.ssresearch.2015.08.005.

Kuc-Czarnecka, M., Lo Piano, S., and Saltelli, A. 2020. "Quantitative Storytelling in the Making of a Composite Indicator." *Social Indicators Research* 149(3): 775-802. https://doi.org/10.1007/s11205-020-02276-0.

Kumar, C. M. 2002. *GIS Methods for Screening Potential Environmental Justice Areas in New England*. M.C.P. Thesis, Massachusetts Institute of Technology. https://dspace.mit.edu/handle/1721.1/68384.

Kuran, C., Morsut, C., Kruke, B., Krüger, M., Segnestam, L., Orru, K., Nævestad, T.-O., Airola, M., Keränen, J., Gabel, F., Hansson, S., and Torpan, S. 2020. "Vulnerability and Vulnerable Groups from an Intersectionality Perspective." *International Journal of Disaster Risk Reduction* 50: 101826. https://doi.org/10.1016/j.ijdrr.2020.101826.

Laidlaw, M. A. S., Filippelli, G. M., Brown, S., Paz-Ferreiro, J., Reichman, S. M., Netherway, P., Truskewycz, A., Ball, A. S., and Mielke, H. W. 2017. "Case Studies and Evidence-Based Approaches to Addressing Urban Soil Lead Contamination." *Applied Geochemistry* 83: 14-30. https://doi.org/10.1016/j.apgeochem.2017.02.015.

Laidlaw, M. A. S., Mielke, H. W., and Filippelli, G. M. 2023. "Assessing Unequal Airborne Exposure to Lead Associated with Race in the USA." *Geohealth* 7(7): e2023GH000829. https://doi.org/10.1029/2023gh000829.

Landry, S., and Chakraborty, J. 2009. "Street Trees and Equity: Evaluating the Spatial Distribution of an Urban Amenity." *Environment and Planning A: Economy and Space* 41: 2651-2670. https://doi.org/10.1068/a41236.

Lane, H. M., Morello-Frosch, R., Marshall, J. D., and Apte, J. S. 2022. "Historical Redlining Is Associated with Present-Day Air Pollution Disparities in U.S. Cities." *Environmental Science & Technology Letters* 9(4): 345-350. https://doi.org/10.1021/acs.estlett.1c01012.

Langowski, J., Berman, W., Brittan, G., LaRaia, C., Lehmann, J.-Y., and Woods, J. 2020. *Qualified Renters Need Not Apply: Race and Voucher Discrimination in the Metro Boston Rental Housing Market*. Boston, MA: The Boston Foundation.

Larrabee Sonderlund, A., Charifson, M., Ortiz, R., Khan, M., Schoenthaler, A., and Williams, N. J. 2022. "A Comprehensive Framework for Operationalizing Structural Racism in Health Research: The Association Between Mass Incarceration of Black People in the U.S. and Adverse Birth Outcomes." *SSM—Population Health* 19: 101225. https://doi.org/10.1016/j.ssmph.2022.101225.

Larsen, K., Gunnarsson-Östling, U., and Westholm, E. 2011. "Environmental Scenarios and Local-Global Level of Community Engagement: Environmental Justice, Jams, Institutions and Innovation." *Futures* 43: 413-423. https://doi.org/10.1016/j.futures.2011.01.007.

Lawrence, Q. 2022. "Black Vets Were Excluded from GI Bill Benefits — A Bill in Congress Aims to Fix That." October 18, 2022, Washington, DC: NPR. https://www.npr.org/2022/10/18/1129735948black-vets-were-excluded-from-gi-bill-benefits-a-bill-in-congress-aims-to-fix-th.

Lee, C. 2020. "A Game Changer in the Making? Lessons from States Advancing Environmental Justice through Mapping and Cumulative Impact Strategies." *Environmental Law Reporter* 50(3): 10203-10215.

Lee, C. 2021. "Another Game Changer in the Making? Lessons from States Advancing Environmental Justice Through Mapping and Cumulative Impact Strategies." *Environmental Law Reporter* 51(8): 10676-10683.

Lee, J. Y., and Van Zandt, S. 2019. "Housing Tenure and Social Vulnerability to Disasters: A Review of the Evidence." *Journal of Planning Literature* 34(2): 156-170. https://doi.org/10.1177/0885412218812080.

Lee, K., Lucia, L., Graham-Squire, D., and Dietz, M. 2019. "Job-Based Coverage Is Less Common Among Workers Who Are Black or Latino, Low-Wage, Immigrants, and Young Adults." *UC Berkeley Labor Center - Rising Health Care Costs in California: A Worker Issue* (blog). https://laborcenter.berkeley.edu/job-based-coverage-is-less-common-among-workers-who-are-black-or-latino-low-wage-immigrants-and-young-adults/.

Lemoigne, Y., and Caner, A. 2008. *Molecular Imaging: Computer Reconstruction and Practice*. Springer Science & Business Media.

Lett, E., Asabor, E., Beltrán, S., Cannon, A. M., and Arah, O. A. 2022. "Conceptualizing, Contextualizing, and Operationalizing Race in Quantitative Health Sciences Research." *Annals of Family Medicine* 20(2): 157-163. https://doi.org/10.1370/afm.2792.

Leung, M. W., Yen, I. H., and Minkler, M. 2004. "Community Based Participatory Research: A Promising Approach for Increasing Epidemiology's Relevance in the 21st Century." *International Journal of Epidemiology* 33(3): 499-506. https://doi.org/10.1093/ije/dyh010.

Levac, D., Colquhoun, H., and O'Brien, K. K. 2010. "Scoping Studies: Advancing the Methodology." *Implementation Science* 5(1): 69. https://doi.org/10.1186/1748-5908-5-69.

Lewis, A. S., Sax, S. N., Wason, S. C., and Campleman, S. L. 2011. "Non-chemical Stressors and Cumulative Risk Assessment: An Overview of Current Initiatives and Potential Air Pollutant Interactions." *International Journal of Environmental Research and Public Health* 8(6): 2020-2073. https://doi.org/10.3390/ijerph8062020.

Li, D., Newman, G. D., Wilson, B., Zhang, Y., and Brown, R. D. 2022. "Modeling the Relationships Between Historical Redlining, Urban Heat, and Heat-Related Emergency Department Visits: An Examination of 11 Texas Cities." *Environment and Planning B: Urban Analytics and City Science* 49 (3): 933-952. https://doi.org/10.1177/23998083211039854.

Li, M., and Yuan, F. 2022. "Historical Redlining and Food Environments: A Study of 102 Urban Areas in the United States." *Health Place* 75: 102775. https://doi.org/10.1016/j.healthplace.2022.102775.

Liboiron, M., Zahara, A., and Schoot, I. 2018. Community Peer Review: A Method to Bring Consent and Self-Determination into the Sciences. *Preprints*. https://doi.org/10.20944/preprints201806.0104.v1.

Licker, R., Dahl, K., and Abatzoglou, J. T. 2022. "Quantifying the Impact of Future Extreme Heat on the Outdoor Work Sector in the United States." *Elementa: Science of the Anthropocene* 10(1): 00048. https://doi.org/10.1525/elementa.2021.00048.

Liévanos, R. S. 2018. "Retooling CalEnviroScreen: Cumulative Pollution Burden and Race-Based Environmental Health Vulnerabilities in California." *International Journal of Environmental Research and Public Health* 15(4): 762. https://doi.org/10.3390/ijerph15040762.

Lindén, D. 2018. "Exploration of Implicit Weights in Composite Indicators: The Case of Resilience Assessment of Countries' Electricity Supply." Master's thesis. Department of Sustainable Development, Environmental Science and Engineering, KTH Royal Institute of Technology. https://kth.diva-portal.org/smash/record.jsf?pid=diva2%3A1266920&dswid=-7180.

Liu, J., Clark Lara, P., Bechle Matthew, J., Hajat, A., Kim, S.-Y., Robinson Allen, L., Sheppard, L., Szpiro Adam, A., and Marshall Julian, D. 2021. "Disparities in Air Pollution Exposure in the United States by Race/Ethnicity and Income, 1990–2010." *Environmental Health Perspectives* 129 (12): 127005. https://doi.org/10.1289/EHP8584.

Locke, D. H., Hall, B., Grove, J. M., Pickett, S. T. A., Ogden, L. A., Aoki, C., Boone, C. G., and O'Neil-Dunne, J. P. M. 2021. "Residential Housing Segregation and Urban Tree Canopy in 37 US Cities." *npj Urban Sustainability* 1 (1): 15. https://doi.org/10.1038/s42949-021-00022-0.

Logan, J. R., Bauer, C., Ke, J., Xu, H., and Li, F. 2020. "Models for Small Area Estimation for Census Tracts." *Geographical Analysis* 52(3): 325-350. https://doi.org/10.1111/gean.12215.

Lohan, T. 2017. "Farmworkers' Dilemma: Affordable Housing, but Undrinkable Water." *The New Humanitarian*, September 26. https://deeply.thenewhumanitarian.org/water/articles/2017/09/26/farmworkers-dilemma-affordable-housing-but-undrinkable-water.

Loidl, M., Wallentin, G., Wendel, R., and Zagel, B. 2016. "Mapping Bicycle Crash Risk Patterns on the Local Scale." *Safety* 2(3): 17. https://www.mdpi.com/2313-576X/2/3/17.

Lu, W., Levin, R., and Schwartz, J. 2022. "Lead Contamination of Public Drinking Water and Academic Achievements Among Children in Massachusetts: A Panel Study." *BMC Public Health* 22(1): 107. https://doi.org/10.1186/s12889-021-12474-1.

Luber, G., and McGeehin, M. 2008. "Climate Change and Extreme Heat Events." *American Journal of Preventive Medicine* 35(5): 429-435. https://doi.org/10.1016/j.amepre.2008.08.021.

Lukachko, A., Hatzenbuehler, M. L., and Keyes, K. M. 2014. "Structural Racism and Myocardial Infarction in the United States." *Social Science & Medicine* 103: 42-50. https://doi.org/10.1016/j.socscimed.2013.07.021.

Luna, M., and Nicholas, D. 2022. "An Environmental Justice Analysis of Distribution-Level Natural Gas Leaks in Massachusetts, USA." *Energy Policy* 162: 112778. https://doi.org/10.1016/j.enpol.2022.112778.

Luxenberg, S. 2019. *Separate: The Story of Plessy v. Ferguson, and America's Journey from Slavery to Segregation*. New York: W. W. Norton & Company.

MacEachren, A. M. 1994. "Visualization in Modern Cartography: Setting the Agenda." In *Visualization in Modern Cartography,* edited by A. M. MacEachren and D. R. Fraser Taylor, 1-12. Academic Press. https://doi.org/10.1016/B978-0-08-042415-6.50008-9.

Mallach, A., 2024. "Shifting the Redlining Paradigm: The Home Owners' Loan Corporation Maps and the Construction of Urban Racial Inequality." *Housing Policy Debate*, pp.1-27. https://doi.org/10.1080/10511482.2024.2321226.

Manware, M., Dubrow, R., Carrión, D., Ma, Y., and Chen, K. 2022. "Residential and Race/Ethnicity Disparities in Heat Vulnerability in the United States." *Geohealth* 6 12): e2022GH000695. https://doi.org/10.1029/2022gh000695.

Marino, E., and Faas, A. J. 2020. "Is Vulnerability an Outdated Concept? After Subjects and Spaces." *Annals of Anthropological Practice* 44(1): 33-46. https://doi.org/10.1111/napa.12132.

Marquez-Velarde, G. 2020. "The Paradox Does Not Fit All: Racial Disparities in Asthma Among Mexican Americans in the U.S." *PLoS ONE* 15(11): e0242855. https://doi.org/10.1371/journal.pone.0242855.

Martinez-Morata, I., Bostick, B. C., Conroy-Ben, O., Duncan, D. T., Jones, M. R., Spaur, M., Patterson, K. P., Prins, S. J., Navas-Acien, A., and Nigra, A. E. 2022. "Nationwide Geospatial Analysis of County Racial and Ethnic Composition and Public Drinking Water Arsenic and Uranium." *Nature Communications* 13(1): 7461. https://doi.org/10.1038/s41467-022-35185-6.

Mascarenhas, M., Grattet, R., and Mege, K. 2021. "Toxic Waste and Race in Twenty-First Century America: Neighborhood Poverty and Racial Composition in the Siting of Hazardous Waste Facilities." *Environment and Society* 12(1): 108-126. https://doi.org/10.3167/ares.2021.120107. Massey, D. S. 1990. "American Apartheid: Segregation and the Making of the Underclass." *American Journal of Sociology* 96(2): 329-357. http://www.jstor.org/stable/2781105.

Massey, D. S. 2001. "The Prodigal Paradigm Returns: Ecology Comes Back to Sociology." In *Does It Take A Village? Community Effects on Children, Adolescents, and Families*, edited by A. Booth and A. C. Crouter, 41-48. Mahwah, NJ: Lawrence Erlbaum Associates.

Massey, D. S. 2020. "Still the Linchpin: Segregation and Stratification in the USA." *Race and Social Problems* 12 (1): 1-12. https://doi.org/10.1007/s12552-019-09280-1.

Maxim, L., and van der Sluijs, J. P., 2011. "Quality in Environmental Science for Policy: Assessing Uncertainty as a Component of Policy Analysis." *Environmental Science & Policy* 14(4): 482-492.

Mazziotta, M., and Pareto, A. 2017. "Synthesis of Indicators: The Composite Indicators Approach." In *Complexity in Society: From Indicators Construction to their Synthesis*, edited by F. Maggino, 159-191. Cham: Springer International. https://doi.org/10.1007/978-3-319-60595-1_7.

McCloskey, D. J., McDonald, M. A., Cook, J., HeurtinRoberts, S., Updegrove, S., Sampson, D., Gutter, S., and Eder, M. 2011. "Community Engagement: Definitions and Organizing Concepts from the Literature." In *Principles of Community Engagement,* 2nd ed. Centers for Disease Control and Prevention and Agency for Toxic Substances and Disease Registry. https://www.atsdr.cdc.gov/communityengagement/index.html.

McDonald, J. A., Llanos, A. A. M., Morton, T., and Zota, A. R. 2022. "The Environmental Injustice of Beauty Products: Toward Clean and Equitable Beauty." *American Journal of Public Health* 112(1): 50-53. https://doi.org/10.2105/AJPH.2021.306606.

McFarland, M. J., Hauer, M. E., and Reuben, A. 2022. "Half of US Population Exposed to Adverse Lead Levels in Early Childhood." *Proceedings of the National Academy of Sciences of the United States of America* 119 (11): e2118631119. https://doi.org/10.1073/pnas.2118631119.

MCRC (Michigan Civil Rights Commission). 2017. *The Flint Water Crisis: Systemic Racism Through the Lens of Flint*. Lansing, MI. https://www.michigan.gov/-/media/Project/Websites/mdcr/mcrc/reports/2017/flint-crisis-report-edited.pdf?rev=4601519b3af345cfb9d468ae6ece9141.

McTarnaghan, S., Junod, A., Shipp, A., Schwabish, J., and Narayanan, A. 2022. "Comment Letter on CEQ's Climate and Economic Justice Screening Tool Beta Version." Washington, DC: Urban Institute.

Meehan, K., Jurjevich, J. R., Chun, N., and Sherrill, J. 2020. "Geographies of Insecure Water Access and the Housing-Water Nexus in US Cities." *Proceedings of the National Academy of Sciences of the United States of America* 117(46): 28700-28707. https://doi.org/10.1073/pnas.2007361117.

MEJ (Mapping for Environmental Justice). 2021. "What Is MEJ?" https://mappingforej.studentorg.berkeley.edu/what-is-mej/ (accessed May 31, 2024).

Méndez, M., Flores-Haro, G., and Zucker, L. 2020. "The (In)visible Victims of Disaster: Understanding the Vulnerability of Undocumented Latino/a and Indigenous Immigrants." *Geoforum* 116: 50-62. https://doi.org/10.1016/j.geoforum.2020.07.007.

Mendez, D. D., Hogan, V. K., and Culhane, J. 2011. "Institutional Racism and Pregnancy Health: Using Home Mortgage Disclosure Act Data to Develop an Index for Mortgage Discrimination at the Community Level." *Public Health Reports* 126(Suppl 3): 102-114. https://doi.org/10.1177/00333549111260s315.

Merz, J. F., Small, M. J., and Fischbeck, P. S. 1992. "Measuring Decision Sensitivity: A Combined Monte Carlo-Logistic Regression Approach." *Medical Decision Making* 12(3): 189-196. https://doi.org/10.1177/0272989x9201200304.

Meschede, T., Eden, M., Jain, S., Jee, E., Miles, B., Martinez, M., Stewart, S., Jacob, J., and Madison, M. 2022. *IERE Research Brief: Final Report from Our GI Bill Study*. Waltham, MA: Institute for Economic and Racial Equity, Brandeis University.

Miller, H. J., Witlox, F., and Tribby, C. P. 2013. "Developing Context-Sensitive Livability Indicators for Transportation Planning: A Measurement Framework." *Journal of Transport Geography* 26(C): 51-64. https://doi.org/10.1016/j.jtrangeo.2012.08.007.

Milton, B., Attree, P., French, B., Povall, S., Whitehead, M., and Popay, J. 2012. "The Impact of Community Engagement on Health and Social Outcomes: A Systematic Review." *Community Development Journal* 47(3): 316-334. https://doi.org/10.1093/cdj/bsr043.

Min, E., Gruen, D., Banerjee, D., Echeverria, T., Freelander, L., Schmeltz, M., Saganić, E., Piazza, M., Galaviz, V. E., Yost, M., and Seto, E. Y. W. 2019. The Washington State Environmental Health Disparities Map: Development of a Community-Responsive Cumulative Impacts Assessment Tool. *International Journal of Environmental Research and Public Health* 16(22): 4470. https://doi.org/10.3390/ijerph16224470.

Ministry of Housing, Communities & Local Government. 2019. *The English Indices of Deprivation 2019—Statistical Release*. London. https://assets.publishing.service.gov.uk/media/5d8e26f6ed915d5570c6cc55/IoD2019_Statistical_Release.pdf.

Mitchell, B., and Chakraborty, J. 2018. "Thermal Inequity: The Relationship Between Urban Structure and Social Disparities in an Era of Climate Change." In *The Routledge Handbook of Climate Justice*, edited by R. Holifield, J. Chakraborty, and G. Walker, 330-346. London: Routledge.

Mitra, S., Jones, K., Vick, B., Brown, D., McGinn, E., and Alexander, M. J. 2013. "Implementing a Multidimensional Poverty Measure Using Mixed Methods and a Participatory Framework." *Social Indicators Research* 110 3): 1061-1081. https://doi.org/10.1007/s11205-011-9972-9.

Mohai, P. 2018. "Environmental Justice and the Flint Water Crisis." *Michigan Sociological Review* 32: 1-41. https://www.jstor.org/stable/26528595.

Mohai, P., and Saha, R. 2007. "Racial Inequality in the Distribution of Hazardous Waste: A National-Level Reassessment." *Social Problems* 54(3): 343-370. https://doi.org/10.1525/sp.2007.54.3.343.

Mohai, P., and Saha, R. 2015. "Which Came First, People or Pollution? Assessing the Disparate Siting and Post-Siting Demographic Change Hypotheses of Environmental Injustice." *Environmental Research Letters* 10(11): 115008. https://doi.org/10.1088/1748-9326/10/11/115008.

Mohai, P., Pellow, D., and Roberts, J. T. 2009. "Environmental Justice." *Annual Review of Environment and Resources* 34(1): 405-430. https://doi.org/10.1146/annurev-environ-082508-094348.

Mohnot, S. 2023. *Broader Coalition Letter on Climate and Economic Justice Screening Tool*. The Greenlining Institute. https://greenlining.org/2023/broader-coalition-letter-on-climate-and-economic-justice-screening-tool/.

Morello-Frosch, R., and Jesdale, B. M. 2006. "Separate and Unequal: Residential Segregation and Estimated Cancer Risks Associated with Ambient Air Toxics in U.S. Metropolitan Areas." *Environmental Health Perspectives* 114(3): 386-393. https://doi.org/10.1289/ehp.8500.

Morello-Frosch, R., and Lopez, R. 2006. "The Riskscape and the Color Line: Examining the Role of Segregation In Environmental Health Disparities." *Environmental Research* 102(2): 181-196. https://doi.org/10.1016/j.envres.2006.05.007.

Morello-Frosch, R., Zuk, M., Jerrett, M., Shamasunder, B., and Kyle, A. D. 2011. "Understanding the Cumulative Impacts of Inequalities in Environmental Health: Implications for Policy." *Health Affairs (Millwood)* 30 (5): 879-887. https://doi.org/10.1377/hlthaff.2011.0153.

Munda, G. 2012. "Intensity of Preference and Related Uncertainty in Non-Compensatory Aggregation Rules." *Theory and Decision* 73(4): 649-669. https://doi.org/10.1007/s11238-012-9317-4.

Munda, G., Albrecht, D., Becker, W., Havari, E., Listorti, G., Ostlaender, N., Paolo, P., and Saisana, M. 2020. "The Use of Quantitative Methods in the Policy Cycle." In *Science for Policy Handbook,* edited by V. Šucha and M. Sienkiewicz, 206-222. Elsevier.

Münzel, T., Sørensen, M., and Daiber, A. 2021. "Transportation Noise Pollution and Cardiovascular Disease." *Nature Reviews Cardiology* 18(9): 619-636. https://doi.org/10.1038/s41569-021-00532-5.

Murray, A., Hall, A., Weaver, J., and Kremer, F. 2021. "Methods for Estimating Locations of Housing Units Served by Private Domestic Wells in the United States Applied to 2010." *Journal of the American Water Resources Association* 57(5): 828-843. https://doi.org/https://doi.org/10.1111/1752-1688.12937.

Nardo, M., Saisana, M., Saltelli, A., Tarantola, S., Hoffman, A., and Giovannini, E. 2005. *Handbook on Constructing Composite Indicators: Methodology and User Guide*. OECD Statistics Working Papers.

NASEM (National Academies of Sciences Engineering and Medicine). 2023a. *Representing Lived Experience in the Climate and Economic Justice Screening Tool: Proceedings of a Workshop–in Brief*, edited by A. DePinto. Washington, DC: The National Academies Press.

NASEM. 2023b. *Using Population Descriptors in Genetics and Genomics Research: A New Framework for an Evolving Field*. Washington, DC: The National Academies Press.

Nayak, S. G., Shrestha, S., Kinney, P. L., Ross, Z., Sheridan, S. C., Pantea, C. I., Hsu, W. H., Muscatiello, N., and Hwang, S. A. 2018. "Development of a Heat Vulnerability Index for New York State." *Public Health* 161: 127-137. https://doi.org/10.1016/j.puhe.2017.09.006.

NEJAC (National Environmental Justice Advisory Council). 2004. *Ensuring Risk Reduction in Communities with Multiple Stressors: Environmental Justice and Cumulative Risks/Impacts*. National Environmental Justice Advisory Council, Cumulative Risks/Impacts Work Group. https://www.epa.gov/sites/default/files/2015-02/documents/nejac-cum-risk-rpt-122104.pdf.

NEJAC. 2011. NEJAC Comments to EPA Plan EJ 2014. National Environmental Justice Advisory Council. https://www.epa.gov/sites/default/files/2015-02/documents/plan-ej-2014-comments-0511.pdf.

Neumayer, E., and Plümper, T. 2007. "The Gendered Nature of Natural Disasters: The Impact of Catastrophic Events on the Gender Gap in Life Expectancy, 1981–2002." *Annals of the Association of American Geographers* 97(3): 551-566. https://doi.org/10.1111/j.1467-8306.2007.00563.x.

Ngo, E. B. 2001. "When Disasters and Age Collide: Reviewing Vulnerability of the Elderly." *Natural Hazards Review* 2(2): 80-89. https://doi.org/10.1061/(ASCE)1527-6988(2001)2:2(80).

NOAA (National Oceanic and Atmospheric Administration). 2022. Climate Mapping for Resilience & Adaptation. [online] resilience.climate.gov. Accessed November 4, 2023. https://resilience.climate.gov.

NRC (National Research Council). 1994. *Ranking Hazardous-Waste Sites for Remedial Action*. Washington, DC: The National Academies Press.

NRC. 1996. *Understanding Risk: Informing Decisions in a Democratic Society*, edited by P. C. Stern and H. V. Fineberg. Washington, DC: The National Academies Press. https://doi.org/10.17226/5138.

NRC. 2009. *Science and Decisions: Advancing Risk Assessment*. Washington, DC: The National Academies Press.

Nyika, L., and Murray-Orr, A. 2017. "Critical Race Theory–Social Constructivist Bricolage: A Health-Promoting Schools Research Methodology." *Health Education Journal* 76(4): 432-441. https://doi.org/10.1177/0017896916689108.

O'Dell, K., Ford, B., Fischer, E. V., and Pierce, J. R. 2019. "Contribution of Wildland-Fire Smoke to US $PM_{2.5}$ and Its Influence on Recent Trends." *Environmental Science & Technology* 53(4): 1797-1804. https://doi.org/10.1021/acs.est.8b05430.

O'Shea, M. J., Toupal, J., Caballero-Gómez, H., McKeon, T. P., Howarth, M. V., Pepino, R., and Gieré, R. 2021. Lead Pollution, Demographics, and Environmental Health Risks: The Case of Philadelphia, USA. *International Journal of Environmental Research and Public Health* 18 (17): 9055. https://doi.org/10.3390/ijerph18179055.

OECD and JRC (Organisation for Economic Cooperation and Development and European Commission Joint Research Centre). 2008. *Handbook on Constructing Composite Indicators: Methodology and User Guide*. Paris: OECD Publishing. https://doi.org/10.1787/9789264043466-en.

Olson, E. D., and Stubblefield, A. 2021. "Lead Pipes Are Widespread and Used in Every State." Natural Resources Defense Council. https://www.nrdc.org/resources/lead-pipes-are-widespread-and-used-every-state (accessed February 17, 2024).

OMB (Office of Management and Budget). 2023. Circular A-4: To the Heads of Executive Agencies and Establishments. Washington, DC: The White House.). https://www.whitehouse.gov/wp-content/uploads/2023/11/CircularA-4.pdf.

Omi, M., and Winant, H. 2014. *Racial Formation in the United States*, 3rd ed. Routledge. https://doi.org/10.4324/9780203076804.

Otoiu, A., Titan, E., and Dumitrescu, R. 2014. "Are the Variables Used in Building Composite Indicators of Well-Being Relevant? Validating Composite Indexes of Well-Being." *Ecological Indicators* 46: 575-585. https://doi.org/10.1016/j.ecolind.2014.07.019.

Papadimitriou, E., Neves, A. R., and Becker, W. 2019. *JRC Statistical Audit of the Sustainable Development Goals Index and Dashboards*. EUR 29776 EN. Luxembourg: Publications Office of the European Union.

Papadimitriou, E., Neves, A. R., and Saisana, M. 2020. *JRC Statistical Audit of the 2020 Environmental Performance Index*. Joint Research Centre, European Commission. Luxembourg: Publications Office of the European Union. https://epi.yale.edu/downloads/jrc120879jrcauditepi2020.pdf.

Paradies, Y., Ben, J., Denson, N., Elias, A., Priest, N., Pieterse, A., Gupta, A., Kelaher, M., and Gee, G. 2015. "Racism as a Determinant of Health: A Systematic Review and Meta-Analysis." *PLoS ONE* 10(9): e0138511. https://doi.org/10.1371/journal.pone.0138511.

Payne-Sturges, D., Turner, A., Wignall, J., Rosenbaum, A., Dederick, E., and Dantzker, H. 2012. "A Review of State-Level Analytical Approaches for Evaluating Disproportionate Environmental Health Impacts." *Environmental Justice* 5(4): 173-187. https://doi.org/10.1089/env.2012.0008.

Payne-Sturges, D. C., Gee, G. C., and Cory-Slechta, D. A. 2021. "Confronting Racism in Environmental Health Sciences: Moving the Science Forward for Eliminating Racial Inequities." *Environmental Health Perspectives* 129(5): 055002. https://doi.org/10.1289/EHP8186.

Peek, L., and Stough, L. M. 2010. "Children with Disabilities in the Context of Disaster: A Social Vulnerability Perspective." *Child Development* 81(4): 1260-1270. https://doi.org/10.1111/j.1467-8624.2010.01466.x.

Perry, A. M. and Harshbarger, D. 2019. "America's Formerly Redlined Neighborhoods Have Changed, and So Must Solutions to Rectify Them." Brookings Institution. https://www.brookings.edu/articles/americas-formerly-redlines-areas-changed-so-must-solutions/.

Phraknoi, N., Sutanto, J., Hu, Y., Goh, Y. S., and Lee, C. E. C. 2023. "Older People's Needs in Urban Disaster Response: A Systematic Literature Review." *International Journal of Disaster Risk Reduction* 96: 103809. https://doi.org/10.1016/j.ijdrr.2023.103809.

Polonik, P., Ricke, K., Reese, S., and Burney, J. 2023. "Air Quality Equity in US Climate Policy." *Proceedings of the National Academy of Sciences of the United States of America* 120(26): e2217124120. https://doi.org/10.1073/pnas.2217124120.

Pontecorvo, E., and Sadasivam, N. 2022. "Three Open Questions About Biden's New Environmental Justice Tool." *Grist Magazine*. https://grist.org/equity/climate-and-economic-justice-screening-tool-biden-comment-period/.

Popovich, N., Figueroa, A., Sunter, D., and Shah, M. 2024. "Identifying Disadvantaged Communities in the United States: An Energy-Oriented Mapping Tool That Aggregates Environmental And Socioeconomic Burdens." *Energy Research & Social Science* 109: 103391. https://doi.org/10.1016/j.erss.2023.103391.

Powers, S. L., Mowen, A. J., and Webster, N. 2023. "Development and Validation of a Scale Measuring Public Perceptions of Racial Environmental Justice in Parks." *Journal of Leisure Research* 55(1): 1-24. https://doi.org/10.1080/00222216.2023.2183369.

Preston, B. L., Yuen, E. J., and Westaway, R. M. 2011. "Putting Vulnerability to Climate Change on the Map: A Review of Approaches, Benefits, and Risks." *Sustainability Science* 6(2): 177-202. https://doi.org/10.1007/s11625-011-0129-1.

Price-Robertson, R. 2011. What Is Community Disadvantage? Understanding the Issues, Overcoming the Problem. Communities and Families Clearinghouse Australia. CAFCA Resource Sheet. https://vdocuments.mx/what-is-community-disadvantage-understanding.html?page=1

Prochaska, J. D., Nolen, A. B., Kelley, H., Sexton, K., Linder, S. H., and Sullivan, J. 2014. "Social Determinants of Health in Environmental Justice Communities: Examining Cumulative Risk in Terms of Environmental Exposures and Social Determinants of Health." *Human and Ecological Risk Assessment: An International Journal* 20(4): 980-994. https://doi.org/10.1080/10807039.2013.805957.

Pullen Fedinick, K., Taylor, S., and Roberts, M. 2019. *Watered Down Justice*. 19-09-A, Natural Resources Defense Council. https://www.nrdc.org/resources/watered-down-justice.

Purnell, B., Theoharis, J., and Woodard, K. 2019. *The Strange Careers of the Jim Crow: North Segregation and Struggle Outside of the South*. New York: NYU Press.

Pursch, B., Tate, A., Legido-Quigley, H., and Howard, N. 2020. "Health for All? A Qualitative Study of NGO Support to Migrants Affected by Structural Violence in Northern France." *Social Science & Medicine* 248: 112838. https://doi.org/10.1016/j.socscimed.2020.112838.

Putnam, R. D. 2000. *Bowling Alone: The Collapse and Revival of American Community*. New York: Simon and Schuster.

Rahman, S., Sunder, P., and Jackson, D. 2022. *A People's History of Structural Racism in Academia: From A(dministration of Justice) to Z(oology)*. https://www.hancockcollege.edu/ccecho/documents/A%20Peoples%20History%20of%20Structural%20Racism%20in%20Academia%20From%20A%20to%20Z.pdf.

Räsänen, A., Heikkinen, K., Piila, N., and Juhola, S. 2019. "Zoning and Weighting in Urban Heat Island Vulnerability and Risk Mapping in Helsinki, Finland." *Regional Environmental Change* 19(5): 1481-1493. https://doi.org/10.1007/s10113-019-01491-x.

Ravalli, F., Yu, Y., Bostick, B. C., Chillrud, S. N., Schilling, K., Basu, A., Navas-Acien, A., and Nigra, A. E. 2022. "Sociodemographic Inequalities in Uranium and Other Metals in Community Water Systems Across the USA, 2006– 2011: A Cross-Sectional Study." *The Lancet Planetary Health* 6(4): e320-e330. https://doi.org/10.1016/S2542-5196(22)00043-2.

Ravichandran, V., Albert, R. M. L., Teirstein, M., Garg, A., Nagovich, J., Wilson, H., and Wilson, S. 2021. *Gaps in Environmental Justice Screening and Mapping Tools and Potential New Indicators*. National Wildlife Federation and Community Engagement, Environmental Justice, and Health Lab. https://www.nwf.org/Home/Educational-Resources/Reports/2021/11-08-21-gaps-in-EJSM-tools.

Razavi, S., Jakeman, A., Saltelli, A., Prieur, C., Iooss, B., Borgonovo, E., Plischke, E., Lo Piano, S., Iwanaga, T., Becker, W., Tarantola, S., Guillaume, J. H. A., Jakeman, J., Gupta, H., Melillo, N., Rabitti, G., Chabridon, V., Duan, Q., Sun, X., Smith, S., Sheikholeslami, R., Hosseini, N., Asadzadeh, M., Puy, A., Kucherenko, S., and Maier, H. R. 2021. "The Future of Sensitivity Analysis: An Essential Discipline for Systems Modeling and Policy Support." *Environmental Modelling & Software* 137: 104954. https://doi.org/10.1016/j.envsoft.2020.104954.

Reardon, S. F., and Bischoff, K. 2011. "Income Inequality and Income Segregation." *American Journal of Sociology* 116(4): 1092-1153. https://doi.org/10.1086/657114.

Reckien, D. 2018. "What Is in an Index? Construction Method, Data Metric, and Weighting Scheme Determine the Outcome of Composite Social Vulnerability Indices in New York City." *Regional Environmental Change* 18(5): 1439-1451. https://doi.org/10.1007/s10113-017-1273-7.

Ren, Q., Panikkar, B., and Galford, G. 2023. "Vermont Environmental Disparity Index and Risks." *Environmental Justice* 16(6): 418-431. https://doi.org/10.1089/env.2021.0063.

Renteria, R., Grineski, S., Collins, T., Flores, A., and Trego, S. 2022. "Social Disparities in Neighborhood Heat in the Northeast United States." *Environmental Research* 203: 111805. https://doi.org/10.1016/j.envres.2021.111805.

Ribeiro, A. I., Amaro, J., Lisi, C., and Fraga, S. 2018. "Neighborhood Socioeconomic Deprivation and Allostatic Load: A Scoping Review." *International Journal of Environmental Research and Public Health* 15(6): 1092. https://doi.org/10.3390/ijerph15061092.

Rice, L. J., Jiang, C., Wilson, S. M., Burwell-Naney, K., Samantapudi, A., and Zhang, H. 2014. "Use of Segregation Indices, Townsend Index, and Air Toxics Data to Assess Lifetime Cancer Risk Disparities in Metropolitan Charleston, South Carolina, USA." *International Journal of Environmental Research and Public Health* 11(5): 5510-5526. https://doi.org/10.3390/ijerph110505510.

Rittel, H. W. J., and Webber, M. M. 1973. "Dilemmas in a General Theory of Planning." *Policy Sciences* 4(2): 155-169. https://doi.org/10.1007/BF01405730.

Roberts, E. M., English, P. B., Wong, M., Wolff, C., Valdez, S., Van den Eeden, S. K., and Ray, G. T. 2006. "Progress in Pediatric Asthma Surveillance II: Geospatial Patterns of Asthma in Alameda County, California." *Preventing Chronic Disease* 3(3): A92. www.cdc.gov/pcd/issues/2006/jul/05_0187.htm.

Romitti, Y., Wing, I. S., Spangler, K. R., and Wellenius, G. A. 2022. "Inequality in the Availability of Residential Air Conditioning Across 115 US Metropolitan Areas." *PNAS Nexus* 1(4): pgac210. https://doi.org/10.1093/pnasnexus/pgac210.

Roth, R. 2013. "Interactive Maps: What We Know and What We Need to Know." *Journal of Spatial Information Science* 6: 59-115. https://doi.org/10.5311/JOSIS.2013.6.105.

Rothbaum, J., Eggleston, J., Bee, A., Klee, M., and Mendez-Smith, B. 2021. *Addressing Nonresponse Bias in the American Community Survey During the Pandemic Using Administrative Data*. U.S. Census Bureau. https://www.census.gov/library/working-papers/2021/acs/2021_Rothbaum_01.html.

Rothstein, R. 2017. *The Color of Law: A Forgotten History of How Our Government Segregated America*. New York: Liveright.

Rowangould, D., Rowangould, G., Craft, E., and Niemeier, D. 2019. "Validating and Refining EPA's Traffic Exposure Screening Measure." *International Journal of Environmental Research and Public Health* 16(1): 3. https://doi.org/10.3390/ijerph16010003.

Rufat, S., Tate, E., Emrich, C. T., and Antolini, F. 2019. "How Valid Are Social Vulnerability Models?" *Annals of the American Association of Geographers* 109(4): 1131-1153. https://EconPapers.repec.org/RePEc:taf:raagxx:v:109:y:2019:i:4:p:1131-1153.

Ryder, S. S. 2017. "A Bridge to Challenging Environmental Inequality: Intersectionality, Environmental Justice, and Disaster Vulnerability." *Social Thought & Research* 34: 85-115. https://doi.org/10.17161/1808.25571.

SAB (EPA Scientific Advisory Board). 2023. *Review of the Updated Methodology of EPA's Environmental Justice Screening and Mapping Tool (EJScreen Version 2.1)*. EPA-SAB-24-002, Office of Environmental Justice and External Civil Rights, US Environmental Protection Agency (Washington, DC). https://sab.epa.gov/ords/sab/r/sab_apex/sab/advisoryactivitydetail?p18_id=2627&clear=18&session=12140937732306.

Sadd, J. L., Pastor, M., Morello-Frosch, R., Scoggins, J., and Jesdale, B. 2011. "Playing It Safe: Assessing Cumulative Impact and Social Vulnerability Through an Environmental Justice Screening Method in the South Coast Air Basin, California." *International Journal of Environmental Research and Public Health* 8(5): 1441-1459. https://doi.org/10.3390/ijerph8051441.

Sadasivam, N. 2023. "Why the White House's Environmental Justice Tool Is Still Disappointing Advocates." Grist. https://grist.org/equity/white-house-environmental-justice-tool-cejst-update-race/.

Sadasivam, N., and Aldern, C. 2022. "The White House Excluded Race from Its Environmental Justice Tool. We Put It Back In." *Grist Magazine*. https://grist.org/equity/climate-and-economic-justice-screening-tool-race/.

Sadd, J. L., Pastor, M., Morello-Frosch, R., Scoggins, J., and Jesdale, B. 2011. "Playing It Safe: Assessing Cumulative Impact and Social Vulnerability Through an Environmental Justice Screening Method in the South Coast Air Basin, California." *International Journal of Environmental Research and Public Health* 8(5): 1441-1459. https://doi.org/10.3390/ijerph8051441.

Sadd, J., Morello-Frosch, R., Pastor, M., Matsuoka, M., Prichard, M., and Carter, V. 2014. "The Truth, the Whole Truth, and Nothing but the Ground-Truth: Methods to Advance Environmental Justice and Researcher-Community Partnerships." *Health Education & Behavior* 41(3): 281-290. https://doi.org/10.1177/1090198113511816.

Sadd, J. L., Hall, E. S., Pastor, M., Morello-Frosch, R. A., Lowe-Liang, D., Hayes, J., and Swanson, C. 2015. "Ground-Truthing Validation to Assess the Effect of Facility Locational Error on Cumulative Impacts Screening Tools." *Geography Journal* 2015: 324683. https://doi.org/10.1155/2015/324683.

Saha, R., and Mohai, P. 2005. "Historical Context and Hazardous Waste Facility Siting: Understanding Temporal Patterns in Michigan." *Social Problems* 52(4): 618-648. https://doi.org/10.1525/sp.2005.52.4.618.

Saisana, M., and Saltelli, A. 2008. *Sensitivity Analysis of the 2008 Environmental Performance Index*. EUR 23485, Joint Research Centre (Luxembourg: Office for Official Publications of the European Communities).

Saisana, M., Saltelli, A., and Tarantola, S. 2005. "Uncertainty and Sensitivity Analysis Techniques as Tools for the Quality Assessment of Composite Indicators." *Journal of the Royal Statistical Society, Series A: Statistics in Society* 168(2): 307-323. https://doi.org/https://doi.org/10.1111/j.1467-985X.2005.00350.x.

Saisana, M., Becker, W., Neves, A. R., Alberti, V., and Dominguez Torreiro, M. 2019. *Your 10-Step Pocket Guide to Composite Indicators & Scoreboards*. JRC118442. Joint Research Centre. https://knowledge4policy.ec.europa.eu/sites/default/files/10-step-pocket-guide-to-composite-indicators-and-scoreboards.pdf.

Saisana, M., Fragoso Neves, A., Nurminen, M., Starnoni, E., Alberti, V., and Tacao Moura, C. J. 2022. *Indices and Scoreboards in EU Policymaking*. JRC131074. European Commission. https://publications.jrc.ec.europa.eu/repository/handle/JRC131074.

Saltelli, A., and Annoni, P. 2010. "How to Avoid a Perfunctory Sensitivity Analysis." *Environmental Modelling & Software* 25(12): 1508-1517. https://doi.org/10.1016/j.envsoft.2010.04.012.

Saltelli, A., and D'Hombres, B. 2010. "Sensitivity Analysis Didn't Help. A Practitioner's Critique of the Stern Review." *Global Environmental Change* 20(2): 298-302. https://doi.org/10.1016/j.gloenvcha.2009.12.003.

Saltelli, A., Ratto, M., Andres, T., Campolongo, F., Cariboni, J., Gatelli, D., Saisana, M., and Tarantola, S. 2008. *Global Sensitivity Analysis. The Primer*, Vol. 304. Hoboken, NJ: John Wiley & Sons.

Saltelli, A., Aleksankina, K., Becker, W., Fennell, P., Ferretti, F., Holst, N., Li, S., and Wu, Q. 2019. "Why So Many Published Sensitivity Analyses Are False: A Systematic Review of Sensitivity Analysis Practices." *Environmental Modelling & Software* 114: 29-39. https://doi.org/10.1016/j.envsoft.2019.01.012.

Salzman, J. 2003. "Methodological Choices Encountered in the Construction of Composite Indices of Economic and Social Well-Being." https://www.csls.ca/events/cea2003/salzman-typol-cea2003.pdf.

Samuels, M. L. 1993. "Simpson's Paradox and Related Phenomena." *Journal of the American Statistical Association* 88(421): 81-88. https://doi.org/10.1080/01621459.1993.10594297.

Santos, K. D. S., Ribeiro, M. C., Queiroga, D. E. U., Silva, I., and Ferreira, S. M. S. 2020. "The Use of Multiple Triangulations as a Validation Strategy in a Qualitative Study." *Ciência & Saúde Coletiva* 25(2): 655-664. https://doi.org/10.1590/1413-81232020252.12302018.

Santos-Hernández, J., and Morrow, B. 2013. "Language and Literacy." In *Social Vulnerability to Disasters*, edited by D. S. K. Thomas, B. D. Phillips, W. E. Lovekamp, and A. Fothergill, 265-280. CRC Press.

Schäfer, R. B., Jackson, M., Juvigny-Khenafou, N., Osakpolor, S. E., Posthuma, L., Schneeweiss, A., Spaak, J., and Vinebrooke, R. 2023. "Chemical Mixtures and Multiple Stressors: Same but Different?" *Environmental Toxicology and Chemistry* 42(9): 1915-1936. https://doi.org/10.1002/etc.5629.

Schaider, L. A., Swetschinski, L., Campbell, C., and Rudel, R. A. 2019. "Environmental Justice and Drinking Water Quality: Are There Socioeconomic Disparities in Nitrate Levels in U.S. Drinking Water?" *Environmental Health* 18(1): 3. https://doi.org/10.1186/s12940-018-0442-6.

Schinasi, L. H., Kanungo, C., Christman, Z., Barber, S., Tabb, L., and Headen, I. 2022. "Associations Between Historical Redlining and Present-Day Heat Vulnerability Housing and Land Cover Characteristics in Philadelphia, PA." *Journal of Urban Health* 99(1): 134-145. https://doi.org/10.1007/s11524-021-00602-6.

Schlake, M. 2015. "Community Engagement: Nine Principles." Cornhusker Economics. http://digitalcommons.unl.edu/cgi/viewcontent.cgi?article=1726&context=agecon_cornhusker.

Schmidtlein, M. C., Deutsch, R. C., Piegorsch, W. W., and Cutter, S. L. 2008. "A Sensitivity Analysis of the Social Vulnerability Index." *Risk Analysis* 28(4): 1099-1114. https://doi.org/10.1111/j.1539-6924.2008.01072.x.

Schneider, J. S. 2023. "Neurotoxicity and Outcomes from Developmental Lead Exposure: Persistent or Permanent?" *Environmental Health Perspectives* 131(8). https://doi.org/10.1289/EHP12371.

Schumann, R., Emrich, C., Butsic, V., Mockrin, M., Zhou, Y., Gaither, C., Price, O., Syphard, A., Whittaker, J., and Aksha, S. 2024. "The Geography of Social Vulnerability and Wildfire Occurrence (1984–2018) in the Conterminous USA." *Natural Hazards* 120: 4297-4327. https://doi.org/10.1007/s11069-023-06367-2.

Schuyler, A. J., and Wenzel, S. E. 2022. "Historical Redlining Impacts Contemporary Environmental and Asthma-Related Outcomes in Black Adults." *American Journal of Respiratory and Critical Care Medicine* 206(7): 824-837. https://doi.org/10.1164/rccm.202112-2707OC.

Schwarz, K., Fragkias, M., Boone, C. G., Zhou, W., McHale, M., Grove, J. M., O'Neil-Dunne, J., McFadden, J. P., Buckley, G. L., Childers, D., Ogden, L., Pincetl, S., Pataki, D., Whitmer, A., and Cadenasso, M. L. 2015. "Trees Grow on Money: Urban Tree Canopy Cover and Environmental Justice." *PLoS ONE* 10(4): e0122051. https://doi.org/10.1371/journal.pone.0122051.

Seto, E., and Huang, C.-H. 2023. "The National Transportation Noise Exposure Map." *medRxiv*: 2023.02.02.23285396. https://doi.org/10.1101/2023.02.02.23285396.

Sevelius, J. M., Gutierrez-Mock, L., Zamudio-Haas, S., McCree, B., Ngo, A., Jackson, A., Clynes, C., Venegas, L., Salinas, A., Herrera, C., Stein, E., Operario, D., and Gamarel, K. 2020. "Research with Marginalized Communities: Challenges to Continuity During the COVID-19 Pandemic." *AIDS and Behavior* 24(7): 2009-2012. https://doi.org/10.1007/s10461-020-02920-3.

Sexton, K., and Linder, S. H. 2011. "Cumulative Risk Assessment for Combined Health Effects from Chemical and Nonchemical Stressors." *American Journal of Public Health* 101(Suppl 1): S81-S88. https://doi.org/10.2105/ajph.2011.300118.

Shaker, Y., Grineski, S. E., Collins, T. W., and Flores, A. B. 2023. "Redlining, Racism and Food Access in US Urban Cores." *Agriculture and Human Values* 40(1): 101-112. https://doi.org/10.1007/s10460-022-10340-3.

Shamasunder, B., Chan, M., Navarro, S., Eckel, S., and Johnston, J. E. 2022. "Mobile Daily Diaries to Characterize Stressors and Acute Health Symptoms in an Environmental Justice Neighborhood." *Health Place* 76: 102849. https://doi.org/10.1016/j.healthplace.2022.102849.

Shepard, P. M. 2002. "Advancing Environmental Justice Through Community-Based Participatory Research." *Environmental Health Perspectives* 110(2): 139.

Sherrieb, K., Norris, F. H., and Galea, S. 2010. "Measuring capacities for community resilience." Springer doi:10.1007/s11205-010-9576-9.

Shertzer, A., Twinam, T., and Walsh, R. 2021. "Zoning and Segregation in Urban Economic History." *Regional Science and Urban Economics* 94: 103652. https://doi.org/10.1016/j.regsciurbeco.2021.103652.

Shindell, D., Zhang, Y., Scott, M., Ru, M., Stark, K., and Ebi, K. L. 2020. "The Effects of Heat Exposure on Human Mortality Throughout the United States." *Geohealth* 4(4): e2019GH000234. https://doi.org/10.1029/2019gh000234.

Shrestha, R., Rajpurohit, S., and Saha, D. 2023. *CEQ's Climate and Economic Justice Screening Tool Needs to Consider How Burdens Add Up*. Washington, DC: World Resource Institute. https://www.wri.org/technical-perspectives/ceq-climate-and-economic-justice-screening-tool-cumulative-burdens.

Siddiqi, S. M., Mingoya-LaFortune, C., Chari, R., Preston, B. L., Gahlon, G., Hernandez, C. C., Huttinger, A., Stephenson, S. R., and Madrigano, J. 2023. "The Road to Justice40: Organizer and Policymaker Perspectives on the Historical Roots of and Solutions for Environmental Justice Inequities in U.S. Cities." *Environmental Justice* 16(5): 340-350. https://doi.org/10.1089/env.2022.0038.

Sims, K. R. E., Lee, L. G., Estrella-Luna, N., Lurie, M. R., and Thompson, J. R. 2022. "Environmental Justice Criteria for New Land Protection Can Inform Efforts to Address Disparities in Access to Nearby Open Space." *Environmental Research Letters* 17(6): 064014. https://doi.org/10.1088/1748-9326/ac6313.

Skelton-Wilson, S., Sandoval-Lunn, M., Zhang, X., Stern, F., and Kendall, J. 2021. *Methods and Emerging Strategies to Engage People with Lived Experience: Improving Federal Research, Policy, and Practice*. Washington, DC: Department of Health and Human Services Office of the Assistant Secretary for Planning and Evaluation. https://aspe.hhs.gov/reports/lived-experience-brief.

Slotterback, C. S., and Lauria, M. 2019. "Building a Foundation for Public Engagement in Planning." *Journal of the American Planning Association* 85(3): 183-187. https://doi.org/10.1080/01944363.2019.1616985.

Smallenbroek, O., Caperna, G., and Papadimitriou, E. 2023. *JRC Audit of the Environmental Performance Index 2022*. Joint Research Centre. Luxembourg: Publications Office of the European Union. https://publications.jrc.ec.europa.eu/repository/handle/JRC131959.

Smedley, A., and Smedley, B. D. 2005. "Race as Biology Is Fiction, Racism as a Social Problem Is Real: Anthropological and Historical Perspectives on the Social Construction of Race." *American Psychologist* 60(1): 16-26. https://doi.org/10.1037/0003-066x.60.1.16.

Smiley, K. T. 2019. "Racial and Environmental Inequalities in Spatial Patterns in Asthma Prevalence in the US South." *Southeastern Geographer* 59(4): 389-402. https://www.jstor.org/stable/26841635.

Solis, R. 1997. "Jemez Principles for Democratic Organizing." SouthWest Organizing Project. Working Group Meeting on Globalization and Trade, Jemez, NM, December 6-8, 1996.

Solomon, G. M., Morello-Frosch, R., Zeise, L., and Faust, J. B. 2016. "Cumulative Environmental Impacts: Science and Policy to Protect Communities." *Annual Review of Public Health* 37: 83-96. https://doi.org/10.1146/annurev-publhealth-032315-021807.

Sotolongo, M. 2022. "Justice40 and Community Definition: How Much of the U.S. Population Is Living in a 'Disadvantaged Community'?" *Initiative for Energy Justice* (blog). https://iejusa.org/justice-40-and-community-definition-blog/.

Southerland, V. A., Anenberg, S. C., Harris, M., Apte, J., Hystad, P., van Donkelaar, A., Martin, R. V., Beyers, M., and Roy, A. 2021. "Assessing the Distribution of Air Pollution Health Risks within Cities: A Neighborhood-Scale Analysis Leveraging High-Resolution Data Sets in the Bay Area, California." *Environmental Health Perspectives* 129 (3): 037006. https://doi.org/10.1289/EHP7679.

Spielman, S. E., Folch, D., and Nagle, N. 2014. "Patterns and Causes of Uncertainty in the American Community Survey." *Applied Geography* 46: 147-157. https://doi.org/10.1016/j.apgeog.2013.11.002.

Spielman, S., Tuccillo, J., Folch, D., Schweikert, A., Davies, R., Wood, N., and Tate, E. 2020. "Evaluating Social Vulnerability Indicators: Criteria and Their Application to the Social Vulnerability Index." *Natural Hazards* 100: 417-436. https://doi.org/10.1007/s11069-019-03820-z.

Spriggs, A., Rotman, R., and Trauth, K. 2024. "Functional Analysis of Web-Based GIS Tools for Environmental Justice Assessment of Transportation Projects." *Transportation Research Part D: Transport and Environment* 128: 104080. https://doi.org/10.1016/j.trd.2024.104080.

Stempowski, D. 2023. "Counting Every Voice: Understanding Hard-to-Count and Historically Undercounted Populations." Random Samplings (blog), U.S. Census Bureau. November 7. https://www.census.gov/newsroom/blogs/random-samplings/2023/10/understanding-undercounted-populations.html.

Steyerberg, E. W., Harrell, F. E., Jr., Borsboom, G. J., Eijkemans, M. J., Vergouwe, Y., and Habbema, J. D. 2001. "Internal Validation of Predictive Models: Efficiency of Some Procedures for Logistic Regression Analysis." *Journal of Clinical Epidemiology* 54(8): 774-781. https://doi.org/10.1016/s0895-4356(01)00341-9.

Stillo, F., and MacDonald Gibson, J. 2017. "Exposure to Contaminated Drinking Water and Health Disparities in North Carolina." *American Journal of Public Health* 107(1): 180-185. https://doi.org/10.2105/ajph.2016.303482.

Stöckl, D., D'Hondt, H., and Thienpont, L. M. 2009. "Method Validation Across the Disciplines—Critical Investigation of Major Validation Criteria and Associated Experimental Protocols." *Journal of Chromatography B: Analytical Technologies in the Biomedical and Life Sciences* 877(23): 2180-2190. https://doi.org/10.1016/j.jchromb.2008.12.056.

Storr, S. 2021. "Environmental Equity and the Cosmetics Industry: The Effect of Class upon Toxic Exposure." Hamilton Digital Commons. https://digitalcommons.hamilton.edu/student_scholarship/64/.

Tarantola, S., Ferretti, F., Lo Piano, S., Kozlova, M., Lachi, A., Rosati, R., Puy, A., Roy, P., Vannucci, G., Kuc-Czarnecka, M., and Saltelli, A. 2024. "An Annotated Timeline of Sensitivity Analysis." *Environmental Modelling & Software* 174: 105977. https://doi.org/10.1016/j.envsoft.2024.105977.

Tarlow, K. R. 2024. "The Colonial History of Systemic Racism: Insights for Psychological Science." *Perspectives on Psychological Science*. https://doi.org/10.1177/17456916231223932.

Tate, E. 2011. "Indices of Social Vulnerability to Hazards: Model Uncertainty and Sensitivity." University of South Carolina.

Tate, E. 2012. "Social Vulnerability Indices: A Comparative Assessment Using Uncertainty and Sensitivity Analysis." *Natural Hazards* 63(2): 325-347. https://doi.org/10.1007/s11069-012-0152-2.

Taylor, N. L., Porter, J. M., Bryan, S., Harmon, K. J., and Sandt, L. S. 2023. "Structural Racism and Pedestrian Safety: Measuring the Association Between Historical Redlining and Contemporary Pedestrian Fatalities Across the United States, 2010-2019." *American Journal of Public Health* 113(4): 420-428. https://doi.org/10.2105/AJPH.2022.307192.

Tee Lewis, P. G., Chiu, W. A., Nasser, E., Proville, J., Barone, A., Danforth, C., Kim, B., Prozzi, J., and Craft, E. 2023. "Characterizing Vulnerabilities to Climate Change Across the United States." *Environment International* 172: 107772. https://doi.org/10.1016/j.envint.2023.107772.

Tessum, C. W., Paolella, D. A., Chambliss, S. E., Apte, J. S., Hill, J. D., and Marshall, J. D. 2021. "$PM_{2.5}$ Polluters Disproportionately and Systemically Affect People of Color in the United States." *Science Advances* 7(18): eabf4491. https://doi.org/10.1126/sciadv.abf4491.

Tester, F., McNicoll, P., and Tran, Q. 2012. "Structural Violence and the 1962-1963 Tuberculosis Epidemic in Eskimo Point, N.W.T." *Études/Inuit/Studies* 36(2): 165-185. https://doi.org/10.7202/1015983ar.

Texas Rising. 2022. "Texas Environmental Justice Explorer." Climate Advocacy Lab. https://climateadvocacylab.org/resource/texas-environmental-justice-explorer (accessed May 31, 2024).

Tishman Environment and Design Center. 2022. "Cumulative Impacts Definitions, Indicators and Thresholds in the US." https://tishmancenter.github.io/CumulativeImpacts/cumulative_impacts.html (accessed December 18, 2023).

Trudeau, C., King, N., and Guastavino, C. 2023. "Investigating Sonic Injustice: A Review of Published Research." *Social Science & Medicine* 326: 115919. https://doi.org/10.1016/j.socscimed.2023.115919.

Tsou, M.-H. 2011. "Revisiting Web Cartography in the United States: The Rise of User-Centered Design." *Cartography and Geographic Information Science* 38(3): 250-257. https://doi.org/10.1559/15230406382250.

Tulve, N., Ruiz, J., Lichtveld, K., Perreault, S., and Quackenboss, J. 2016. "Development of a Conceptual Framework Depicting a Child's Total (Built, Natural, Social) Environment in Order to Optimize Health and Well-Being." *Journal of Environment and Health Science* 2: 1-8. https://doi.org/10.15436/2378-6841.16.1121.

Turner, S. E., and Bound, J. 2002. "Closing the Gap or Widening the Divide: The Effects of the G.I. Bill and World War II on the Educational Outcomes of Black Americans. National Bureau of Economic Research Working Paper 9044. https://doi.org/10.3386/w9044.

UNDP (United Nations Development Programme). 1990. *Human Development Report 1990: Concept and Measurement of Human Development*. New York and Oxford: Oxford University Press.

University of California Hastings College of Law. 2010. *Environmental Justice for All: A Fifty-State Survey of Legislation, Policies, and Initiative*, 4th ed.

U.S. Congress, House. 2021. "H.R.2021—Environmental Justice for All Act." 117 Cong., 1st sess., December 30. https://www.congress.gov/bill/117th-congress/house-bill/2021 (accessed December 19, 2023).

U.S. Congress, Senate. 2021. "S.2630— Environmental Justice Act of 2021." 117 Cong., 1st sess., August 5. https://www.congress.gov/bill/117th-congress/senate-bill/2630 (accessed December 19, 2023).

USGCRP (U.S. Global Change Research Program). 2023. *The Fifth National Climate Assessment*. Washington, DC. http://nca2023.globalchange.gov/.

van Donkelaar, A., Hammer, M. S., Bindle, L., Brauer, M., Brook, J. R., Garay, M. J., Hsu, N. C., Kalashnikova, O. V., Kahn, R. A., Lee, C., Levy, R. C., Lyapustin, A., Sayer, A. M., and Martin, R. V. 2021. "Monthly Global Estimates of Fine Particulate Matter and Their Uncertainty." *Environmental Science & Technology* 55(22): 15287-15300. https://doi.org/10.1021/acs.est.1c05309.

Vermont State Legislature. 2022. "An Act Relating to Environmental Justice in Vermont." S.148 (Act 154) § 1 (2). https://legislature.vermont.gov/bill/status/2022/S.148 (accessed December 22, 2023).

Verdin, A., Funk, C., Peterson, P., Landsfeld, M., Tuholske, C., and Grace, K. 2020. "Development and Validation of the CHIRTS-Daily Quasi-Global High-Resolution Daily Temperature Data Set." *Scientific Data* 7(1): 303. https://doi.org/10.1038/s41597-020-00643-7.

Vinyeta, K., Whyte, K., and Lynn, K. 2015. *Climate Change Through an Intersectional Lens: Gendered Vulnerability and Resilience in Indigenous Communities in the United States*. General Technical Report PNW-GTR-923. U.S. Department of Agriculture. https://www.fs.usda.gov/pnw/pubs/pnw_gtr923.pdf.

Volin, E., Ellis, A., Hirabayashi, S., Maco, S., Nowak, D. J., Parent, J., and Fahey, R. T. 2020. "Assessing Macro-Scale Patterns in Urban Tree Canopy and Inequality." *Urban Forestry & Urban Greening* 55: 126818. /https://doi.org/10.1016/j.ufug.2020.126818.

Wackernagel, M., and Rees, W. 1996. *Our Ecological Footprint: Reducing Human Impact on Earth*, Vol. 9. Gabriola Island, BC and Philadelphia, PA: New Society.

Wang, Y., Apte, J. S., Hill, J. D., Ivey, C. E., Patterson, R. F., Robinson, A. L., Tessum, C. W., and Marshall, J. D. 2022. "Location-Specific Strategies for Eliminating US National Racial-Ethnic $PM_{2.5}$ Exposure Inequality." *Proceedings of the National Academy of Sciences of the United States of America* 119(44): e2205548119. https://doi.org/10.1073/pnas.2205548119.

Wang, Y., Apte, J. S., Hill, J. D., Ivey, C. E., Johnson, D., Min, E., Morello-Frosch, R., Patterson, R., Robinson, A. L., Tessum, C. W., and Marshall, J. D. 2023. "Air Quality Policy Should Quantify Effects on Disparities." *Science* 381(6655): 272-274. https://doi.org/10.1126/science.adg9931.

Washington State Legislature. 2021. "The Healthy Environment for All Act (HEAL)," SB 5141, codified as RCW 70A.02. https://lawfilesext.leg.wa.gov/biennium/2021-22/Pdf/Bills/Session%20Laws/Senate/5141-S2.SL.pdf (accessed December 22, 2023).

Wei, J., Wang, J., Li, Z., Kondragunta, S., Anenberg, S., Wang, Y., Zhang, H., Diner, D., Hand, J., Lyapustin, A., Kahn, R., Colarco, P., da Silva, A., and Ichoku, C. 2023. "Long-term Mortality Burden Trends Attributed to Black Carbon and PM2•5 from Wildfire Emissions Across the Continental USA from 2000 to 2020: a Deep Learning Modelling Study." *The Lancet Planetary Health* 7 (12): e963-e975. https://doi.org/10.1016/S2542-5196(23)00235-8.

Weinberger, K. R., Harris, D., Spangler, K. R., Zanobetti, A., and Wellenius, G. A. 2020. "Estimating the Number of Excess Deaths Attributable to Heat in 297 United States Counties." *Environ Epidemiol* 4 (3): e096. https://doi.org/10.1097/ee9.0000000000000096.

West Virginia Development Office. 2020. *West Virginia Community Development Block Grant, Amendment 6 to Disaster Recovery Action Plan: For the Use of CDBG-DR Funds in Response to the Floods and Severe Storms of June 2016*. https://wvfloodrecovery.com/wp-content/uploads/actionplan/SubstantialAmendement6.pdf.

WHEJAC (White House Environmental Justice Advisory Council). 2021. *Final Recommendations: Justice40 Climate and Economic Justice Screening Tool & Executive Order 12898 Revisions*. https://www.epa.gov/sites/default/files/2021-05/documents/whiteh2.pdf.

WHO/ILO/UNEP (World Health Organization, International Labour Organization, and United Nations Environment Programme). 2008. *Uncertainty and Data Quality in Exposure Assessment*. Geneva, Switzerland: World Health Organization. https://www.who.int/publications/i/item/9789241563765.

Wien, S., Miller, A. L., and Kramer, M. R. 2023. "Structural Racism Theory, Measurement, and Methods: A Scoping Review." *Frontiers in Public Health* 11. https://www.frontiersin.org/journals/public-health/articles/10.3389/fpubh.2023.1069476.

Wilkins, D., and Schulz, A. J. 2023. "Antiracist Research and Practice for Environmental Health: Implications for Community Engagement." *Environmental Health Perspectives* 131(5): 55002. https://doi.org/10.1289/ehp11384.

Wilkinson, M. D., Dumontier, M., Aalbersberg, I. J., Appleton, G., Axton, M., Baak, A., Blomberg, N., Boiten, J. W., da Silva Santos, L. B., Bourne, P. E., Bouwman, J., Brookes, A. J., Clark, T., Crosas, M., Dillo, I., Dumon, O., Edmunds, S., Evelo, C. T., Finkers, R., Gonzalez-Beltran, A., Gray, A. J., Groth, P., Goble, C., Grethe, J. S., Heringa, J., 't Hoen, P. A., Hooft, R., Kuhn, T., Kok, R., Kok, J., Lusher, S. J., Martone, M. E., Mons, A., Packer, A. L., Persson, B., Rocca-Serra, P., Roos, M., van Schaik, R., Sansone, S. A., Schultes, E., Sengstag, T., Slater, T., Strawn, G., Swertz, M. A., Thompson, M., van der Lei, J., van Mulligen, E., Velterop, J., Waagmeester, A., Wittenburg, P., Wolstencroft, K., Zhao, J., and Mons, B. 2016. "The FAIR Guiding Principles for Scientific Data Management and Stewardship." *Scientific Data* 3: 160018. https://doi.org/10.1038/sdata.2016.18.

Williams, D. R., and Rucker, T. D. 2000. "Understanding and Addressing Racial Disparities in Health Care." *Health Care Financing Review* 21(4): 75-90.

Williams, D. R., and Collins, C. 2001. "Racial Residential Segregation: A Fundamental Cause of Racial Disparities in Health." *Public Health Reports* 116(5): 404-416. https://doi.org/10.1093/phr/116.5.404.

Williams, E., Polsky, D., Archer, J.-M., Rodriguez, A., Han, R., Stewart, K., and Wilson, S. 2022. "MD EJSCREEN v2.0: Visualizing Overburdening of Environmental Justice Issues Using the Updated Maryland Environmental Justice Screening Tool." *Environmental Justice* 15(6). https://doi.org/10.1089/env.2020.0055.

Wilson, B. 2020. "Urban Heat Management and the Legacy of Redlining." *Journal of the American Planning Association* 86(4): 443-457. https://doi.org/10.1080/01944363.2020.1759127.

Winstead, E. 2023. "Cancer and Climate Change: The Health Threats of Unnatural Disasters." *Cancer Currents Blog*, NIH National Cancer Institute. April 5. https://www.cancer.gov/news-events/cancer-currents-blog/2023/cancer-climate-change-impact.

Wisdom, J., and Creswell, J. 2013. *Mixed Methods: Integrating Quantitative and Qualitative Data Collection and Analysis While Studying Patient-Centered Medical Home Models*. AHRQ Publication No. 13-0028-EF, Rockville, MD: Agency for Healthcare Research and Quality.

Wisner, B., Blaikie, P., Cannon, T., and Davis, I. 2004. *At Risk: Natural Hazards, People's Vulnerability and Disasters*, 2nd ed. Routledge.

Wisner, B., Gaillard, J. C., and Kelman, I. 2012. *Handbook of Hazards and Disaster Risk Reduction*, 1st ed. Routledge.

Wispelwey, B., Tanous, O., Asi, Y., Hammoudeh, W., and Mills, D. 2023. "Because Its Power Remains Naturalized: Introducing the Settler Colonial Determinants of Health." *Frontiers in Public Health* 11. https://www.frontiersin.org/journals/public-health/articles/10.3389/fpubh.2023.1137428.

Woo, B., Kravitz-Wirtz, N., Sass, V., Crowder, K., Teixeira, S., and Takeuchi, D. T. 2019. "Residential Segregation and Racial/Ethnic Disparities in Ambient Air Pollution." *Race and Social Problems* 11(1): 60-67. https://doi.org/10.1007/s12552-018-9254-0.

Woods, L. L., II. 2012. "The Federal Home Loan Bank Board, Redlining, and the National Proliferation of Racial Lending Discrimination, 1921–1950." *Journal of Urban History* 38(6): 1036-1059. https://doi.org/10.1177/0096144211435126.

Woods, L. L., II. 2013. "Almost 'No Negro Veteran … Could Get a Loan': African Americans, the GI Bill, and the NAACP Campaign Against Residential Segregation, 1917–1960." *Journal of African American History* 98: 392. https://doi.org/10.5323/jafriamerhist.98.3.0392.

World Economic Forum. 2020. *Global Gender Gap Report 2020*. Geneva. https://www.weforum.org/reports/gender-gap-2020-report-100-years-pay-equality/in-full.

Wu, S.-Y., Yarnal, B., and Fisher, A. 2002. "Vulnerability of Coastal Communities to Sea-Level Rise: A Case Study of Cape May County, New Jersey, USA." *Climate Research* 22: 255-270. https://doi.org/10.3354/cr022255.

Xu, C., and Gertner, G. 2008. "A General First-Order Global Sensitivity Analysis Method." *Reliability Engineering & System Safety* 93: 1060-1071. https://doi.org/10.1016/j.ress.2007.04.001.

Xu, P., Huang, H., and Dong, N. 2018. "The Modifiable Areal Unit Problem in Traffic Safety: Basic Issue, Potential Solutions and Future Research." *Journal of Traffic and Transportation Engineering* (English ed.) 5(1): 73-82. https://doi.org/https://doi.org/10.1016/j.jtte.2015.09.010.

Yitshak-Sade, M., Lane, K. J., Fabian, M. P., Kloog, I., Hart, J. E., Davis, B., Fong, K. C., Schwartz, J. D., Laden, F., and Zanobetti, A. 2020. "Race or Racial Segregation? Modification of the $PM_{2.5}$ and Cardiovascular Mortality Association." *PLoS ONE* 15(7): e0236479. https://doi.org/10.1371/journal.pone.0236479.

Yonto, D., and Schuch, C. 2020. "Developing and Ground-Truthing Multi-Scalar Approaches to Mapping Gentrification." *Papers in Applied Geography* 6 (4): 352-368. https://doi.org/10.1080/23754931.2020.1789499.

Yudell, M., Roberts, D., DeSalle, R., and Tishkoff, S. 2016. "Taking Race Out of Human Genetics." *Science* 351(6273): 564-565. https://doi.org/10.1126/science.aac4951.

Zartarian, V. G., Xue, J., Gibb-Snyder, E., Frank, J. J., Tornero-Velez, R., and Stanek, L. W. 2023. "Children's Lead Exposure in the U.S.: Application of a National-Scale, Probabilistic Aggregate Model with a Focus on Residential Soil and Dust Lead (Pb) Scenarios." *Science of the Total Environment* 905: 167132. https://doi.org/10.1016/j.scitotenv.2023.167132.

Zeng, A. 2022. "Justice40: The Complicated Business of Defining Disadvantaged Communities." Synapse Energy Economics, Inc. August 19. https://www.synapse-energy.com/justice40-complicated-business-defining-disadvantaged-communities.

Appendix A

Biographical Sketches of Committee Members

Harvey J. Miller (*Co-Chair*), is the Bob and Mary Reusche Chair in Geographic Information Science, professor of geography, courtesy professor of city and regional planning, director of the Center for Urban and Regional Analysis at The Ohio State University. His research and teaching activities include geospatial data analytics applied to questions and challenges facing sustainable mobility, equitable and resilient communities, and the relationships between human mobility and health. Previously, Dr. Miller served as a member and chair of the National Academies of Sciences, Engineering, and Medicine (National Academies) Mapping Science Committee, co-chair of the National Academies Geographical and Geospatial Sciences Committee and member of the National Academies Board on Earth Sciences and Resources. He serves on the Regional Data Advisory Committee of the Mid-Ohio Regional Planning Commission in Columbus, Ohio, and is the president of the University Consortium for Geographic Information Science, a not-for-profit organization that creates and supports communities of practice for geographic information science research, education, and policy in higher education and allied institutions in the public and private sectors. Dr. Miller is an elected fellow of the American Association of Geographers and the American Association for the Advancement of Science. He has a PhD in geography from The Ohio State University.

Eric Tate (*Co-Chair*) is a professor in the Center for Policy Research on Energy and the Environment in the School of Public and International Affairs at Princeton University. His research and teaching examine intersections of environmental hazards and society, primarily using geospatial models of flood hazards, vulnerability, and risk. His research focuses on themes of social vulnerability indicators, flood adaptation, and uncertainty analysis. He currently serves on the boards of directors of the Anthropocene Alliance and the Gulf Environmental Protection and Stewardship Board at the National Academies; on the Resilient America Roundtable of the National Academies; and was a co-author of the Adaptation chapter of the Fifth National Climate Assessment. Tate earned a Ph.D. in geography from the University of South Carolina, an M.S. in environmental and water resources engineering from the University of Texas, and a B.S. in environmental engineering from Rice University.

Susan Anenberg is an associate professor of environmental and occupational health and of global health at the George Washington University (GW) Milken Institute School of Public Health. She is also the director of the GW Climate and Health Institute. Anenberg's research focuses on the health implications of air pollution and climate change, from local to global scales. Previously, Anenberg was a co-founder at Environmental Health Analytics, LLC, the deputy managing director for recommendations at the U.S. Chemical Safety Board, an environmental scientist at the U.S. Environmental Protection Agency (EPA), and a senior advisor for clean cookstove initiatives at the U.S. State Department. She received her Ph.D. in environmental science and engineering, environmental policy from the University of North Carolina. Anenberg currently serves on the U.S. Environmental Protection Agency's Science Advisory Board and Clean Air Act Advisory Committee, the World Health Organization's Global Air Pollution and Health Technical Advisory Group, and as president of the GeoHealth section of the American Geophysical Union. She has written public comments to EPA on the importance of including environmental justice analysis in regulatory impact analyses and has chaired an EPA Science Advisory Board committee providing advice to EPA on including distributional analyses in air quality regulations. Anenberg currently serves on the National Academies Committee to Advise the U.S. Global Change Research Program.

Lauren Bennett is the group product engineering lead and program manager of spatial analysis and science at Esri, Inc. In her 15 years at Esri, she has also worked as a solution engineer for the federal sciences team, as well as a lead product engineer on the Spatial Statistics software development team. Bennett's research has focused on spatial statistics and spatiotemporal analysis, especially their application to human geography problems including public health, social equity, and urban planning. Bennett received a B.A. in geography from McGill University, an M.S. in geographic and cartographic science from George Mason University, and a Ph.D. in information systems and technology from Claremont Graduate University.

Jayajit Chakraborty is a professor and Mellichamp Chair in Racial Environmental Justice at the Bren School of Environmental Science and Management at the University of California, Santa Barbara. He currently serves as a member of the U.S. Environmental Protection Agency (EPA) Science Advisory Board, Work Group for Review of Science Supporting EPA Decisions, and the EPA Environmental Justice Science Committee. He is chairing the Scientific Review Panel for the EPA's EJScreen Mapping and Screening Tool and serving as a member of the EPA Environmental Justice Science and Analysis Review Panel. Dr. Chakraborty's research and outreach activities encompass a wide range of concerns related to the social dimensions of climate and environmental change, with an emphasis on environmental justice and community vulnerability to hazards and disasters. He is particularly interested in applying geospatial tools and spatial statistical techniques for analyzing environmental and social injustices. Dr. Chakraborty has published 4 books and more than 120 articles/chapters, including *The Routledge Handbook of Environmental Justice* and a chapter for the US Government's Fifth National Climate Assessment (NCA5). He has been a principal/co-principal investigator for over 30 sponsored projects, which include grants from the EPA, National Science Foundation, National Aeronautics and Space Administration, US Department of Transportation, US Department of Treasury, and other agencies. Dr. Chakraborty has a Ph.D. in geography and M.S. in urban and regional planning, both from the University of Iowa.

Ibraheem Karaye is assistant professor of population health and director of the Health Science Program at Hofstra University. His research broadly focuses on the physical, mental, and environmental health impacts of disasters and mass trauma on socially vulnerable populations, including racial and ethnic minorities and older adults. He also examines health disparities and the distribution of health outcomes globally and within the United States. Karaye uses large secondary datasets and novel statistical and spatial analytic methods to study social variables. His publication "The Impact of Social Vulnerability on COVID-19 in the U.S.: An Analysis of Spatially Varying Relationships" was recognized as a finalist for the 2020 Article of the Year by the *American Journal of Preventive Medicine*. He currently serves as an academic editor for the journal, *PLoS ONE*. Karaye received his medical degree from Bayero University Kano in Nigeria and holds a master of public health degree in epidemiology and a doctorate in public health (epidemiology and environmental health) from Texas A&M University.

Marcos Luna is a professor of geography and sustainability and coordinator of the graduate Geo-Information Science program at Salem State University in Salem, Massachusetts. His research focus is on environmental justice and applications of geospatial analytic techniques to social and environmental inequities, particularly around energy and climate change. He has published research on the inequity of natural gas leaks, urban noise, transit efficiency and equity, energy, air pollution, and environmental policy. Luna holds an M.A. in geography from the California State University, Los Angeles, and a Ph.D. in urban affairs and public policy from the University of Delaware. In addition to academic research, he works with community organizations and policy makers on issues including residential housing and segregation, transportation equity, voter mapping and outreach, and climate change adaptation. He is a member of the board of directors for GreenRoots, Inc., an environmental justice organization based in Chelsea, Massachusetts, and he is a governor-appointed member of the Massachusetts Environmental Justice Advisory Council, which is charged with (re)assessing the appropriateness of the state's definition of "environmental justice communities."

Bhramar Mukherjee, NAM, is the University of Michigan (UM) John D. Kalbfleisch Collegiate Professor and Chair, Department of Biostatistics; professor, Department of Epidemiology, professor, Global Public Health, UM School of Public Health; research professor and core faculty member, Michigan Institute of Data Science; and founding director of the UM Summer Institute on Big Data. She is also the associate director for quantitative data sciences, UM Rogel Cancer Center, and the associate workgroup director for cohort development for UM Precision Health. Her research interests include statistical methods for analysis of electronic health records, studies of gene–environment interaction, and analysis of multiple pollutants, and she collaborates in research related to cancer, cardiovascular diseases, reproductive health, exposure science and environmental epidemiology. Mukherjee is a fellow of the American Statistical Association and the American Association for the Advancement of Science and is the recipient of many awards for scholarship, service, and teaching. Mukherjee has an M.S. in applied statistics and data analysis from the Indian Statistical Institute, an M.S. in mathematical statistics from Purdue University, and a Ph.D. in statistics from Purdue University. She serves on the National Academies Committee on Applied and Theoretical Statistics and has served on National Academies committees on the Reassessment of the Department of Veterans Affairs Airborne Hazards and Open Burn Pit Registry and on the Rising Midlife Mortality Rates and Socioeconomic Disparities.

Kathleen Segerson, NAS, is a Board of Trustees Distinguished Professor of Economics at the University of Connecticut. Her research focuses on the incentive effects of alternative environmental policy instruments, including applications in the following areas: groundwater contamination, hazardous waste management, land use regulation, climate change, and nonpoint pollution from agriculture. In addition, she has worked on valuing ecosystem services and the protection of marine species. Segerson is a fellow of the Association of Environmental and Resource Economists and of the American Agricultural Economics Association. Segerson holds a Ph.D. from Cornell University and a B.A. from Dartmouth College. She has served or is currently serving on a number of advisory boards, including the U.S. Environmental Protection Agency's Science Advisory Board (SAB) and the Committee on Valuing the Protection of Ecological Systems and Services, the National Academy of Sciences Advisory Committee for the U.S. Global Change Research Program and the Review Panel on the National Climate Assessment, the National Academies Board on Agriculture and Natural Resources, the U.S. National Member Organization of the International Institute of Applied Systems Analysis, and the Advisory Board of the Beijer Institute of Ecological Economics in Stockholm.

Monica E. Unseld is the founder and executive director of the nonprofit Until Justice Data Partners, utilizing her experience as a subject-matter expert in environmental and public health, and believes that science should be accessible to all. The organization partners with marginalized communities nationwide and internationally to apply research methods to environmental and social justice issues, through her specializations in endocrine disruption, environmental signaling, and public health. Prior to her nonprofit work, she was an assistant professor and worked at a data think tank in Louisville, Kentucky. She has almost 15 years of volunteer environmental justice work experience as a subject-matter expert in science and general research methodology, working with predominantly Black- and Brown-led groups and coalitions to help normalize the use of research and data. In December 2022, she was named a Science Defender by the Union of Concerned Scientists for her efforts to democratize data. She obtained her doctorate in biology in 2008 from the University of Louisville and her masters in public health in 2018 from Benedictine University.

Walker Wieland is an environmental program manager with the Office of Environmental Health Hazard Assessment (OEHHA), at the California Environmental Protection Agency. Wieland is leading the development of CalHeatScore, an extreme heat ranking system for California. Wieland has 12 years of experience in planning and conducting research studies to characterize the distribution of environmental pollutants or sources of pollution to support the development of screening cumulative impact analysis. He is co-author of each version of CalEnviroScreen, California's pioneering environmental justice screening tool. He has held multiple leadership positions throughout state government in geographic information systems (GIS) and open data and has formerly received certification as a GIS professional. Wieland is an award-winning public speaker and routinely provides training to agencies on considering cumulative impacts and environmental justice mapping in their policies and programs. He also has consulted with agencies across the United States and internationally in implementing their own environmental health screening tools. Wieland received his B.A. in environmental studies from California State University, Sacramento, and his A.S. in GIS from American River College.

Appendix B

Public Meeting Agendas

The committee's deliberations were informed by formal presentations, formal and informal discussions, and written input from several individuals. This appendix includes the agendas from the committee's formal information-gathering sessions and lists those individuals invited to make presentations. Many more individuals (not listed) asked questions and offered their expertise in commentary to help enrich the discussions in both closed and open sessions.

FEBRUARY 13, 2023

HYBRID MEETING

2:00 pm EST	Welcomes and Introduction
2:05 pm	Introduction to Climate and Economic Justice Screening Tool (CEJST) **Natasha DeJarnett**, *Deputy Director of Environmental Justice Data and Evaluation*
2:35 pm	Clarifying Questions from Committee for CEQ
2:45 pm	Break
3:00 pm	Orientation from Sponsor **Cecilia Martinez**, Principal Advisor, *Bezos Earth Fund*

3:25 pm Clarifying Questions from Committee for Sponsor

3:35 pm General Discussion Between Committee and Sponsor

4:00 pm Opportunity for Public Comment (*Advance Registration Required*)
- Email Oshane Orr for advanced registration at oorr@nas.edu
- Public comment registration closes at 11 am EST, February 13th
- Members of the public will be called on in the order that they registered and have 2 minutes to provide their comments. At the end of 2 minutes, the next commentor will be called.

4:30 pm End of Open Session

FEBRUARY 23, 2023

VIRTUAL MEETING

1:00 pm EST Welcomes and Introduction

1:10 pm CDC's Environmental Justice Index (EJI) Explorer
Benjamin McKenzie, *Epidemiologist, CDC*
Sharunda Buchanan, *Director, Office of Priority Projects, Innovation and Environmental Justice, CDC*

1:30 pm Clarifying Questions from Committee for CDC

1:35 pm EPA's EJScreen: Environmental Justice Screening and Mapping Tool
Tai Lung, *Environmental Protection Specialist, U.S. EPA*
Matthew Lee, *Environmental Protection Specialist, U.S. EPA*

1:55 pm Clarifying Questions from Committee for U.S. EPA

2:00 pm Break

2:20 pm Panel Discussion
Benjamin McKenzie, *Epidemiologist, CDC*
Sharunda Buchanan, *Director, CDC*
Tai Lung, *Environmental Protection Specialist, U.S. EPA*
Matthew Lee, *Environmental Protection Specialist, U.S. EPA*
Lucas Brown, *Senior Advisor for Justice40 at CEQ*

4:00 pm End of Open Session

MAY 12, 2023

VIRTUAL MEETING

3:00 pm EST	Welcomes and Introduction
3:05 pm	Presentation on U.S. DOT Equitable Transportation Community Explorer **Kristin Wood**, *Justice40 Program Lead in the U.S. DOT Secretary's Office of Policy Development and Coordination*
3:25 pm	Clarifying Questions from Committee
3:35 pm	General Discussion Between Committee and Presenter
4:00 pm	End of Open Session

MAY 16, 2023

VIRTUAL MEETING

1:00 pm EST	Welcomes and Introduction
1:05 pm	Presentation on mapping present and future vulnerability **Weihsueh Chiu**, *Professor in the Department of Veterinary Integrative Biosciences at Texas A&M University*
1:25 pm	Clarifying Questions from Committee
1:35 pm	General Discussion Between Committee and Presenter
2:00 pm	End of Open Session

JUNE 1, 2023

HYBRID COMMUNITY WORKSHOP

This workshop of the Committee on Utilizing Advanced Environmental Health and Geospatial Data and Technologies to Inform Community Investment will explore how well data used within the Climate and Economic Justice Screening Tool (CEJST) represent lived experiences of different kinds of communities across the nation. CEJST, a geospatial tool, was developed by the White House Council on Environmental Quality to identify communities experiencing climate and economic burdens. Workshop participants will discuss (a) how information obtained using CEJST reflects conditions within communities, (b) gaps in the data used within CEJST, and (c) how cumulative burdens are experienced and represented in CEJST. The information gathered during this workshop will inform the study committee's deliberations and recommendations regarding a future data strategy for CEJST. View the full study description at https://www.nationalacademies.org/our-work/utilizing- advanced-environmental-health-and-geospatial-data-and-technologies-to-inform-community-investment.

Purpose	**To understand connections between the lived experiences of historically marginalized and overburdened populations and:** • the representativeness of the current measures in the tool. • measures currently absent from the tool that relate thematically to the Justice40 categories. • how cumulative burdens are experienced.
10:00 am EST	Welcomes and introduction, background about project and CEJST tool, objectives and logistics for the day
10:20 am	Presentation regarding the categories and indicators of burden in CEJST and the data used to represent them
10:50 am	Break
11:00 am	Flash talks from experts on understanding the lived experience of communities: **Michelle Alvarado,** Executive Director, Resiliency Law Center **Loka Ashwood,** Associate Professor of Sociology, University of Kentucky **Nayamin Martinez,** Executive Director, Central California Environmental Justice Network **Vi Waghiyi,** Native of Village of Savoonga, Physician, Environmental Health and Justice Program Director with Alaska Community Action on Toxics (ACAT), WHEJAC **Neza Xiuhtecutli,** General Coordinator, Farmworker Association of Florida
11:55 am	Plenary: Concluding the morning sessions
12:00 pm	Lunch
12:45 pm	CEJST demonstration to provide a base level understanding of each component of the tool to prepare participants for hands-on exercise

1:00 pm	Hands on exercise: How well does CEJST represent conditions in your community? The exercise focuses on the representation of conditions through CEJST categories of burden. Exercise worksheet: https://forms.gle/23LxjfHK8ravhAgL9
2:00 pm	Break
2:15 pm	Plenary reports from rapporteurs followed by group discussion on common observations about data gaps and unique observations
3:30 pm	Panel discussion on defining disadvantaged communities and approaches for how CEJST can better identify these communities: **Jack Ding**, Environmental Justice Technology Fellow, Anthropocene Alliance/Environmental Policy Innovation Center **Rep. Larry Lambert**, State Representative, 7th District, Delaware General Assembly, House of Representatives, **Marisa Sotolongo,** PhD candidate in the School of Public Policy and Urban Affairs, Northeastern University; Research Assistant, Initiative for Energy Justice **Mathy Stanislaus, Esq.**, Vice Provost & Executive Director, the Environmental Collaboratory, Drexel University **Sacoby Wilson**, Professor and Director, Center for Community Engagement, Environmental Justice and Health (CEEJH) Lab, Maryland Institute for Applied Environmental Health, School of Public Health, University of Maryland, College Park
4:10 pm	Open plenary discussion: Committee member opens questions to the participants to reflect on the data needs identified throughout the day's sessions and how they can be addressed Form for sending other useful datasets for committee consideration: https://forms.gle/LF1wYwMnbHUVn5yf6
4:45 pm	Closing Remarks
5:00 pm	End of Open Session

JULY 20, 2023

VIRTUAL MEETING

2:00 pm EST	Welcomes and Introduction
2:05 pm	Presentation on composite indicators **Michaela Saisana**, *Head of the Monitoring, Indicators and Impact Evaluation Unit, lead European Commission's Competence Centre on Composite Indicators and Scoreboards (COIN), Joint Research Centre*
2:30pm	Clarifying Questions from Committee
2:45 pm	General Discussion Between Committee and Presenter
3:30 pm	End of Open Session

Appendix C

Screening Tools Examined by the Committee

The White House Council on Environmental Quality's (CEQ) Climate and Economic Justice Screening Tool (CEJST)[1] was designed to screen for communities that qualify for extra consideration for investment under the Justice40 Initiative. On the basis of surveys, information gathered during committee meetings, and the knowledge from members of the committee, the committee chose a subset of 12 tools from which to highlight key features of geographically based EJ tools. To facilitate comparison of the tools, the committee created a matrix to summarize information about each tool (e.g., purpose of tool; geographic resolution; methodology employed to rank, compare, or score the index; categories or themes with corresponding indicators used; and data sources.)

[1] See the Climate and Economic Justice Screening Tool (CEJST) at https://screeningtool.geoplatform.gov/en/ (accessed December 15, 2023).

TABLE C.1 Properties of the White House Council on Environmental Quality (CEQ) Climate and Economic Screening Tool (December 2022, Version 1.0)

Purpose

Used by federal agencies to help them identify disadvantaged communities that can benefit from the Justice40 Initiative.

Geography

2010 Census Tracts.
All 50 states, the District of Columbia, and the U.S. territories (Puerto Rico, American Samoa, the Northern Mariana Islands, Guam, and the U.S. Virgin Islands).
Data chosen based on availability for 50 states + DC, alternative data selected for territories when needed.

Method

Threshold Approach-Iterative Tool
An area is considered burdened:

- If a census tract meets the criteria of burdened for one category, then it is considered burdened. Tracts that meet the criteria in multiple categories are denoted.
- If a tract is not considered burdened by the above criteria but is at or above the 50th percentile for low income AND is completely surrounded by burdened tracts, then it is considered burdened.
- If the tract is 99.5% covered by federally recognized tribal land.

Categories/Themes	Indicators
Climate Change	Expected agriculture loss rate ≥ 90th percentile OR Expected building loss rate ≥ 90th percentile OR Expected population loss rate ≥ 90th percentile OR Projected flood risk ≥ 90th percentile OR Projected wildfire risk ≥ 90th percentile
Energy	Energy cost ≥ 90th percentile OR $PM_{2.5}$ in the air ≥ 90th percentile
Health	Asthma ≥ 90th percentile OR, Diabetes ≥ 90th percentile OR, Heart disease ≥ 90th percentile OR, Low life expectancy ≥ 90th percentile
Categories/Themes	**Indicators**
Housing	Historic underinvestment = Yes OR, Housing cost ≥ 90th percentile OR, Lack of green space ≥ 90th percentile OR, Lack of indoor plumbing ≥ 90th percentile OR, Lead paint ≥ 90th percentile
Legacy Pollution	Abandoned mine land present = Yes OR Formerly used defense site present = Yes OR Proximity to hazardous waste facilities ≥ 90th percentile (treatment, storage and disposal facilities) OR, Proximity to Superfund or National Priorities List sites ≥ 90th percentile OR, Proximity to Risk Management Plan sites ≥ 90th percentile.

TABLE C.1 Continued

Transportation	Diesel particulate matter ≥ 90th percentile OR, Transportation barriers ≥ 90th percentile OR, Traffic proximity and volume ≥ 90th percentile.
Water and Wastewater	Underground storage tanks and releases ≥ 90th percentile OR, Wastewater discharge ≥ 90th percentile.
Workforce Development	Linguistic isolation ≥ 90th percentile OR, Low median income ≥ 90th percentile OR, Poverty ≥ 90th percentile OR, Unemployment ≥ 90th percentile
Socioeconomic	Low income ≥ 65th percentile (coupled with all but the workforce development indicators) High school education < 90th percentile (coupled with the workforce development indicators).

Data Sources

- Census, American Community Survey
- Centers for Disease Control and Prevention (CDC) PLACES: Local Data for Better Health
- CDC U.S. Small-Area Life Expectancy Estimates Project
- U.S. Department of Energy (DOE) Low-Income Energy Affordability Data (LEAD) Tool
- U.S. Department of the Interior (DOI)/Bureau of Indian Affairs (BIA) Land Area Representation
- DOI Abandoned Mine Lands Inventory System
- U.S. Department of Transportation (DOT) Transportation Access Disadvantage
- U.S. Environmental Protection Agency (EPA) EJScreen
- Federal Emergency Management Agency (FEMA) National Risk Index
- First Street Foundation Climate Risk Data Access
- Home Owner's Loan Corporation (HOLC) Historic Redlining Scores
- Housing and Urban Development (HUD) Comprehensive Housing Affordability Strategy
- Multi-Resolution Land Characteristics (MRLC) consortium Percent Developed Imperviousness (CONUS)
- U.S. Army Corps of Engineers (USACE) formerly used defense sites

TABLE C.2 Properties of the Centers for Disease Control and Prevention and Agency for Toxic Substances and Disease Registry Social Vulnerability Index (2020 Version)

Purpose

To assist public health officials and emergency response planners in identifying and mapping the communities that are most likely to require support before, during, and after a hazardous event.

Geography

U.S. Census Tracts

Method

Percentile-Based Ranking. The SVI ranks tracts based on 16 social factors, including unemployment, racial and ethnic minority status, and disability. These tracts are further categorized into four related themes. As a result, each tract is assigned a ranking for each Census variable and for each of the four themes, along with an overall ranking. In addition to tract-level rankings, SVI 2010, 2014, 2016, 2018, and 2020 also provide corresponding rankings at the county level. Tract rankings are determined by percentiles, with values ranging from 0 to 1, where higher values indicate greater vulnerability. For each tract, we calculated its percentile rank among all tracts for (1) the 16 individual variables, (2) the four themes, and (3) its overall position.

TABLE C.2 Continued

Categories/Themes	Indicators
Socioeconomic Status	Below 150% poverty Unemployed Housing cost burden No high school diploma No health insurance
Household Characteristics	Ages 65 & older Ages 17 & younger Civilian with a disability Single-parent households English language proficiency.
Racial & Ethnic Minority Status	Hispanic or Latino (of any race); Black and African American, not Hispanic or Latino; American Indian and Alaska Native, not Hispanic or Latino; Asian, not Hispanic or Latino; Native Hawaiian and Other Pacific Islander, not Hispanic or Latino; Two or more races, not Hispanic or Latino; Other races, not Hispanic or Latino
Housing Type & Transportation	Multiunit structures Mobile homes Crowding No vehicle Group quarters
Adjunct Variables	An estimate of daytime population derived from LandScan 2020 estimates 2016–2020 ACS estimates for households without a computer with a broadband Internet subscription 2016–2020 ACS estimates for Hispanic/Latino persons, not Hispanic or Latino Black/African American persons, not Hispanic or Latino Asian persons, not Hispanic or Latino American Indian and Alaska Native persons, not Hispanic or Latino Native Hawaiian and Other Pacific Islander persons, not Hispanic or Latino persons of two or more races, and not Hispanic or Latino persons of some other race

Data Sources

- American Community Survey (ACS), 2016–2020 (5-year) data

TABLE C.3 Properties of FEMA National Risk Index (March 2023 Release)

Purpose

Designed to depict the communities in the United States and territories that are most vulnerable to 18 different natural hazards. These hazards encompass: Avalanche, Coastal Flooding, Cold Wave, Drought, Earthquake, Hail, Heat Wave, Hurricane, Ice Storm, Landslide, Lightning, Riverine Flooding, Strong Wind, Tornado, Tsunami, Volcanic Activity, Wildfire, and Winter Weather.
The National Risk Index offers Risk Index values, scores, and ratings derived from data related to Expected Annual Loss caused by natural hazards, Social Vulnerability, and Community Resilience. Additionally, distinct values, scores, and ratings are available for Expected Annual Loss, Social Vulnerability, and Community Resilience. Both the Risk Index and Expected Annual Loss encompass the option to view composite scores for all hazards collectively, or separately for each of the 18 hazard types.

Geography

U.S. Census Tracts/County

Method

Definitions
Risk: The National Risk Index defines risk as the possibility of negative outcomes due to natural hazards.
Risk Components: The risk equation in the National Risk Index has three main parts: natural hazards risk, consequence enhancement, and consequence reduction.
Natural Hazards Risk (EAL): This component calculates the expected loss each year in terms of building, population, and agriculture value caused by natural hazards.
Consequence Enhancement (Social Vulnerability): This factor analyzes demographic characteristics to measure how susceptible different social groups are to the negative effects of natural hazards.
Consequence Reduction (Community Resilience): This factor uses demographic attributes to gauge a community's ability to prepare for, adapt to, withstand, and recover from the impacts of natural hazards.
Combination: Social Vulnerability and Community Resilience are combined into a single factor called Community Risk Factor (CRF).
Final Risk Calculation: The CRF is multiplied by the Expected Annual Loss (EAL) to calculate the overall risk score.

Categories/Themes	**Indicators**
Expected Annual Loss	The EAL for each census tract or county is the average economic loss in dollars resulting from natural hazards each year. EAL is quantified—in dollar amounts—for each of the 18 hazard types
Social Vulnerability	Below 150% poverty Unemployed Housing cost burden No high school diploma No health insurance Ages 65 & older Ages 17 & younger Civilian with a disability Racial & ethnic minority status Multiunit structures Mobile homes Crowding No vehicle Group quarters Single-parent households English language proficiency.

TABLE C.3 Continued

Community Resilience	This is derived from the University of South Carolina's Hazards and Vulnerability Research Institute's (HVRI) Baseline Resilience Indicators for Communities (BRIC). The HVRI BRIC dataset includes a set of 49 indicators that represent six types of resilience: social, economic, community capital, institutional capacity, housing/infrastructure, and environmental.

Data Sources

- Sources for EAL data include:
  - Alaska Department of Natural Resources,
  - Arizona State University's Center for Emergency Management and Homeland Security,
  - California Department of Conservation,
  - California Office of Emergency Services
  - California Geological Survey,
  - Colorado Avalanche Information Center,
  - CoreLogic's Flood Services,
  - Federal Emergency Management Agency (FEMA) National Flood Insurance Program,
  - Humanitarian Data Exchange,
  - Iowa State University's Iowa Environmental Mesonet,
  - Multi-Resolution Land Characteristics Consortium,
  - National Aeronautics and Space Administration's (NASA's) Cooperative Open Online Landslide Repository,
  - National Earthquake Hazards Reduction Program,
  - National Oceanic and Atmospheric Administration's (NOAA's) National Centers for Environmental Information,
  - NOAA's National Hurricane Center, NOAA's National Weather Service,
  - NOAA's Office for Coastal Management,
  - NOAA's National Geophysical Data Center,
  - NOAA's Storm Prediction Center,
  - Oregon Department of Geology and Mineral Industries,
  - Pacific Islands Ocean Observing System,
  - Puerto Rico Seismic Network,
  - Smithsonian Institution's Global Volcanism Program,
  - State of Hawaii's Office of Planning's Statewide GIS Program,
  - U.S. Army Corps of Engineers' Cold Regions Research and Engineering Laboratory,
  - U.S. Census Bureau, U.S. Department of Agriculture's (USDA) National Agricultural Statistics Service),
  - U.S. Forest Service's Fire Modeling Institute's Missoula Fire Sciences Lab,
  - U.S. Forest Service's National Avalanche Center,
  - U.S. Geological Survey's Landslide Hazards Program,
  - United Nations Office for Disaster Risk Reduction,
  - University of Alaska Fairbanks' Alaska Earthquake Center, University of Nebraska–Lincoln's National Drought Mitigation Center,
  - University of Southern California's Tsunami Research Center,
  - Washington State Department of Natural Resources.
- Social Vulnerability data are provided by the Centers for Disease Control and Prevention (CDC) Agency for Toxic Substances and Disease Registry Social Vulnerability Index
- Community Resilience data are provided by University of South Carolina's Hazards and Vulnerability Research Institute's (HVRI) 2020 Baseline Resilience Indicators for Communities. https://www.sc.edu/study/colleges_schools/artsandsciences/centers_and_institutes/hvri/index.php/bric
- The source of the boundaries for counties and census tracts are based on the U.S. Census Bureau's 2021 TIGER/Line shapefiles.

continued

TABLE C.3 Continued

Data Sources

- Building value and population exposures for communities are based on FEMA's Hazus 6.0.
- Agriculture values are based on the USDA 2017 Census of Agriculture.
- Sources for Expected Annual Loss data include:
  - Alaska Department of Natural Resources,
  - Arizona State University's Center for Emergency Management and Homeland Security,
  - California Department of Conservation,
  - California Office of Emergency Services
  - California Geological Survey,
  - Colorado Avalanche Information Center,
  - CoreLogic's Flood Services,
  - FEMA National Flood Insurance Program,
  - Humanitarian Data Exchange,
  - Iowa State University's Iowa Environmental Mesonet,

Data Sources

  - Multi-Resolution Land Characteristics Consortium,
  - NASA's Cooperative Open Online Landslide Repository (COOLR),
  - National Earthquake Hazards Reduction Program (NEHRP),
  - National Oceanic and Atmospheric Administration's National Centers for Environmental Information (NCEI),
  - National Oceanic and Atmospheric Administration's National Hurricane Center, National Oceanic and Atmospheric Administration's National Weather Service (NWS),
  - National Oceanic and Atmospheric Administration's Office for Coastal Management,
  - NASA's National Geophysical Data Center,
  - NOAA's Storm Prediction Center,
  - Oregon Department of Geology and Mineral Industries,
  - Pacific Islands Ocean Observing System,
  - Puerto Rico Seismic Network,
  - Smithsonian Institution's Global Volcanism Program,
  - State of Hawaii' Office of Planning's Statewide GIS Program,
  - U.S. Army Corps of Engineers' Cold Regions Research and Engineering Laboratory,
  - U.S. Census Bureau
  - USDA National Agricultural Statistics Service,
  - U.S. Forest Service's Fire Modeling Institute's Missoula Fire Sciences Lab,
  - U.S. Forest Service's National Avalanche Center,
  - U.S. Geological Survey's Landslide Hazards Program,
  - United Nations Office for Disaster Risk Reduction,
  - University of Alaska Fairbanks' Alaska Earthquake Center
  - University of Nebraska–Lincoln's National Drought Mitigation Center,
  - University of Southern California's Tsunami Research Center,
  - Washington State Department of Natural Resources.
- Social Vulnerability data are provided by the CDC Agency for Toxic Substances and Disease Registry Social Vulnerability Index
- Community Resilience source data are provided by HVRI Baseline Resilience Indicators for Communities (https://www.sc.edu/study/colleges_schools/artsandsciences/centers_and_institutes/hvri/index.php/bric)
- Boundaries for counties and Census tracts are based on the U.S. Census Bureau's 2021 TIGER/Line shapefiles.
- Building value and population exposures for communities are based on FEMA's Hazus 6.0.

Agriculture values are based on the USDA 2017 Census of Agriculture.

TABLE C.4 Properties of the Environmental Protection Agency's Environmental Justice Screening and Mapping Tool (EJSCREEN) (2019 Version)

Purpose

EPA characterizes EJSCREEN as a pre-decisional screening tool not designed for decision making or determinations regarding the existence or absence of environmental justice concerns.

Geography

2011–2017 American Community Survey Block Groups

50 States + District of Columbia, Puerto Rico. Does not include Virgin Islands, American Samoa, Northern Mariana Islands, Guam

Method

Multiple Indexes. No single score

Demographic Index = (% minority + % low-income)/2

EJ Index. Users can construct their own environmental justice index by combining a single environmental indicator with the demographic index)

EJ Index = (Environmental Indicator) × (Demographic Index for Block Group –Demographic Index for U.S.) × (Population Count for Block Group)

Categories/Themes	**Indicators**
Environmental indicators Air Pollution Traffic Proximity Lead Paint Waste/Hazardous Materials Proximity Wastewater Discharge	National Air Toxics Assessment (NATA) Air Toxics Cancer Risk—Lifetime inhalation cancer risk NATA Respiratory Hazard Index— Ratio of exposure concentration to reference concentration NATA diesel particulate matter Particulate matter ($PM_{2.5}$) Ozone (summer seasonal 8-hour max average) Lead paint (% housing units built before 1960) Traffic proximity and volume (count of vehicles at major roads within 500 m) Proximity to Risk Management Plan (RMP) sites (count of facilities within 5 km) Proximity to treatment, storage and disposal facilities (count of facilities within 5 km) Proximity to National Priorities List sites (count of facilities within 5 km) Wastewater discharge (toxicity weighted stream concentrations)
Social Indicators (A&B are used for index development, other indicators are included though) Low Income Minority	Low income Minority Less than high school education Linguistic isolation Individuals under 5 years Individuals over 64 years

continued

TABLE C.4 Continued

Data Sources

- EPA NATA
- EPA Office of Air and Radiation $PM_{2.5}$/ozone monitor data
- U.S. Department of Transportation traffic data
- Census ACS 2013–2017 data
- EPA RMP database
- EPA RCRAInfo database
- EPA Comprehensive Environmental Response, Compensation, and Liability Information System (CERCLIS/Superfund) database
- EPA Risk-Screening Environmental Indicators model

TABLE C.5 Properties of the Department of Transportation's Data on Transportation Disadvantaged Communities (2022 Data)

Purpose

Created a mapping tool to help grant applicants understand if their community is disadvantaged.

Geography

Census tracts (year not denoted)

Includes geographies not denoted in methodology, but provides shapefile; includes 50 states + District of Columbia

Geography of territories in shapefile but no data provided (listed as 0)

Method

Index. Percentile rank of all 22 indicators, 99th percentile = "most disadvantaged." For each category: if average of percentile of indicators = 50th percentile, then the census tract is considered disadvantaged for that tract (except resilience category = 75th percentile).

- If a tract is considered disadvantaged for a category, it receives a "1" for the category. If not, it receives a "0." Scores are summed. Tracts with 4 or higher are considered "Transportation Disadvantaged."

TABLE C.5 Continued

Categories/Themes	Indicators
Transportation Access Disadvantage Health Disadvantage Environmental Disadvantage Economic Disadvantage Resilience Disadvantage, Equity Disadvantage	>30 min commute No vehicle Walkability Transportation burden Population 65+ Uninsured Disability Homes built before 1960 Diesel Cancer Traffic proximity $PM_{2.5}$ Ozone Less high school education Renters Unemployment Gini Index (endemic inequality), Low income Poverty Housing costs Climate hazards Linguistic isolation

Data Sources

- American Community Survey 5-year est 2015–2019
- Centers for Disease Control and Prevention Social Vulnerability Index (2018)
- U.S. Environmental Protection Agency (EPA) Smart Location Database v2.0
- EPA EJSCREEN
- Federal Emergency Management Agency (FEMA) National Risk Index
- Housing and Urban Development Exchange
- FEMA Resilience Analysis and Planning Tool (GINI Index)

TABLE C.6 Properties of the Department of Health and Human Services Environmental Justice Index (EJI) (influenced by Centers for Disease Control and Prevention)

Purpose

A national, place-based tool designed to measure the cumulative impacts of environmental burden through the lens of human health and health equity. The EJI delivers a single score for each community so that public health officials can identify and map areas most at risk for the health impacts of environmental burden.

Geography

Census tracts (year not denoted)
EJI 2022 includes only the continental United States (48 states plus the District of Columbia)
Does not include Alaska, Hawaii, or U.S. territories and dependencies due to a lack of data for these states/territories.

Method

Index
Groups 36 indicators into environmental, social, and health "modules."
Overall EJI score = sum of three modules (percentile ranked)

- Notes that EJI ranking is for identifying areas needing special attention or to characterize local factors driving cumulative impacts on health to inform policy. Develops "EJI SER" for secondary outcome analysis.

Categories/Themes	**Indicators**
Social Vulnerability Module	Minority status poverty No high school diploma Unemployment Housing tenure Housing-burdened-lower income house household Lack of health insurance, Broadband access Age 65+ Age 17 under Disability Speaks English less than well Mobile homes Group quarters.
Health Vulnerability Module	High blood pressure Asthma Cancer Poor mental health Diabetes

TABLE C.6 Continued

Environmental Burden Module	Ozone $PM_{2.5}$ Diesel particulate matter Air toxics cancer risk National Priority List sites Toxic release inventory sites Treatment, storage, disposal sites Risk Management Plan sites Coal mines Lead mines Lack of recreational parks Houses built pre-1980 Walkability High-volume roads Railways, airports Impaired surface water

Data Sources

- U.S. Centers for Disease Control and Prevention PLACES estimates
- U.S. Environmental Protection Agency (EPA) Air Quality System
- EPA National Air Toxics Assessment
- EPA Facility Registry Service
- U.S. Mine Safety and Health Administration Mine Data Retrieval System
- TomTom MultiNet® Enterprise Dataset
- U.S. Census Bureau American Community Survey 2015–2019
- EPA National Walkability Index
- EPA Watershed Index Online

TABLE C.7 Properties of the Department of Energy's Justice40 Disadvantaged Communities Energy Justice Mapping Tool (2022 Data)

Purpose

Data used to define the U.S. Department of Energy's working definition of disadvantaged communities (DACs) as pertaining to Executive Order 14008, or the Justice40 Initiative. The dataset provides the 36 inputs to the index at the census-tract level as well as the classification of each census tract as disadvantaged or not disadvantaged.

Geography

Census tracts (year not denoted)
Documentation does not describe what geographies are included.
50 states + District of Columbia
Shapefile includes Puerto Rico & Virgin Islands, but no data.

Method

Percentile values of each "indicator of burden" are calculated for each census tract and then summed. Equal weighting. Final scores range from 0 to 36, with 36 being the most disadvantaged. The top 20% of census tracts for each state was selected to be representative. Tracts are excluded if do not have 30% or more of households within the tract at or below 200% of the federal poverty line and/or considered below low-income households as defined by Housing and Urban Development (HUD). All tribal lands are included, per the Office of Management and Budget interim guidance. (This is not defined.)

Categories/Themes	**Indicators**
Energy Burden	Energy burden (energy housing costs)
	Non-grid-connected heating fuel
	Outage duration
	Outage events
	Transportation costs
Environmental and Climate Hazards	Cancer risk
	Climate hazard loss of life estimates
	Diesel
	Homes built before 1960
	National Priority List proximity
	$PM_{2.5}$
	Risk Management Plan site proximity
	Traffic proximity
	Treatment, storage and disposal facilities proximity
	Water discharge
Socioeconomic Vulnerabilities	30-minute commute
	Disabled population
	Food desert
	Homelessness

TABLE C.7 Continued

Socioeconomic Vulnerabilities *(continued)*	Housing costs Incomplete plumbing Internet access Job access Less high school education Linguistic isolation Low-income population Mobile homes No vehicle Parks Population 65+ years Renters Single parent Unemployed Uninsured
Fossil Dependence	Coal employment Fossil energy employment

Data Sources

- American Community Survey (ACS) 5-year estimates, 2015–2019
- Longitudinal Employer-Household Dynamics Survey
- EPA Environmental Justice Screening Tool (EJScreen)
- Census Tract–City Place Relationship Table
- Low-Income Energy Affordability Data Tool
- Current Employment Statistics
- Electric Emergency Incident and Disturbance Report
- Food Access Research Atlas Data
- Housing and Transportation Affordability Index (H+T® Index)
- Indicators of Broadband Need Map
- USA Parks geodatabase
- Continuum of Care GIS Tools, 2019
- Low- And Moderate-Income Summary Data—All Block Groups, based on 2011–2015 ACS
- Federal Emergency Management Agency National Risk Index, 2020

TABLE C.8 Properties of the California Environmental Protection Agency's Office of Environmental Health Hazard Assessment's California EnviroScreen 4.0

Purpose

The tool analyzes the cumulative effects of pollution burden and additional socioeconomic and health factors to identify which communities might need policy, investment, or programmatic interventions. CalEnviroScreen is a screening tool used to help identify communities disproportionately burdened by multiple sources of pollution and with population characteristics that make them more sensitive to pollution.

Geography

Census tracts (California)

Method

Index showing overall percentile ranks. Assigns scores for 21 indicators in each geographic area. Percentiles are averaged for each of the four subcomponents. More weight is given to exposure factors. CalEnviroScreen Score is a product of pollution subcomponent multiplied by population characteristics. Model's components that contribute to cumulative impacts include Pollution Burden with subcomponents Exposures, Environmental Impacts; Population Characteristics with subcomponents Sensitive Populations and Socioeconomic Factors.

Categories/Themes	**Indicators**
Exposure	Ozone concentration in air, $PM_{2.5}$ concentration in air, diesel particulate matter in air, drinking water contaminants, children's lead risk from housing, use of high-hazard, high-volatility pesticides, toxic releases from facilities
Environmental Impacts; Population Characteristics with Subcomponents	Toxic cleanup sites, groundwater threats from leaving underground storage sites and cleanup sites, hazardous waste facilities and generators, impaired water bodies, solid waste sites and facilities
Sensitive Populations	Asthma emergency department visits, cardiovascular diseases (emergency department visits for heart attacks), low-birth-weight infants.
Socioeconomic Factors	Educational attainment, housing-burdened low-income households, linguistic isolation, poverty, unemployment

TABLE C.8 Continued

Data Sources
• California Air Resources Board • California Emissions Inventory Development and Reporting System • Water Boundary Tool–Tracking California • California Residential Parcel Data–Digital Map Products • California Department of Pesticide Regulation • EPA Toxics Release Inventory • TomTom Find/Route/Display • EnviroStor Cleanup Sites Database • Department of Toxic Substances Control • State Water Resources Control Board • EnviroStor Hazardous Waste Facilities Database • Solid Waste Information System • California Department of Resources Recycling and Recovery • California Office of Statewide Health Planning and Development • California Department of Public Health • American Community Survey–US Census Bureau • Comprehensive Housing Affordability Strategy–Housing and Urban Development

TABLE C.9 New Jersey EJMap (Beta) (Centers for Disease Control and Prevention Agency for Toxic Substances and Disease Registry 2020 version)

Purpose

Facilities seeking permits/renewals in an overburdened community (OBC) must analyze their potential contribution to environmental and public health stressors. The department (1) identified justifiable and quantifiable environmental and public health stressors in overburdened communities, (2) designated a geographic unit of analysis for comparison, and (3) developed a methodology for determining whether an OBC is currently subject to adverse cumulative stressors.

Geography

2020–Block groups

Method

Creates two summary maps: OBC & Environmental Stressors. Defines overburdened communities as block groups that meet at least one of the following: (1) at least 35% low-income households; (2) at least 40% of the residents identify as minority or as members of a state-recognized tribal community; and/or (3) at least 40% of the households have limited English proficiency. Additional label "adjacent" provided to describe block groups next to an OBC, or a block group with 0 population.
Identifies Core Environmental and Social Stressors (stressors) by including 26 stressors.

Categories/Themes	Indicators
Concentrated Areas of Air Pollution	Ground-level ozone Fine particulate matter Cancer risk from diesel particulate matter Cancer risk from air toxics excluding diesel particulate matter Noncancer risk from air toxics
Mobile Sources of Air Pollution	Traffic—cars, light- and medium-duty trucks Traffic—heavy-duty trucks Railways
Contaminated Sites	Known contaminated sites Soil contamination deed restrictions Groundwater classification exception areas/currently known extent restrictions
Transfer Stations	Solid waste facilities Scrap metal facilities
Point Sources of Water Pollution	Surface water Combined sewer overflows
Factors with Potential to Cause Public Health Issues	Drinking water Potential lead exposure Lack of recreational open space Lack of tree canopy Impervious surface Flooding (urban land cover)
Density/Proximity Stressors	Emergency planning sites Permitted air sites New Jersey Pollutant Discharge Elimination System (NJPDES) sites
Social Determinants of Health	Unemployment Education

TABLE C.9 Continued

Data Sources
• Air Quality Index 3-year average 2018–2020 • U.S. Environmental Protection Agency National Air Toxics Assessment—2017 Air Tox Screen • Federal Highway Administration Highway Performance Monitoring System (HPMS) Annual Daily Traffic • New Jersey Department of Transportation • New Jersey's Known Contaminated Sites List • New Jersey Deed Permits (NJ Open Data Portal) • New Jersey Solid and Hazardous Waste Facilities (NJ Open Data Portal) • Scrap metal facilities (NJ Open Data Portal) • NJ USGS-based method of identifying watersheds • NJ Surface Water Non-attainment file (open data portal) • Combined Sewer Overflow (NJ Open Data Portal) • NJ Private Well Data • American Community Survey • NJ Recreational Open Spaces Inventory Database • U.S. Forest Service "Analytical" Tree Canopy Cover (TCC) Dataset • NJPDES Residual Category V sludge processing facilities.

TABLE C.10 Properties of the Census Community Resilience Estimates (U.S. Census Bureau, 2019 Estimates, updated August 10, 2021)

Purpose

Community resilience refers to a community's ability to handle the pressures of a disaster. The 2019 Community Resilience Estimates (CRE) are created using data from the 2019 American Community Survey and the Census Bureau's Population Estimates Program about individuals and households. Local leaders, policy makers, public health authorities, and community members can utilize these estimates to evaluate how well communities might cope with challenges and to devise strategies for lessening the impact and facilitating recovery.

Geography

Census tract

Method

An index is generated that produces aggregate-level (tract, county, and state) small-area estimates: the CRE. The CRE provide an estimate of the number of people with a specific number of risks. In its current data file layout form, the estimates are categorized into three groups: zero risks, 1–2 risks, and 3+ risks.

Categories/Themes	Indicators
No broad categories/themes. There are 10 risk factors (RFs) for households and individuals.	**RF 1:** Income-to-poverty ratio (IPR) < 130% percent (HH). **RF 2:** Single- or zero-caregiver household—only one or no individuals living in the household who are 18–64 years (HH). **RF 3:** Unit-level crowding defined as >0.75 persons per room (HH) **RF 4:** Communication barrier defined as either 1. Limited English-speaking households (HH) 2. No one in the household over age 16 with a high school diploma (HH) **RF 5:** No one in the household is employed full time, year-round. The flag is not applied if all residents of the household are ages 65 years or older (HH). **RF 6:** Disability posing constraint to significant life activity. Persons who report having any one of the six disability types (I): hearing difficulty, vision difficulty, cognitive difficulty, ambulatory difficulty, self-care difficulty, and independent living difficulty. **RF 7:** No health insurance coverage (I) **RF 8:** Being age 65 years or older (I) **RF 9:** Households without a vehicle (HH) **RF 10:** Households without broadband Internet access (HH)

Data Sources

- American Community Survey
- Census Bureau's Population Estimates Program: https://www.census.gov/programs-surveys/popest.html

TABLE C.11 Properties of the Climate Mapping for Resilience and Adaptation (CMRA) (National Oceanic and Atmospheric Administration and U.S. Department of the Interior, August 2022)

Purpose

The CMRA Assessment Tool offers condensed overviews of reliable datasets for counties, census tracts, and tribal lands. These overviews offer a uniform perspective on diverse spatial data, enabling users to explore the convergence of climate data with other federal informational resources such as the Climate and Economic Justice Screening Tool and the Tracking of Building Code Adoption.

Geography

Census tract/county

Method

N/A

Categories/Themes	Components of Mapping Tool
Climate Data Summaries, Climate and Economic Justice Screening Tool, FEMA's Nationwide Building Code Adoption Tracking, Coastal Inundation, Current Flooding Hazards, Census TIGER/Lines	Drought Inland flooding Coastal flooding Extreme heat Wildfire Coastal inundation Demographics

Data Sources

- U.S. Census Bureau, 2019 version (county and census tract boundaries)
- 2019 American Community Survey (population estimates)
- Scenarios for the National Climate Assessment: https://scenarios.globalchange.gov/loca-viewer/
- Localized Constructed Analogs
- U.S. Geological Survey THREDDS
- Applied Climate Information System
- National Oceanic and Atmospheric Administration Coastal Digital Elevation Models
- 2022 Interagency Sea Level Rise Technical Report Data Files
- Federal Emergency Management Agency's National Hazard Flood Layer
- Climate and Economic Justice Data
- Building Code Adoption Tracking

TABLE C.12 Properties of the U.S. Department of Transportation Equitable Transportation Community (ETC) Explorer (2023)

Purpose

The USDOT's ETC Explorer is intended to enhance the capabilities of the Council on Environmental Quality's Climate & Economic Justice Screening Tool (CEJST). The aim of the ETC Explorer is to offer users a more comprehensive insight into a community's exposure to transportation challenges. This understanding helps ensure that investments effectively target the transportation-related issues causing disadvantage, thereby ensuring that the benefits are appropriately addressing these concerns.

Geography

Census tract

Method

Composite index
Percentile-based ranking
Measures cumulative burden

Categories/Themes	Indicators
Environmental Burdens	Ozone $PM_{2.5}$ Diesel particulate matter Air toxics cancer risk Hazardous site proximity Toxic release site proximity Treatment and disposal proximity Risk management plan sites Coal mine proximity Lead mine proximity Impaired surface water High-volume road proximity Railway proximity Airport proximity Port proximity Pre-1980 housing Percent over 65 years Percent under 17 years Percent disabled Limited English proficiency Percent mobile homes 200% poverty line High school graduation status Unemployment House tenure Housing cost burden Percent uninsured
Social Vulnerabilities	Percent lacking Internet Endemic inequality

TABLE C.12 Continued

Categories/Themes	Indicators
Climate and Disaster Risk	Annualized hazard losses Impervious surfaces Future extreme weather risk
Health Vulnerabilities	Asthma Cancer High blood pressure Diabetes Low mental health
Transportation Insecurity	Transportation access Transportation cost burden Fatalities per 100,000 people

Data Sources

- U.S. Census Bureau
- USDOT
- U.S. Environmental Protection Agency (EPA)
- Bureau of Labor Statistics
- EPA Smart Location Database
- Highway Performance Monitoring System
- EPA's Environmental Justice Screening Tool (EJScreen) 2022
- Point data retrieved for nonabandoned coal mine and nonabandoned lead mine locations from the Mine Safety Administration
- Point data for hazardous sites, toxics release sites, treatment, storage and disposal sites, and Risk Management Plan sites from the EPA Facility Registry Service.
- Data on high-volume roads sourced from the Highway Performance Monitoring System
- Railway and airport locations obtained from the Bureau of Transportation Statistics Geospatial dataset
- Impaired water data were from EPA Watershed Index Online
- Health Vulnerability indicator data obtained from the Centers for Disease Control and Prevention PLACES 2020
- Hospital proximity data from the American Community Survey (ACS) and Department of Homeland Security
- Social vulnerability data from ACS 2015–2020
- Data on individuals without insurance and Internet access were directly obtained from the U.S. Census Bureau
- Federal Emergency Management Agency National Risk Index

TABLE C.13 Properties of the Massachusetts Department of Public Health Environmental Justice Tool (Massachusetts Executive Office of Energy and Environmental Affairs)

Purpose

The purpose of the MA-DPH-EJ Tool is to support the application of the Massachusetts Executive Office of Energy and Environmental Affairs environmental justice policy, improve inclusive community planning for environmental assessment, and provide insights for various tasks such as siting, permitting, Brownfields cleanup, Massachusetts Environmental Policy Act review, grant applications, transportation projects, and evaluations of community, health, or climate impacts.

Geography

Census block group

Method

Environmental justice communities refer to Census block groups that fulfill one or more EJ criteria. Vulnerable Health EJ Criteria indicate communities that satisfy a minimum of 1 EJ criterion AND at least 1 health indicator criterion.

Categories/Themes	Indicators
State-Designated Environmental Justice Categories	The annual median household income is 65% or less of the statewide annual median household income, OR Minorities make up 40% or more of the population, OR 25% or more of households identify as speaking English less than "very well," OR Minorities make up 25% or more of the population and the annual median household income of the municipality in which the neighborhood is located does not exceed 150% of the statewide annual median household income
Vulnerable Health Environmental Justice Criteria	Heart attack Childhood blood lead level ≥5 μg/dL Low birth weight Asthma

Data Sources

- Massachusetts Center for Health Information Analysis
- Massachusetts Registry of Vital Records and Statistics
- Massachusetts Department of Public Health Childhood Lead Poisoning Prevention Program
- Decennial Census and American Community Survey 5-year estimate

Appendix D

Example Datasets for Consideration for EJ Tool Indicators

This appendix includes a list of datasets that could be considered for inclusion in tools such as CEJST. Identifying appropriate indicators and datasets through an inclusive, community-informed process could lead to improved and more informed identification of representative data, and scientific and technical advances may drive improvements in data quality and completeness. To select appropriate indicators, it is necessary to understand the relationships between burden, indicator, and dataset. These relationships are described in Chapter 3 as part of the discussion on the conceptual foundation for constructing composite indicators. Table D.1 is organized into 10 categories of what the Council on Environmental Quality (CEQ) labels "burdens" in CEJST—Climate Change, Energy, Health, Housing, Legacy Pollution, Transportation, Water and Wastewater, Workforce Development, Racial Segregation, and Structural Racism. Each of these categories contains a set of indicators that are intended to represent the burden category. The spatial resolution of each indicator and the year of the latest version of the dataset(s) used are also provided. The committee does offer suggestions of other datasets that could be considered at a variety of scales. The column labeled "Spatial Resolution" in Table D.1 represents the minimum resolution or spatial unit at which data or measure is currently available or can be computed. This may or may not be the most appropriate scale depending upon for the tool objective or measure. Data selection criteria, including such characteristics as scale, need to be established by tool developers based on a structured tool development process, the objectives of the tool, and the concepts to be measured.

TABLE D.1 Example Datasets for Consideration for Environmental Justice Tool Indicators

Potential Indicators	Dataset Name, Source (Date Accessed)	Latest Year	Spatial Resolution
Climate Change			
Deaths from climate disasters	U.S. Billion-Dollar Weather and Climate Disasters National Oceanic and Atmospheric Administration National Centers for Environmental Information, https://www.ncei.noaa.gov/access/billions/ (March 15, 2024)	2021	County
Days with maximum temperature above 35°C	Repository supporting the implementation of FAIR principles in the Intergovernmental Panel on Climate Change Working Group 1 (IPCC-WG1) Atlas. Iturbide M. et al., https://github.com/IPCC-WG1/Atlas (March 15, 2024)	2021	Tract
Days with maximum temperature above 40°C	Repository supporting the implementation of FAIR principles in the IPCC-WG1 Atlas. Iturbide M. et al., https://github.com/IPCC-WG1/Atlas (March 15, 2024)	2021	Tract
Frost days	Repository supporting the implementation of FAIR principles in the IPCC-WG1 Atlas. Iturbide M. et al., https://github.com/IPCC-WG1/Atlas (March 15, 2024)	2021	Tract
Maximum of maximum temperatures	Repository supporting the implementation of FAIR principles in the IPCC-WG1 Atlas. Iturbide M. et al., https://github.com/IPCC-WG1/Atlas (March 15, 2024)	2021	Tract
Mean temperature	Repository supporting the implementation of FAIR principles in the IPCC-WG1 Atlas. Iturbide M. et al., https://github.com/IPCC-WG1/Atlas (March 15, 2024)	2021	Tract
Urban Heat Island Extreme Heat Days	Repository supporting the implementation of FAIR principles in the IPCC-WG1 Atlas. Iturbide M. et al., https://github.com/IPCC-WG1/Atlas (March 15, 2024)	2021	Tract
Sea level rise (meters)	Repository supporting the implementation of FAIR principles in the IPCC-WG1 Atlas. Iturbide M. et al., https://github.com/IPCC-WG1/Atlas (March 15, 2024)	2021	Tract
Heat wave—annualized frequency	National Risk Index Federal Emergency Management Agency (FEMA), https://hazards.fema.gov/nri/ (March 15, 2024)	2022	Tract

TABLE D.1 Continued

Potential Indicators	Dataset Name, Source (Date Accessed)	Latest Year	Spatial Resolution
Cold wave—annualized frequency	National Risk Index FEMA, https://hazards.fema.gov/nri/ (March 15, 2024)	2022	Tract
Drought—annualized frequency	National Risk Index FEMA, https://hazards.fema.gov/nri/ (March 15, 2024)	2022	Tract
Coastal flooding—annualized frequency	National Risk Index FEMA, https://hazards.fema.gov/nri/ (March 15, 2024)	2022	Tract
Riverine flooding—annualized frequency	National Risk Index FEMA, https://hazards.fema.gov/nri/ (March 15, 2024)	2022	Tract
Hurricane—annualized frequency	National Risk Index FEMA, https://hazards.fema.gov/nri/ (March 15, 2024)	2022	Tract
Tornado—annualized frequency	National Risk Index FEMA, https://hazards.fema.gov/nri/ (March 15, 2024)	2022	Tract
Winter weather—annualized frequency	National Risk Index FEMA, https://hazards.fema.gov/nri/ (March 15, 2024)	2022	Tract
Increased $PM_{2.5}$ mortality—cardiovascular disease (ages 65+)	*Climate Change and Social Vulnerability in the US,* EPA 430-R-21-003, https://www.epa.gov/cira/technical-appendices-and-data (March 15, 2024)	2021	Tract
Increased ozone mortality (all ages)	*Climate Change and Social Vulnerability in the US,* EPA 430-R-21-003, https://www.epa.gov/cira/technical-appendices-and-data (March 15, 2024)	2021	Tract
Various	Climate-related hazards U.S. Climate Resilience Toolkit/Global Change.gov, https://resilience.climate.gov/ (March 15, 2024)	Various	Various
Energy			
Ozone	EPA model-monitor fusion dataset EPA's Office of Air and Radiation, https://cfpub.epa.gov/ols/catalog/catalog_full_record.cfm?&FIE LD4=CALLNUM&INPUT4=454%2FS%2D15%2D001&LIBCODE=&COLL=&SORT_TYPE=YRDESC&item_count=1 (March 15, 2024)	2019	
Other $PM_{2.5}$	Satellite-derived $PM_{2.5}$ van Donkelaar et. al. (2021), https://pubs.acs.org/doi/abs/10.1021/acs.est.1c05309 (March 15, 2024)	2022	0.01 x 0.01 degree

continued

TABLE D.1 Continued

Potential Indicators	Dataset Name, Source (Date Accessed)	Latest Year	Spatial Resolution
Energy			
Electric vehicle charging station proximity	Alternative Fuels Data Center U.S. Department of Energy, https://afdc.energy.gov/stations/#/analyze?country=US&fuel= ELEC (March 15, 2024)	2022	Tract
Health			
Health care access indicators	PLACES: Local Data for Better Health Center for Disease Control and Prevention (CDC), https://www.cdc.gov/places/about/index.html (March 15, 2024)	Various	Multiple
Self-reported physical health	PLACES: Local Data for Better Health CDC, https://www.cdc.gov/places/about/index.html (March 15, 2024)	Various	Multiple
Self-reported mental health	PLACES: Local Data for Better Health CDC, https://www.cdc.gov/places/about/index.html (March 15, 2024)	Various	Multiple
Lack of health insurance	PLACES: Local Data for Better Health CDC, https://www.cdc.gov/places/about/index.html (March 15, 2024)	Various	Multiple
Alchohol abuse	PLACES: Local Data for Better Health CDC, https://www.cdc.gov/places/about/index.html (March 15, 2024)	Various	Multiple
Food access	Food Access Research Atlas U.S. Department of Agriculture Economic Research Service, https://www.ers.usda.gov/data-products/food-access-research-atlas/download-the-data/ (March 15, 2024)	2019	Tract
Medically underserved areas	HRSA Data Explorer Health Resources and Services Administration (HRSA), https://data.hrsa.gov/tools/data-explorer (March 15, 2024)	2020	Tract
Proximity to hospitals	Homeland Infrastructure Foundation-Level Data (HIFLD) U.S. Department of Homeland Security (DHS), https://hifld-geoplatform.opendata.arcgis.com/datasets/geoplatform::nursing-homes/about (March 15, 2024)	2017	Tract
Proximity to nursing homes	HIFLD, DHS, https://hifld-geoplatform.opendata.arcgis.com/datasets/geoplatform::nursing-homes/about (March 15, 2024)	2017	Tract
Location of hospital and nonhospital healthcare facilities	Provider of Services File—Hospital & Non- Hospital Facilities Centers for Medicare and Medicaid Services, https://data.cms.gov/provider-characteristics/hospitals-and-other-facilities/provider-of-services-file-hospital-non-hospital (March 15, 2024)	2023	Address
Locations of doctors and clinicians	Doctors and Clinicians national downloadable file Centers for Medicare and Medicaid Services, https://data.cms.gov/provider-data/dataset/mj5m-pzi6 (March 15, 2024)	2023	Address

TABLE D.1 Continued

Potential Indicators	Dataset Name, Source (Date Accessed)	Latest Year	Spatial Resolution
Housing			
HUD Public Housing units	Real Estate Assessment Center (REAC) Physical Inspection Scores U.S. Department of Housing and Urban Development (HUD), https://www.huduser.gov/portal/datasets/pis.html (March 15, 2024)	2021	Tract
Housing affordability (renters): HUD criteria	Office of Policy Development and Research Consolidated Planning/CHAS Data HUD, https://http://www.huduser.gov/portal/datasets/cp.html#2006- 2018_query (March 15, 2024)	2018	Tract
Housing affordability (owners): HUD criteria	Office of Policy Development and Research Consolidated Planning/CHAS Data HUD, https://http://www.huduser.gov/portal/datasets/cp.html#2006- 2018_query (March 15, 2024)	2018	Tract
Number and area of parks	National Neighborhood Data Archive Inter-university Consortium for Political and Social Research, https://doi.org/10.3886/E117921V1 (March 15, 2024)	2018	Tract
Legacy Pollution			
Hazardous waste landfills	Hazardous Waste Incinerators/Landfills Locations EPA, https://app.box.com/s/h4zayqq6lwsli7b5a6basp59za-1z10i7 (March 15, 2024)	Various	
GHG emissions	Facility Level Information on GHG Tool EPA, https://ghgdata.epa.gov/ghgp/main.do#/facility/ (March 15, 2024)	2021	State
Agricultural pesticides	Estimated Annual Agricultural Pesticide Use for Counties U.S. Geological Survey, https://doi.org/10.5066/P9F2SRYH (March 15, 2024)	2020	County
Transportation			
NO_2	TROPOMI NO_2 in the United States Goldberg, D. L. et al., https://doi.org/10.1029/2020EF001665 (March 15, 2024)	2021	Tract
Noise pollution	National Transportation Noise Map Bureau of Transportation Statistics, https://www.bts.gov/geospatial/national-transportation-noise-map (March 15, 2024)	2020	30 meters
Railroads, marine, transit, roads, aviation	Bureau of Transportation Statistics Open Data Site Geospatial at the Bureau of Transportation Statistics (arcgis.com) (March 15, 2024)	Current	Tract

continued

TABLE D.1 Continued

Potential Indicators	Dataset Name, Source (Date Accessed)	Latest Year	Spatial Resolution
Delay (congestion) per capita/	Urban Mobility Report Texas A&M Transportation Institute, https://mobility.tamu.edu/umr/ (March 15, 2024)	2019	Tract
Road quality and maintenance	International Roughness Index, U.S. Department of Transportation, https://www.fhwa.dot.gov/policyinformation/hpms.cfm (March 15, 2024)	2018	Tract
Walkability	Score of walkability and bikeability from 0 to 100 Walk Score, https://www.walkscore.com (March 15, 2024)	2022	Tract
Bikability	Score of walkability and bikeability from 0 to 101 Walk Score, https://www.walkscore.com (March 15, 2024)	2022	Tract
Water/Wastewater			
Drinking water violations or enforcement	Safe Drinking Water Information System EPA, https://sdwis.epa.gov/ords/sfdw_pub/r/sfdw/sdwis_fed_reports_public/200 (March 15, 2024)	1st quarter, 2023	Quarterly summary
Facilities with enforcement or violation	Significant Non-Compliance for National Pollutant Discharge Elimination System permits EPA—Enforcement and Compliance History Online, https://echo.epa.gov/ (March 15, 2024)	Current	National
CCR (Consumer Confidence Report) Compliance	Safe Drinking Water Information System (SDWIS) Federal Reporting Services EPA, https://sdwis.epa.gov/ords/sfdw_pub/r/sfdw/sdwis_fed_reports_public/200 (March 15, 2024)	2023	Community water system
Private domestic wells	Well density EPA U.S. Private Domestic Wells, https://experience.arcgis.com/experience/be9006c30a2148f595 693066441fb8eb (March 15, 2024)	2020	Census-block group
Lead service lines	Drinking Water Infrastructure Needs Survey and Assessment EPA Drinking Water State Revolving Fund, https://www.epa.gov/dwsrf/epas-7th-drinking-water- infrastructure-needs-survey-and-assessment (March 15, 2024)	2023	State
Workforce Development			
Median earnings of full-time, year-round workers	American Community Survey U.S. Census Bureau, https://data.census.gov/table (March 15, 2024)	2020	National

TABLE D.1 Continued

Potential Indicators	Dataset Name, Source (Date Accessed)	Latest Year	Spatial Resolution
Percent of working-age adults (19–64 years) with employer-based health insurance	American Community Survey U.S. Census Bureau, https://data.census.gov/table (March 15, 2024)	2020	National
Racial Segregation			
Dissimilarity Index	Decennial Census or American Community Survey U.S. Census Bureau Housing Patterns: Appendix B: Measures of Residential Segregation, https://www.census.gov/topics/housing/housing-patterns/guidance/appendix-b.html (March 15, 2024)	Revised 2021	Census block
Gini Coefficient	Decennial Census or American Community Survey https://www.census.gov/topics/housing/housing-patterns/guidance/appendix-b.html (March 15, 2024)	Revised 2021	Census block
Entropy Index/ Information Index	Decennial Census or American Community Survey https://www.census.gov/topics/housing/housing-patterns/guidance/appendix-b.html (March 15, 2024)	Revised 2021	Census block
Atkinson Index	Decennial Census or American Community Survey https://www.census.gov/topics/housing/housing-patterns/guidance/appendix-b.html (March 15, 2024)	Revised 2021	Census block
Isolation Index	Decennial Census or American Community Survey https://www.census.gov/topics/housing/housing-patterns/guidance/appendix-b.html (March 15, 2024)	Revised 2021	Census block
Delta Index	Decennial Census or American Community Survey https://www.census.gov/topics/housing/housing-patterns/guidance/appendix-b.html (March 15, 2024)	Revised 2021	Census block
Structural Racism			
Index of Concentration at the Extremes (ICE)	American Community Survey Krieger, N. et al. 2017. https://doi.org/10.1007/s11524-016-0116-z. (March 15, 2024)	2017	Census-block group
Index of Disproportionality	American Community Survey Chantarat, T. et al. 2022. https://doi.org/10.1111/1475-6773.13976. (March 15, 2024); Dougherty, G. et al. 2020. https://doi.org/10.1016/j.amepre.2020.05.019. (March 15, 2024)	2022	Public Use Micro-data Area (PUMA) or County
Other			
Prison population	American Community Survey U.S. Census Bureau, https://data.census.gov/cedsci/ (March 15, 2024)	2021	County

continued

TABLE D.1 Continued

Potential Indicators	Dataset Name, Source (Date Accessed)	Latest Year	Spatial Resolution
Undocumented population	Unauthorized Immigrant Population Profiles Migration Policy Institute, https://www.migrationpolicy.org/programs/us-immigration-policy-program-data-hub/unauthorized-immigrant-population-profiles#SD (March 15, 2024)	2019	County
School children	Elementary/Secondary Information System National Center for Education Statistics, https://nces.ed.gov/ccd/elsi/ (March 15, 2024)	2022	School